BRWM - Tech Services

ENVIRONMENTAL ENGINEERING P.E. EXAMINATION GUIDE & HANDBOOK

W. Christopher King, Ph.D., P.E., DEE

AN
AMERICAN ACADEMY OF
ENVIRONMENTAL ENGINEERS®
PUBLICATION

The contents of this publication are not intended to be and should not be construed to be a standard of the American Academy of Environmental Engineers (The Academy) and are not intended for use as a reference in purchase specificaitons, contracts, regulations, statutes, or any other legal document.

No reference made in this publication to any specific method, product, process or service constitutes or implies an endorsement, recommendation, or warranty thereof by The Academy.

The Academy makes no representation or warranty of any kind, whether express or implied, concerning the accuracy, completeness, suitability or utility of any information, apparatus, product, or process discussed in this publication, and assumes no liability thereof. Anyone utilizing this information assumes all liability arising from such use, including but not limited to infrigement of any patent or patents.

© 1996 by American Academy of Envronmental Engineers.

All rights reserved.

Printed in the United States of America.

Library of Congress Cataloging-in-Publication Data

Environmental engineering P.E. examanation guide & handbook.
    p. cm.
  Includes bibliographical references and index.
  ISBN 1-883767-13-X
    1. Environmental engineering -- Problems, excercises, etc.
2. Environmental engineering -- Outlines, syllabi, etc.   I. American Academy of Environmental Engineers.
TD157.15.E54   1996
628'.076 -- dc20   96-33784
                                                                                CIP

ABOUT THE AUTHORS

W. Christopher King, Ph.D., P.E., DEE is a Colonel is serving as Professor and Program Director of Environmental Engineering at the United States Military Academy. Colonel King earned his bachelors degree in chemical engineering and masters degree in civil engineering at Tennessee Technological University in 1972 and 1974. He completed his Ph.D. in environmental engineering at the University of Tennessee in 1988. In 22 years with the Army, Colonel King has worked in a broad range of environmental media including water quality analysis, wastewater engineering, air pollution engineering, and hazardous waste management. He serves as the chairman of the American Society of Civil Engineering committee on waste minimization. Colonel King is a licensed professional engineer in Minnesota and is AAEE board-certified in hazardous waste. In 1992 he won the Environmental Engineering Excellence Honor Award for planning for his work on the Health Risk Assessment of the Kuwait Oil Fires, and the Army Science Conference Award for research in geophysical remote sensing.

William C. Anderson, P.E., DEE is the Executive Director of the American Academy of Environmental Engineers. He received his baccalaureate degree in civil engineering from Iowa State University, is a registered engineer in five states, and a Diplomate of AAEE. In addition to his service to the Academy, he was Chairman of the American Society of Civil Engineer's Environmental Engineering Division. He is a member of the Air & Waste Management Association, American Society of Civil Engineers, American Water Works Association, American Environmental Engineering Professors, National Society of Professional Engineers, Water Environment Federation, and is Vice Chairman of the Environmental Engineering Subcommittee of the Engineering Professional Examination Committee of the National Council of Examiners for Engineering and Surveying. He was awarded the Paul F. Morgan Medal by WPCF and the Lewis Van Carpenter Award by New York Water Pollution Control Association. He is the author of over 100 technical papers and publications and is the editor-in-chief of *Environmental Engineer*®.

James F. Braithwaite, P.E., DEE is a Vice-President and Program Manager for RMT, Inc., an international environmental engineering and consulting firm specializing in solid and hazardous waste management, air quality engineering, remedial investigation, closure/clean up activities, and strategic environmental management. Prior to merging with RMT, Inc. in 1994, Mr. Braithwaite was President of Braithwaite Consultants, Inc., a midwestern environmental engineering and science consulting firm formed in 1983. He received his Bachelors Degree in Mechanical Engineering in 1972 from Michigan State University and pursued graduate studies at the University of Michigan in solid waste management and hydrology. Mr. Braithwaite is a licensed professional engineer in several states, and has been appointed as a Subject Matter Expert in solid waste management and has been a member of the Environmental Subcommittee the National Council of Examiners for Engineer and Surveyors since 1993. As a Diplomate of the American Academy of Environmental Engineers, Mr. Braithwaite is board-certified as a specialist in solid waste management, and has recently completed a 3-year term as Chair of the Academy's Solid Waste Subcommittee.

John D. Dietz, Ph.D., P.E. joined the faulty at the University of Central Florida in 1982 where he has conducted research and instruction in municipal water and wastewater treatment, industrial waste management, and process control systems. Prior to entering academia, Dr. Dietz obtained experience in the private sector with design of municipal and industrial wastewater treatment systems. He holds bachelors and masters Degrees in Civil Engineering from the University of Illinois and a Doctor of Philosophy in Environmental Systems Engineering from Clemson University. He is a registered professional engineer in Florida and Mississippi. Dr. Dietz is a member of the Water Environment Federation, the American Water Works Association, the American Society of Civil Engineers, the American Electroplaters and Surface Finishers Society, and the International Association on Water Quality.

About the Authors

Richard F. Dominguez, Ph.D., P.E. is Visiting Professor of Environmental Engineering at the United States Military Academy at West Point, New York. He is also the President of E_2O Consultants, Inc. a engineering consulting firm specializing in environmental, water resources, and ocean engineering which he co-founded while on the faculty of Texas A & M University.

Dr. Dominguez received his Bachelors Degree from Worcester Polytechnic Institute in civil engineering. He holds a Masters Degree from Colorado State University in the field of hydraulics and fluid mechanics and a Doctor of Philosophy from Oregon State University in ocean engineering. His 30-year professional career has been divided between university teaching and administration, and consulting engineering practice. He is a registered professional engineer in New York and a member of the National Society of Professional Engineers, American Society of Civil Engineers and the American Society for Engineering Education.

Ronald L. Kathren, P.E., DEE is Professor of Health Physics and Director of the United States Transuranium and Uranium Registries at Washington State University, and Affiliate Professor Environmental Health at the University of Washington. A registered Professional Engineer in California, he holds degrees from UCLA and the University of Pittsburgh and is a Diplomate of both the American Academy of Environmental Engineers and the American Board of Health Physics. During a career spanning more than 35 years, he has been the recipient of numerous awards including the Radiology Centennial Hartman Medal and the Founders Award and Elda E. Anderson Award of the Health Physics Society. He is past-president of the Health Physics Society, and the author or co-author of more than 150 scientific and technical articles in the peer reviewed literature as well as several books including Radioactivity in the Environment (1984), Radiation Protection (1985), and the Plutonium Story (1994).

Katherine R. King, P.E. DEE is currently the Manager of Michigan Operations of RMT, Inc., a national environmental engineering and consulting firm specializing in solid and hazardous waste management, air quality engineering, remedial investigation, closure/clean up activities, and strategic environmental management. Ms. King is a registered Professional Engineer in Michigan, Illinois, and Indiana and is a Diplomate in the American Academy of Environmental Engineers (AAEE) with certification in Hazardous Waste Management. She received her Bachelors degree in Environmental Engineering from Michigan Technological University where she currently serves on the Civil and Environmental Engineering Professional Advisory Committee. Ms. King is also a member of AAEE Admissions Committee and is active in the Water Environment Federation, Air & Waste Management Association, America Society of Civil Engineers and the Engineering Society of Detroit.

Jon M. Rueck, P.E., DEE received both his B.S. in Nuclear Engineering and M.S. in Mechanical Engineering degrees from Kansas State University. He is the recepient of ASHRAE's Homer Addams Award. He has worked with the Kansas Division of Environment in Industrial and Municipal Waste Water Treatment and Water Quality Planning. He has also served the Air Quality Bureau in Source Enforcement as a continuous emission monitoring specialist, and as air quality program analyst. He is a Diplomate of the American Academy of Environmental Engineers with specialty certification in Air Pollution Control Engineering, the National Society of Professional Engineers, and is Chairman of the Midwest Section of the Air & Waste Management Association. He is currently president of Rueck Associates, a firm which sells and services continuous emission monitoring systems.

TABLE OF CONTENTS

Chapter 1 — INTRODUCTION **13**
  1.1 Environmental Engineering History 13
  1.2 Engineering Credentials 24
  1.3 Principles and Practice (P.E.) Examination 27
  1.4 Environmental Engineering Examination 29
  1.5 References 30

Chapter 2 — COMMON PRINCIPLES **33**
  2.1 Introduction 33
  2.2 Math Basics 33
  2.3 Units of Environmental Engineering 34
  2.4 Mass Balance 37
  2.5 Engineering Economics 39
  2.5 Standard Notations 47
  2.6 References 47

Chapter 3 — ESSENTIAL CHEMISTRY **49**
  3.1 Introduction 49
  3.2 Definitions 49
  3.3 Equivalent Weight 52
  3.4 The Basic Chemical Reaction 52
  3.5 Stoichiometric Calculations 56
  3.6 Reaction Equilibrium 59

3.7 Reaction Kinetics 60
3.8 Acids and Bases 62
3.9 Standard Chemistry Notations (Chemical symbols are listed in Table 3-1) 64
3.10 References 65

Chapter 4 — FLUID MECHANICS 67
4.1 Introduction 67
4.2 Fluid and Flow Properties 67
4.3 Hydrostatics 69
4.4 Conservation Principles 70
4.5 Conduit Flow 74
4.7 Standard Fluids Notations 86
4.8 References 87

Chapter 5 — HYDRAULICS & HYDROLOGY 89
5.1 Introduction 89
5.2 Hazen-Williams pipe flow analysis 89
5.3 Pipe networks 91
5.4 Pumps 92
5.5 Free Surface Flow or Open Channel Flow 102
5.6 Hydrology 123
5.7 Groundwater Hydrology 135
5.8 Standard Hydraulics/ Hydrology Notations 142
5.9 References 144

Chapter 6 — WATER TREATMENT 147
6.1 Introduction 147
6.2 Water Quality 148
6.3 Basic Water Treatment 154
6.4 Settling 155
6.5 Coagulation/Flocculation 165
6.6 Filtration 167
6.7 Disinfection 169
6.8 Solids Management 172
6.9 Additional Water Treatment Processes 174
6.10 Operations 179
6.11 Standard Water Treatment Notations 179
6.12 References 180

Chapter 7 — WASTEWATER TREATMENT 181
7.1 Introduction 181
7.2 Treatment Standards 182
7.3 Receiving Stream Water Quality 184
7.4 Wastewater Treatment 190
7.5 Preliminary Treatment 199
7.6 Sedimentation 202
7.7 Biological Treatment 207
7.8 Gas Transfer - Aeration 228
7.9 Filtration 234
7.10 Disinfection 236
7.11 Waste Solids Management 239
7.12 Physical Chemical Treatment 255

7.12.1 Oxidation 255
7.13 Standard Wastewater Treatment Notations 257
7.14 References 259

Chapter 8 — AIR POLLUTION 261
8.1 Introduction 261
8.2 Air Pollution Regulations 264
8.3 Ambient Air Quality 267
8.4 Stationary Source Air Pollution 269
8.5 Particulate Control 273
8.6 Scrubbers 285
8.7 Control of Nitrogen from Stationary Sources 292
8.8 Treatment of Volatile Organic Gases 293
8.9 References 298

Chapter 9 — SOLID AND HAZARDOUS WASTE MANAGEMENT 301
9.1 Introduction 301
9.2 Regulations 301
Operating Criteria 309
9.3 Basic Solid Waste Management 312
9.4 Landfilling 321
9.5 Incineration/Energy Recovery 343
9.6 Resource Recovery Systems 350
9.7 Hazardous Waste Treatment 359
9.8 Site Remediation 372
9.9 Standard Solid and Hazardous Waste Notations 384
9.10 References 390

Chapter 10 — HEALTH, SAFETY, AND ENVIRONMENTAL PROTECTION 393
10.1 Introduction 393
10.2 Legal Bases and References 394
10.3 Indoor Air Quality 395
10.4 Noise 409
10.5 Ionizing Radiation 417
10.5 Risk Assessment 425
10.6 Standard Notations 427
10.7 References 428

Appendix — CONVERSION FACTORS 431

INDEX 435

PREFACE

The inspiration for this book came as a result of my work as a volunteer for the National Council of Examiners for Engineering and Surveying (NCEES). Viewing the results of those taking the environmental engineering principles and practice examination, it became clear to me that many were not familiar with some of the basics applicable to the various environmental engineering specialties. This is not an unusual condition since most degree programs emphasize the water and wastewater specialties and those specialties are where most environmental engineers practice. Also, it was apparent that many had not acquired proficiency in using the chemical and biological analytical tools on which environmental management processes and technologies are based.

Environmental engineering has come to be a relatively comprehensive discipline encompassing many specialties — those included on the examination in particular. Therefore, I decided to write this book to provide, first a guide to the basics of all that environmental engineering encompasses and secondly, as a long-term reference for practicing engineers. Typically, most practicing environmental engineers concentrate in a particular specialty. But, from time-to-time, questions may arise in another specialty for which colleagues and friends expect the "environmental engineer" to have the answer. The information in this book will enable you to address those inquiries generally and to identify references which can provide more detailed information.

I want to acknowledge the assistance of those who assisted in or prepared individual chapters in this book without whom it would not have been completed. They are described in About the Authors. I particularly want to acknowledge the assistance of William C. Anderson, P.E., DEE whose continuing encouragement and editing of the entire manuscript made this book possible. For the chapter on Solid and Hazardous Waste, the authors were aided by Sharon Bailey, P.E. (composting and resource recovery), Steven Wittman (landfill gas), Charles O. Velzy, P.E., DEE (waste-to-energy), and to John Wolfe, Ph.D. and Eugene A. Glysson, Ph.D., P.E., DEE (complete chapter review). Equally important was the publications staff of the American Academy of Environmental Engineers, including John M. Buterbaugh, Manager of Publications and Communications, Susan C. Zarriello, Production Manager, and Catherine L. Schultz, Production Assistant.

W. Christopher King, P.E., DEE
March 1996

Chapter 1

INTRODUCTION

by William C. Anderson, P.E., DEE

This chapter addresses three separate, but important, topics for all environmental engineers and for licensure candidates taking the environmental engineering discipline principles and practice of engineering written examination or, as it is more commonly known, the P.E. examination. In Section 1.1, the rich heritage of environmental engineering, from its beginning in the 1830s to the late 20th century is summarized. The types of credentials available to environmental engineers and their role in the practice of the profession is discussed in Section 1.2. Finally, the development and scope of the environmental engineering P.E. examination is described to inform licensure candidates what the examination covers and why.

1.1 Environmental Engineering History

Environmental engineering is a relatively new name for a type of engineering that began in the United States in the 1830s. Under different names, it continued to evolve to satisfy environmental challenges posed by urbanization, suburbanization, and the other needs of the nation during the industrial revolution in the late 1800s through the information revolution of the 1990s.

Until 1970, when Earth Day captured the public's attention leading to a concentrated effort to clean up environment, the profession was practiced by but a few.

Yet, the pioneers — Mills, Cheseborough, Sedgewick, Hazen, Metcalf, Eddy, Camp, Fair, Wolman, to name a few — blazed a trail establishing design protocols still in use today. Of necessity, this section is but a brief overview of the rich heritage on which modern environmental engineering is founded.

1.1.1 The Beginning Hydraulic engineering best describes environmental engineering at its birth. Early communities were usually located on or adjacent to plentiful sources of fresh water. As these communities grew and people were forced to live farther and farther from the water source, private companies formed to convey water to the outlying areas. By 1800, there were 18 private waterworks in the U.S. (McKinney 1994).

The inability of these private water companies to meet the water needs of rapidly-growing cities forced the larger municipalities such as New York, Boston and Chicago to consider public water systems. Colonel De Witt Clinton, Jr., an Army engineer and son of the former governor of New York, was retained to examine New York's water supply needs in 1832. He recommended that water be obtained from the Croton River and conveyed to New York City through a 40-mile aqueduct. In 1836, John B. Jervis, a self-educated hydraulic engineer, who had learned his engineering on the Erie Canal and the Mohawk and Hudson Railroad, began construction of a dam across the Croton River to provide storage for the aqueduct (Jervis 1876). By 1842, fresh water began flowing from the Croton Reservoir to New York City through the aqueduct which carried 95 Mgal/day (4.16 m^3/sec) (Hazen 1907).

In Boston, city officials, noting the success of the Croton Aqueduct in New York, retained Jervis and Professor Walter R. Johnson of Philadelphia to find a new water supply for Boston in 1844 (Brandlee 1868). The team identified Lake Cochituate as the best source of water and Jervis designed the Cochituate Aqueduct. At the same time, Ellis S. Cheseborough was retained as Chief Engineer of the West Division of the Boston Water Works (Cain 1991). Under Cheseborough, construction of the Cochituate Aqueduct was completed in 1848 and Boston established its public water system. He became the Commissioner of the Boston Water Works and Boston's first City Engineer.

As City Engineer, Cheseborough was responsible for managing the stormwater sewer system. Boston had taken over the storm sewers in 1823 to ensure that they were properly maintained and that future sewers were built to proper specifications (MSBH 1903). Storm sewers were designed to flow by gravity to the nearest stream and eventually to the ocean. Because Boston prohibited the dumping of human sewage into storm sewers, sanitary sewage was collected in privies and vaults which were pumped out at regular intervals. This septage was carried in special wagons to farms outside the City. But, with the increased water supplies, more water closets and indoor bathtubs increased the flow of wastewaters and quickly filled the sewage storage vaults. A few home owners solved their problem by building their own sewers and connecting them to the storm sewers without official permission. In addition to domestic sewage pollution of storm water, horse manure from the streets ended up in storm sewers when it rained. While smaller communities typically constructed two sets of sewers, a storm sewer and a sanitary sewer, Boston, like New York and other large cities, combined the sanitary sewers and the storm sewers in one pipe for economic reasons. Under either approach, sewer design required hydraulic engineering.

After serving Boston as its City Engineer for seven years, Cheseborough resigned in 1855 and moved to Chicago. There, he developed its sewerage system, the accomplishment for which he is most remembered. Cheseborough developed an innovative (for the time) plan to employ combined sewers to drain the city's waste into the Chicago River which required raising the elevation of most of the downtown area. The new sewerage system contaminated Lake Michigan, the city's water supply. To correct this problem, Cheseborough developed and implemented an intake works offshore which was connected to the city by a 2 mile-long tunnel.

1.1.2 A Discipline Takes Shape As the fruits of hydraulic engineering, water systems serving the public, became more common and concerns about the quality of the water used for public water supplies grew, Sanitary Chemistry emerged as a new facet of environmental engineering. In 1873, the Massachusetts State Board of Health asked Professor William Ripley Nichols, who was in charge of Chemistry at the then infant Massachusetts Institute of Technology (M.I.T.) to analyze the water quality of the major rivers used for public water supplies in Massachusetts (MSBH 1874). Nichols agreed and set up the first Sanitary Chemistry Laboratory in 1874 to perform water analyses. Ellen Richards, then Ellen Swallow, the first coed at M.I.T., had just graduated with her SB degree in Chemistry from M.I.T. She was named Professor Nichols' assistant and did most of the analyses at the new Sanitary Chemistry Laboratory. She went on to become one of the foremost sanitary chemists in the United States.

Water pollution problems also increased as populations rose and industrial and water system development advanced. Research by the Sanitary Chemistry Laboratory at M.I.T. into the chemical quality of Massachusetts' river water showed that pollution was becoming significant and needed to be reduced before wastewaters contaminated public water supplies. Because there was no viable system for treating sewage, in 1887 the Massachusetts State Board of Health established the Lawrence Experiment Station, the first of its kind, to do the necessary research. Hiram F. Mills, a hydraulic engineer from Lawrence, Massachusetts, who also was a member of the Board of Health, was chosen as the station's first director. Mills recognized that research on wastewater treatment required not only engineers, but also chemists and biologists. As a member of the M.I.T. Board of Trustees, Mills did not hesitate to draw upon M.I.T. to supplement the staff at the experiment station. Professor William T. Sedgewick was appointed as Consulting Biologist and

Dr. Thomas M. Brown, Professor of Chemistry was appointed Consulting Chemist to the State Board of Health (McCracken and Sebian 1988).

In the beginning, environmental research at the Lawrence Experiment Station relied heavily on British research on intermittent sand filtration. In Britain, Sir Edward Frankland had demonstrated that intermittent sand filtration was a viable method for treating municipal sewage. Picking up on his work, Allen Hazen, George Fuller, and Harry Clark, the staff of the Lawrence Experiment Station, demonstrated that wastewater treatment was a biochemical process in addition to a physical process. The station's results, published in two large volumes in 1890 (MSBH 1890), were carefully studied across the world and stimulated considerable research on municipal wastewater treatment. These reports, *Purification of Sewage and Intermittent Filtration of Water* and *Examination of Water Supplies and Inland Waters of Massachusetts*, contained over 1,500 pages because of Mills' insistence that all the data collected be shared with others (McCracken and Sebian, 1988). Using the Lawrence Experiment Station results, the British developed trickling filters. By 1900, biological wastewater treatment concepts had become well established.

The findings by the Lawrence Experiment Station were made possible by the newly-discovered discipline of bacteriology. The discipline rapidly advanced after Dr. Robert Koch, a German physician and self-educated bacteriologist, demonstrated that bacteria could cause diseases (Collard 1976). Koch became a university professor in Berlin and under his leadership, research in the fledgling field moved quickly from growing bacteria on potato slices to gelatin and agar during the 1880s. This development enabled Professor Sedgwick and Edwin O. Jordan, a recent graduate of Sedgwick's, to conduct research on sewage bacteriology.

Their efforts focused on isolating and identifying different bacteria in pure cultures. They discovered that bacteria isolated from intermittent sand filters used to treat municipal sewage were non-pathogenic. However, isolating pathogenic bacteria was difficult, even from contaminated wastewaters. Professor Sedgwick, Jordan, and Ellen Richards demonstrated that nitrification was a biological process that should be caused by bacteria, but they were unable to isolate the nitrifying bacteria in pure culture. Nitrate formation was the primary indicator for the stability of wastewaters after filtration. Allen Hazen, the first Chemist-in-Charge at the Lawrence Experiment Station, suggested using a liquid medium containing only inorganic materials rather than using gelatin. After several attempts, Jordan developed a nitrifying culture in an organic-free medium. At the same time, in Russia, Serge Winogradsky, a soil microbiologist, successfully applied the same approach. Winogradsky was able to isolate nitrifying bacteria on silica gel media.

Professor Sedgwick's studies also included algae and protozoa found in surface waters. Sedgwick developed a quantitative method for counting these large microorganisms that was modified by George W. Rafter. The *Sedgwick-Rafter Concentrator and Counting Cell* became standard equipment for evaluating algae and protozoa in water supplies. The work at M.I.T. and the Lawrence Experiment Station established Water Microbiology as a major area for environmental engineering.

The concept that bacteria cause disease was not readily accepted in 1890, despite its wide acceptance in the scientific community. Even the isolation of several disease-producing bacteria in pure cultures was not considered adequate proof. A typhoid fever epidemic in Lowell and Lawrence, Mass., in December 1890 and early 1891 provided the impetus for a concerted research effort.

By using engineering techniques and scientific logic, Sedgwick developed a form to collect data on a house-to-house basis, to determine the source of infections. Together with George V. McLauthlin, Sedgwick began collecting data from every house reporting cases of typhoid fever and those in the immediate vicinity. Each house with a reported death was marked on a map of Lowell along with the houses with reported typhoid cases. This mapping showed that some areas of the city were affected more severely than others. This data, when compared with a map of Lowell's five different water systems, clearly demonstrated that the areas that obtained drinking water from the Merrimack River were most affected by the typhoid fever epidemic. Applying their methodology in Lawrence, Sedgewick and McLauthlin found that people who used Merrimack River water were most affected. The documentation of these epidemics of typhoid fever, together with several other smaller epidemics transmitted by water, milk, and direct contact, provided irrefutable proof that typhoid fever could be spread by polluted water.

Formal engineering design evolved slowly as engineers learned the best design concepts, but it eventually became the backbone for environmental engineering. Early design efforts were focused on water distribution and sewers. After the Civil War, there was a decades-long debate between proponents of "separate" versus "combined" sewers. Some engineers favored building separate sewers for stormwater and sanitary wastewater. Others favored combining stormwater and wastewater into a single sewer. George E. Waring was among the proponents of separate sewers. Waring, a self-educated engineer, became interested in sewerage systems in the 1870s. He was retained by the city of Memphis, Tenn., as a consulting engineer after epidemics broke out there in 1878 and 1879. Waring recommended that the city construct separate sewers, using small-diameter pipes with automatic flush tanks (Odell 1880). After the Memphis project, Waring worked on the Buffalo, NY, trunk sewer and was also appointed to the National Board of Health. In 1895, he was named Commissioner of Sanitation in New York City. In three short years, Waring improved the city's solid waste collection and processing which had been one of the worst of the major cities.

James P. Kirkwood was one of the first American engineers to design a slow sand filter for water treatment based on data collected in Europe as part of a study of water treatment for St. Louis, Mo., published in 1869 (Baker 1948). Yet, Kirkwood's initial design of slow sand filters for St. Louis was ignored. It was not until 1871 that Poughkeepsie, NY, constructed a slow sand filter based on his design.

In response to the typhoid epidemics of 1890 and 1891, Hiram Mills designed an intermittent sand filter to treat Merrimack River water in Lawrence. The sand filter was placed into operation in 1893 and removed 98 percent of the bacteria from the polluted river water (MSBH 1894). Deaths from typhoid fever took a dramatic drop, clearly demonstrating the value of sand filtration for purifying polluted water. With slow sand filters proven, considerable effort was directed toward developing mechanical, rapid sand filters. However, the interest in patenting such devices slowed their development.

Rudolph Hering made perhaps the most significant early contributions to the development of engineering design for water supply and sewage treatment. Hering was born in Philadelphia and educated in Germany. He returned to the United States as an engineer, eventually becoming the Assistant City Engineer of Philadelphia in 1873. In 1880, Hering was commissioned by the National Board of Health to go to Europe and study the latest methods for sewage treatment. He presented a report on his findings to the American Society of Civil Engineers in 1881 (Hering 1881). In 1889, he was appointed by President Harrison to make plans for sewerage and drainage in Washington, D.C. Over the years, Hering prepared water supply and treatment studies for 150 cities. It is not surprising that he became known as the "Dean of Sanitary Engineering" (ASCE 1972).

While working at the Lawrence Experiment Station, Allen Hazen had examined chemical precipitation and the sedimentation processes. Because of this expertise, Hazen was invited to Chicago in 1893 to operate the wastewater treatment plant constructed for the Columbian Exposition, a showcase for modern technology. A sewerage system connected each of the exposition's major buildings to the wastewater treatment plant enclosed in a separate building. Because of the flat terrain, each building was equipped with an ejector to lift the sewage into the sewers and convey it to the treatment plant. Unfortunately, the multitude of ejectors and their random operation created a number of problems such as periodic sewer ruptures caused by system pressures exceeding pipe capacity and peak flows disrupting the settling characteristics in the 30 foot high Dortmund settling tanks. Coping with the Columbian Exposition wastewater treatment plant operations showed Hazen that there was a major difference between operating pilot plants at the Lawrence Experiment Station and actual full-scale plants. When the Columbian Exposition closed, Hazen returned to Boston to become a consulting civil engineer, eventually locating his practice in New York City. His paper, "On Sedimentation", was presented before the American Society of Civil Engineers in 1904 and became one of the classic papers on sedimentation theory and design (Hazen 1904). In 1907, he summarized water treatment plant design in his book, *Clean Water and How to Get It* (Hazen 1907). Hazen became one of the most successful and respected water treatment plant design engineers in the early 1900s.

As Hazen had demonstrated, design engineering grew out of need and experience rather than from theory. This was true for both sewage treatment plants and water treatment plants, although there was more interest in water treatment plants. Most design engineers depended on information published in the latest engineering magazines. In the United States, engineers quickly adopted trickling filter designs and activated sludge technology from the British and Imhoff tanks from the Germans. As each new plant was constructed, design engineers learned what worked and what did not. Leonard Metcalf and Harrison P. Eddy, consulting engineers in Boston, brought the best of American wastewater technology together for all engineers in 1915 with the publication of *American Sewerage Practice* (Metcalf and Eddy 1915). They set a standard for professionalism while demonstrating their knowledge for future customers. They also provided texts for teaching future generations of design engineers.

A discipline is born when the development of knowledge and its application evolves from individual experimentation into a formal course of study. In 1889, M.I.T. established the first program in Sanitary Engineering (Wylie 1975). It was designated Course XI and incorporated courses in sanitary chemistry and sanitary biology into the Civil Engineering Department. The new department was named Civil and Sanitary Engineering and degrees were offered at the undergraduate level in Civil Engineering and in Sanitary Engineering. By 1893, when engineering faculty from across the country gathered at the Columbian Exposition in Chicago to organize a professional society to represent engineering educators, the only other school that had a degree program in Sanitary Engineering was the University of Illinois, which offered a degree in Municipal and Sanitary Engineering. A survey of engineering education in 1899 by Ira O. Baker, President of the Society for the Promotion of Engineering Education (SPEE) (Baker 1900), indicated that there were 110 engineering schools, but only 89 were active and only two schools offered degrees in Sanitary Engineering. There were 9,679 students enrolled in engineering, with 19 in Sanitary Engineering. Of the 1,413 engineering degrees awarded in 1898-1899, only one was in Sanitary Engineering (McKinney 1994).

1.1.3 The 20th Century Before World War II Environmental engineers entered the 20th century with hope and aspiration of the opportunities that lay ahead. M.I.T.'s Sedgwick was confident that Sanitary Engineers had a special place in the future of technology, even though most Course XI graduates at M.I.T. were still concentrating on hydraulic engineering. Environmental engineering was water-oriented. The need for safe water supplies for a dynamic, growing nation occupied many. Additionally, with connection between safe drinking water and polluted waters firmly established, substantial effort was focused on the abatement of water pollution. Interestingly, even air pollution, a post World War II focus, attracted some interest.

While these various facets of environmental engineering are interconnected, technology development and applications were pursued separately. Any interchanges between the specialties existed primarily at universities teaching sanitary engineering, in the Public Health Service, and associated state departments of health.

In 1911, George C. Whipple, who had been a biologist at the Boston Water Works for eight years after graduating and a member of Hazen, Whipple and Fuller, water and wastewater consultants in New York City since 1903, was appointed Professor of Sanitary Engineering at Harvard University. After his appointment, he joined forces with his professor Sedgwick at M.I.T and in 1913 the M.I.T.-Harvard School of Public Health was established with Sedgwick as the Program Head and Whipple as Secretary. This association enabled the Harvard Sanitary Engineering program to maintain a focus on Public Health Engineering in addition to customary civil engineering.

The Harvard Sanitary Engineering faculty was joined in 1918 by one of its recent graduates, an immigrant from South Africa via Germany, Gordon Maskew Fair. This inauspicious beginning launched a career that would have profound impact on the profession. A survey taken in 1949 showed that about half of all American doctorate degrees in Sanitary Engineering up to that time had been earned at Harvard and over half of the State Sanitary Engineers had received advanced degrees under Fair's direction (Anderson, 1986).

When Fair began work, the technology of most processes and practices for environmental control were characterized by a high degree of empiricism. The remedies prescribed for freeing air and water of pollutants were derived from workaday experience rather than from scientific observation and analysis. His research in environmental science was motivated by the belief that a set of theoretical principles governed a wide range of artificial and natural purification processes - and that these could be specified in mathematical language so that engineers could use an orderly process of calculation in designing treatment works for water and air. His success in developing the theory of purification kinetics was embodied in widely read books and papers including his well-known textbook, *Water Supply and Waste Water Disposal* (with J.C. Geyer). He provided additional contributions in limnology, the broad application of the principles of physical chemistry to complex processes of water purification, to specific problems of quantitive measurement of tastes and odors, and mechanisms of biodegradation of certain organic compounds in streams. Perhaps his greatest achievements stemmed from his ability as a theorist to deploy the methods of science and the techniques of mathematical analysis, a precursor of today's emphasis of mathematical modeling and the key to computerization, to a discipline that had evolved for many centuries as a practical art.

Fair's grasp of environmental engineering was visionary and prophetic. Years ago he understood that environmental control is a multi-media challenge.

The impact of sewage on water quality and the need for safe drinking water identified in Massachusetts and other major urban centers in the U.S. gave impetus for the creation of a sanitary engineering component within the US Public Health Service (PHS) in 1890. At the start of the new century, this predecessor of EPA came to be a dominating force. In 1901, Congress authorized the construction of a PHS

Hygienic Laboratory, "for the investigation of infectious and contagious diseases." This was followed by commissions in 1908-1909 to study water pollution and protect water supplies in Lake Michigan and Lake Erie. A major re-organization in 1912 gave the PHS a broad mandate to study the diseases of man and conditions influencing their propagation.

In 1913, a group of medical officers, engineers, and scientists took over the laboratories at an abandoned Marine Hospital in Cincinnati, Ohio, with a mission to control water pollution. This center produced much of the fundamental research on which the control of water pollution is founded including:

- definition of the Oxygen Sag Equation by Streeter and Phelps;
- confirmation of the rate of oxygenation of polluted waters by Theriault;
- confirmation of the rate of atmospheric reaeration by Streeter;
- definition of the elements of bacterial pollution by Hoskins;
- development of major elements of stream biology by Purdy; and
- initiation of studies on industrial wastes (Dworsky, 1990).

For nearly thirty years, between 1913-1938, many of the Cincinnati group, augmented by a second, but still small, wave of engineers and scientists, structured and implemented plans which carried the nation rationally toward its goal of water pollution control. Some of these include:

- a strategic selection of rivers to understand the properties of their differences (1914);
- the initiation of a comprehensive survey of stream pollution (1915);
- support for the growth and improved capacity of the states to participate in efforts to control stream pollution (1920);
- public education efforts focusing on the importance of water pollution control measures as an aspect of comprehensive water resources development (1936); and
- increased technical assistance to states through the creation of a separate PHS Office of Stream Sanitation (1932) (Dworsky, 1990).

Their efforts culminated in Public Law 845, enacted by the 80th Congress on June 30, 1948, the Nation's first comprehensive Water Pollution Control Act.

While educators and government officials labored to understand the science of water pollution control in the first half of the century, wastewater was being treated in urban areas of all sizes. Technology developments continued to flow back and forth across the Atlantic. The studies at Lawrence Experiment Station fueled the creation of the trickling filter in England and the Imhoff Tank in Germany. These developments were applied in many U.S. communities; Chicago used Imhoff tanks to treat over 500 million gallons per day by the mid 1920s. In 1914, the activated sludge method of wastewater treatment, today's most commonly-used wastewater treatment technology, was developed in Manchester, England. Shortly thereafter, in 1918, Houston, Texas, placed the first large-scale activated sludge plant in the U.S. in operation.

The work at the Lawrence Experiment Station fostered not only wastewater treatment, but water treatment as well. The station's research proved the efficacy and value of sand filtration to protecting human health. First with slow sand filters and then into the 20th century with rapid sand filters, water filtration became the preferred treatment technology for all but those with a pure, upland, source of supply.

The Interstate Quarantine Act in 1893 laid the foundation for the control of communicable disease and the regulation of its carriers, such as water. A regulation pursuant to this act compelled trains and other interstate carriers to use waters of known quality which had been certified by the local health authority. The nation's first drinking water standards were adopted in 1914 as an aid to the enforcement of the Quarantine Act. These were implemented by state agencies with support from the PHS.

Arguably, the most significant development in water treatment in the first half of the 20th century occurred in 1923. That was the year Abel Wolman, who had recently been appointed Chief Engineer of the Maryland Department of Health, developed and perfected techniques for controlled chlorination of water. These techniques made possible the prescription of chlorine feed rates for water leaving the treatment plant sufficient to provide for effective and reliable disinfection of the water supplied. With filtration and/or reliable disinfection, the major public health threat of water-borne disease ceased to exist in the U.S. by World War II.

Clearly, the first half of the 1900s was occupied with water-related issues. However, air quality concerns began to receive attention shortly after the turn of the century. The major atmospheric concern stemmed from the presence of smoke. Up to the late 1940's, most American urban centers had smoke abatement agencies. The transfer of interest from solely smoke to more comprehensive air quality issues came about as a result of developments in the field of Industrial Hygiene. Early atmospheric pollution studies were an extension of the science of industrial hygiene and included:

- a PHS and US Bureau of Mines study of silicosis (1910).
- PHS organizes a Division of Industrial Hygiene (1912).

- studies of air in industrial workshops were well underway, e.g., carbon monoxide from the use of gas- heated equipment (1916); and

- industrial hygiene activities of the PHS included studies of municipal dust, the radium dial painting industry, and a comparative study of air pollution in fourteen of the largest cities (1931) (Dworsky 1990).

Not withstanding the foregoing, the seminal event in the specialty of air pollution control was triggered by an air pollution episode in Donora, Pennsylvania, in 1944; twenty persons died and 5,910 became ill. It triggered the first major comprehensive study of air pollution by the PHS. The resulting Surgeon General's Report noted:

- "This study is the opening move in what may develop into a major field of operation in improving the nation's health.

- We have realized, during our growing impatience with the annoyance of smoke, that pollution from gases, fumes and microscopic particles was also a factor to be reckoned with.

- The Donora report has completely confirmed two beliefs we held at the outset of the investigation:

    1. how little fundamental knowledge exists regarding the effects of atmospheric pollution on health, and

    2. how long range and complex is the job of overcoming air pollution."

1.1.4 Post World War II The industrial explosion that accompanied World War II had two significant impacts in environmental engineering. On the one hand, the richness and diversity of technological exploitation resulted in a plethora of chemical and other industrial wastes discharged to the water and air and deposited on the land. At the time, these were of little concern to a public tired from war and anxious to enjoy the fruits of peace. They systematically ignored the concerns and warnings from environmental engineers of the day. But the day-by-day, year in and year out callous disregard of wasteful production sowed the seeds for an environmental revolution triggered by Rachel Carlson's 1964 book, *Silent Spring* and culminated with Earth Day 1970. That event unleashed a torrent of regulations which has dominated the profession ever since.

The immediate post World War II period was also marked by a significant increase in environmental engineering research, made possible by significant federal grants to universities. This established a pattern that continues to this day that has significantly increased the knowledge of the science underlying environmental engineering. It also produced increasing numbers of environmental engineers with masters and doctorate degrees and spawned a new class of environmental engineer — the academic-researcher — which replaced, for the most part, the academic-practitioner that was the norm before the War.

As the next millennium approaches, environmental engineers practice in a world of increasing complexity as they attempt to work in harmony with a most complex system - nature - and a rapidly expanding knowledge base made possible by the "information age". Yet, it is meaningful to reflect upon the profession's history and to take note that what environmental engineers do today and tomorrow is grounded on fundamental principles developed and perfected 50, 100, and 150 years ago. These principles remain those which today's environmental engineers *must* know to be deemed minimally competent and, therefore, entitled to a license to practice the profession.

1.2 Engineering Credentials

1.2.1 Definitions There are 4 terms that need definition to provide the foundation for discussing credentialling of engineers - licensing, registration, certification, and accreditation.

> *Licensing* — A license is authorization granted by a government to an individual or entity to perform a function or service, e.g., driver's license, engineer's license, etc. The root of licensing lies in the police powers embodied in governments to protect the health, safety, and welfare of its constituents.
>
> *Registration* — Registration is listing with and by some body. It can be a governmental or a non-governmental entity that does the registration. Properly applied, registration is the listing of an individual, e.g., registering with Selective Service. It grants no authority nor does it address qualifications.
>
> *Certification* — A voluntary act which, in some organized fashion, measures an individual's qualifications to perform a specialized function. Because it is voluntary, it conveys no authority or privilege, i.e., one does not need to possess the certificate to perform a function or service, albeit custom or market forces may require it. Certification exists in many professions and trades.
>
> *Accreditation* — It is like certification, except that it applies to institutions and programs, not individuals. A familiar example is the accreditation of college curricula.

Confusion in credentialling is rampant because these four words are seldom correctly applied in society. Engineers are familiar with the word "registration" which is often used when the correct term that should be used is "license." Yet, engineers have no lock on the misapplication of these four words. Throughout American society the misapplication of these terms abound. And, their misapplication has been institutionalized by encoding them in laws and regulations. Many states continue to call their engineering licensing programs *engineering registration*, a misnomer. In 1995, National Council of Examiners for Engineering and Surveying

(NCEES) adopted a policy requesting all licensing boards to use the term "licensing" instead of "registration."

1.2.2 Licenses The first law governing the practice of engineering in the United States was passed by Wyoming in 1907. This law was created because many non-professionals were practicing engineering and surveying. By 1920 there were 13 states that required engineers and surveyors to be licensed to practice. That same year, Iowa called a meeting and representatives from seven of the thirteen states met in Chicago, Illinois. They founded what is today the National Council of Examiners for Engineering and Surveying (NCEES). It currently consists of licensing boards from all fifty states and five jurisdictions (District of Columbia, Puerto Rico, Virgin Islands, Guam, and Northern Mariana Islands) that license engineers.

NCEES provides an organization through which state boards act and counsel together to better discharge their responsibilities in regulating the practice of engineering and land surveying to protect the welfare of the public by safeguarding life, health, and property. During the early years of the Council, its activities focused on the education and experience requirements for licensing as well as the development of a model law which could be used by states as they enacted legislation requiring licensing and/or modifying existing legislation.

The requirements for licensing in all states and jurisdictions include education, experience, and examination. The NCEES started developing uniform examinations with the offering of the Fundamentals of Engineering (FE) examination for the first time in 1965. The Principles and Practice (P.E.) examinations were first offered in 1966. Since then, use of the uniform national examinations by the Member Boards has steadily increased. By 1984, all 55 states and jurisdictions were using these examinations. Further, uniformity is achieved by all states using the same passing point developed and recommended by the Council.

The FE examination tests those basic engineering science knowledges which are normally acquired through a bachelors engineering degree. Successful completion of the FE examination is a preresquite to take the P.E. examination. For licensing as a professional engineer, most states require four years experience and passing of the P.E. examination. Currently the P.E. examination is offered in sixteen disciplines: chemical, civil, electrical, mechanical, and special structural examinations offered twice a year in April and October; and aeronautical/aerospace, agricultural, control systems, fire protection, industrial, manufacturing, metallurgical, mining/mineral, nuclear, and petroleum offered once a year in October.

1.2.3 Specialty Certification Specialty certification is certification of a particular specialty of a discipline. Generically, every profession becomes more and more specialized as its technology advances and the understanding of the forces at play increases. Specialization requires specialists. At some point in time either a general license or certification is no longer sufficient. Specialty licensing and specialty certification is a mechanism used to identify those with special capabilities, be it the large truck driver as opposed to the auto driver or the neurosurgeon as opposed to the general practitioner.

Specialty certification in the learned professions was introduced in medicine by ophthalmologists in 1917. Between then and 1971, 23 medical specialty boards were created that grant certificates in 83 specialties. Quoting from the presidential address of Dr. Derrick T. Vail as he took office in the American Academy of Ophthalmology and Otolaryngology in 1908:

> "I hope to see the time when ophthalmology will be taught in this country as it should be taught. That day will come when we as oculists, demand that certain amount of preliminary education and training be enforced before a man may be licensed to practice ophthalmology. It should no longer be possible for a man to be called an oculist, by himself or by the laity, after he has spent a month or six weeks in some postgraduate school or after serving as an assistant for six months or a year in some oculist's office. After a sufficiently long time in an ophthalmic institution in America or abroad, he should be permitted to appear before a proper examining board for examination, similar to any State Board for Examination and Registration, and if he is found competent let him then be permitted and licensed to practice ophthalmology."

That statement encompasses the ideals of a proper specialty certification program. To wit, formalized training in the specialty, experience in the specialty, and an examination to prove that the training and experience have taken root.

The environmental engineering profession pioneered specialty certification in engineering in 1955. This occured for three reasons:

- Environmental engineers had a long history of working with the medical profession. They saw first hand the benefits specialty certification brought to the medical field.

- Due to the public nature of their work, most environmental engineers in 1955 were licensed.

- At the time specialty certification began, environmental engineers were a relatively small cohesive group of professionals who shared similar experiences and institutions. Environmental engineering was a far different profession than today in its composition.

The Academy's program embraces the same principles found in the medical specialties — formal training at the college level, a state license to practice engineering, a prescribed amount of practical experience, a review of qualifications by peers, and examinations; in its program both written and oral examinations are used.

1.2.4 Relationship of Licensing and Specialty Certification The examination for licensure as a professional engineer is designed to differentiate between the person who has the minimum qualifications (minimum competency) to practice without harm to the public and those who do not. Accordingly, the environmental licensing examination covers the broad spectrum of environmental practice. Those individuals that wish to document their competence in a particular specialty of environmental engineering can seek certification from the American Academy of Environmental Engineers. That process provides an indepth evaluation for each of seven specialty areas of a person's ability at a level above the minimally competent.

1.3 Principles and Practice (P.E.) Examination

1.3.1 Development The specifications for all of the P.E. examinations are based upon a Professional Activities and Requirements (PAR) Analysis. This analysis, done by discipline, queries individuals practicing at the entry level, usually individuals who have been licensed less than six years, as to what professional activities they perform. The most recent PAR survey was conducted by the NCEES between 1987 and 1989 with the final report produced in 1989.

The last step in the PAR Analysis process is to convene a panel of discipline subject matter experts who review the survey data and develop the specifications for the examinations to include not only the professional activities, but the professional requirements as well. Professional requirements are those knowledges, skills, and abilities which are necessary in order to perform a specific professional activity. The examinations are based upon these specifications and are knowledge driven. Each of the problems on the examination are developed to test certain of the knowledges.

1.3.2 Scoring The P.E. examination is scored using a criterion referenced system. The criteria are established to define what knowledges must be demonstrated by the minimally competent candidate at the entry level. One of the most critical considerations in developing and administering examinations concerns the establishment of passing scores which reflect a standard of minimum competency. Minimum competency, as measured by the examination component of the licensing process, is defined as:

> *The lowest level of knowledge at which a person can practice professional engineering in such a manner that will safeguard life, health and property and promote the public welfare.*

The concept of minimal competency is the prime consideration in selecting the problems used on the examinations. Independent audits of this process are conducted periodically in the form of passing score workshops. At these workshops, a panel of engineers who have not been involved with the preparation of the examinations review a series of actual candidates' solutions and decide whether they believe that the candidates have demonstrated minimum competence and should be licensed. These decisions are compared to the actual assigned scores using a mathematical model developed to assess both the consistency of individual judgments and the panelists' opinion of the score needed to demonstrate competence.

Each problem is developed so that a candidate who is minimally qualified in that subject matter should score six points. In the case of essay problems, where partial credit is given, a scoring plan is developed which defines each of the score levels in problem-specific terms. Only scores of 0, 2, 4, 6, 8, or 10 points are awarded. The scoring plan defines the amount of knowledge shown or the errors which will result in each of these points. Multiple-choice problems are composed of ten one-point questions. There is no penalty for incorrect responses, but no credit is given where two or more responses are marked. Particular emphasis is placed on defining the characteristics of a response that will receive six points.

The complete eight-hour examination is composed of four essay and four multiple choice problems and has a maximum score of 80 points. The examination is compensatory - poor scores on some problems can be offset by superior performance elsewhere. To pass, NCEES has determined that a candidate must score at least 48 points or an average of six points per problem.

1.3.3 Discipline Specific Examinations In the early years, several states developed the practice of allowing individuals to work problems in any of the disciplines examined. This became known as a combined examination. Other states, however, required candidates to take discipline-specific examinations. This condition created problems in comity licensing. At its 1990 Annual Meeting, the NCEES decided that all individuals must take discipline-specific examinations beginning January 1, 1994.

It must be noted that discipline-specific examination component of the licensing process in no way dictates to the Member Boards whether they should license by discipline or issue general licenses as a professional engineer. The two issues are separate. Currently only eleven states license by discipline, the remainder issue general licenses with the expectation that individuals will practice in their area of expertise.

1.3.4 Examination Format To improve examination accuracy and reliability, the environmental engineering examination has only eight test items. This format requires all candidates to work the same problems, a practice similar to the Fundamentals of Engineering examination. As a result, statistical methods can be employed to assess test performance and adequacy.

The eight problems are comprised of four essay (free response) problems in the four-hour morning session and four multiple-choice problems, each containing ten items, in the four-hour afternoon session.

1.4 Environmental Engineering Examination

1.4.1 Background   The 1989 PAR Analysis indicated that most environmental engineers had bachelor degrees in civil engineering and spent most of their time practicing in the water, wastewater, solid and hazardous waste areas. Additionally, individuals with bachelor degrees in mechanical and chemical engineering were also found to be practicing environmental engineering. While the specifications for the civil, mechanical, and chemical examinations currently contain environmental engineering problems, engineers and Member Boards determined a stand-alone environmental engineering examination was needed. That examination was offered for the first time in October of 1993.

1.4.2 Examination Development   The 1989 PAR Analysis was completed by 500 individuals who were civil, mechanical, and chemical engineers and working in the environmental area. The analysis by that group of individuals was reviewed in 1992 by a panel of fourteen experts provided by American Academy of Environmental Engineers. These individuals were selected so as to represent all of the seven specialty practice areas of environmental engineering recognized by the Academy. Additionally, these panel members were selected to be representative of geographical regions across the United States and the distribution of practicing engineers as to gender, ethnicity, age, and varying academic degrees. After reviewing the professional activities, the panel determined the knowledges, skills, and abilities necessary to perform those professional activities.

1.4.3 Scope of The Examination   The areas of practice to be covered by the environmental engineering examination are as follows:

1. *Planning, Research, Development & Design of Water Systems* including water supply, drainage and flood control, wetlands, and collection/distribution.

2. *Project Implementation, Operations & Monitoring of Water Systems* including water supply, drainage and flood control, wetlands, and collection/distribution.

3. *Planning, Research, Development & Design of Wastewater Systems* including domestic and process (industrial/ commercial) wastewaters.

4. *Project implementation, Operations & Monitoring of Wastewater Systems* including domestic and process (industrial/commercial) wastewaters.

5. *Planning, Research, Development & Design of Solid Waste & Hazardous Materials Systems* including domestic and process (industrial/commercial) sources.

6. *Project Implementation, Operations & Monitoring of Air Quality Systems* including stationary, mobile and indoor air sources.

7. *Planning, Research, Development & Design for Health, Safety & Environmental Protection* including emergency response, risk analysis, radiation protection, noise, toxicology and industrial hygiene.

8. *Project Implementation, Operations & Monitoring for Health, Safety & Environmental Protection* including emergency response, risk analysis, radiation protection, noise, toxicology and industrial hygiene.

The eight problems included in the examination, one for each of the practice areas listed above require a variety of approaches and methodologies including design, analysis, application, and operations. Problems may include economic aspects.

1.5 References

American Society of Civil Engineers, Committee on History and Heritage. 1972. *A Biographical Dictionary of American Civil Engineers*, ASCE Historical Publication No. 2.

Anderson, W.C. 1986. "Dr. Gordon Maskew Fair - The Master", *The Diplomate* vol. 22 no. 3, pp 6-8.

Baker, I.O. 1900. "Engineering Education in the United States at the End of the Century," *Proc. 8th Annual Meeting of Soc. for Prom. of Engr. Educ.*, vol. 8, pp. 11-27.

Baker, M.N. 1892. *Third Annual Issue of the Manual of American Water Works*, Engr. News Publishing Co., New York, NY.

Baker, M.N. 1948. *The Quest for Pure Water*, Amer. Water Works Assoc.

Bradlee, N.J. 1868. *History of the Introduction of Pure Water into the City of Boston*, Alfred Mudge & Sons, Boston, Mass.

Cain, L.P. 1991. "Raising and Watering a City: Ellis Sylvester Cheseborough and Chicago's First Sanitation System," *The Engineer in America*, pp. 69-88, Univ. of Chicago Press, Chicago, Ill.

Collard, P. 1976. *The Development of Microbiology*, Cambridge Univ. Press, Cambridge, Mass.

Dworsky L. B. 1990. "The United States Public Health Sciences", *The Diplomate*, vol. 26, no. 4, pp 7-12, 31.

Hazen, Allen. 1904. "On Sedimentation," *Trans. Amer. Soc. Civil Engrs.*, vol. 53.

_____. 1907. *Clean Water and How to Get It*, John Wiley & Sons, New York, N.Y.

Hering, Rudolph. 1881. "Sewerage Systems," *Trans. Amer. Soc. Civil Engrs.*, vol. 10.

Hunt, C.L. 1958. *The Life of Ellen H. Richards, 1842-1911*, Amer. Home Economics Assoc., Washington, D.C.

Jervis, J.B. 1876. "A Memoir of American Engineering," *Trans. Amer. Soc. Civil Engrs.*, vol. 5, pp. 39-67.

Koeppel, Gerard. 1994. "A Struggle for Water," *Invention & Technology*, vol. 9, no. 3, pp. 18-31.

Massachusetts State Board of Health. 1871. *Second Annual Report of the State Board of Health.*

_____. 1874. *Fifth Annual Report of the State Board of Health.*

_____. 1890. *Twenty-First Annual Report of the State Board of Health.*

_____. 1893. *Twenty-Fourth Annual Report of the State Board of Health.*

_____. 1894. *Twenty-Fifth Annual Report of the State Board of Health.*

_____. 1903. *Twenty-Fourth Annual Report of the State Board of Health.*

McCracken, R.A. and D. Sebian. 1988. "Lawrence Experiment Station," *The Diplomate*, vol. 11, no. 2, pp 12-18.

McKinney, A. 1994. "A Stroll Through Time in Environmental Engineering", *Environmental Engineer*, vol. 31, no. 1, pp 12-13, 25, 33.

Metcalf, L. and H.P. Eddy. 1915. *American Sewerage Practice*, McGraw-Hill, New York, N.Y.

Odell, F.S. 1881. "The Sewerage of Memphis," *Trans. Amer. Soc. Civil Engrs.*, vol. 10, pp. 23-52.

Prescott, S.C. 1954. *When M.I.T. Was "Boston Tech," 1861-1916*, The Technology Press, Cambridge, Mass.

Schrenk, H.H., H. Heimann, G.D. Clayton, W.M. Gafafer, and H. Wexler. 1949. *Air Pollution in Donora, Pa.*, Public Health Bull. No. 306, U.S. Public Health Service, Washington, D.C.

Wood, DeVolson, I.O. Baker, and J.B. Ju-nson. 1984. *Engineering Education, Proc. Section E, World Engr. Congress, Chicago, Ill.*, vol. 1, Soc. Prom. Engr. Educ.

Wylie, F.E. 1975. *M.I.T. in Perspective*, Little, Brown & Co., Boston, Mass.

Chapter 2

COMMON PRINCIPLES

by W. Christopher King, P.E., DEE

2.1 Introduction

A core of engineering and science knowledge, common to all environmental engineering disciplines, is the essential foundation for the profession. Chemistry, fluids, hydrology, and hydraulics are discussed in depth in subsequent chapters. This chapter focuses on mathematics, conversion of units of measurements, and engineering economics.

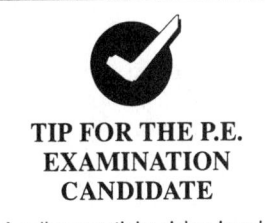

TIP FOR THE P.E. EXAMINATION CANDIDATE

A college math book is a handy reference to take to the examination.

2.2 Math Basics

Several math theorems are presented that routinely appear in solving simple environmental engineering problems. They are reviewed in brief; details can be found in any algebra text.

Exponentials

$a^x a^y = a^{x+y}$ [2-4]

$(a^x)^y = a^{xy}$ [2-5]

$a^x/a^y = a^{x-y}$ [2-6]

Logarithms

$\ln_e(x) = 2.303 \log_{10}(x)$ [2-7]

$\log(MN) = \log M + \log N$ [2-8]

$\log(M/N) = \log M - \log N$ [2-9]

$\log M^p = p(\log M)$ [2-10]

2.3 Units of Environmental Engineering

A wide variety of units are used in environmental engineering practice. Generally, there are common units of expressions in each specialty, e.g., water supply, solid waste management, etc., which are often different, even if applied to the same operation, media, or process unit. Also different professions involved, e.g., chemists, engineers, operators, etc. may use different units to quantify the same phenomena. This lack of standardization in units has been compounded by regulatory standards which often contain two different units to express the same standard, e.g., wastewater effluent standards are expressed in concentration (mg/L) and mass (lb/day) and retain a historical foundation, e.g., in air pollution grains are still used in some standards (7,000 grains = 1 lb).

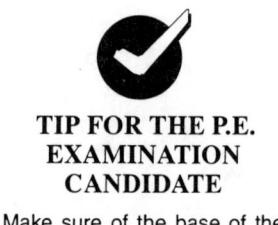

TIP FOR THE P.E. EXAMINATION CANDIDATE

Make sure of the base of the logrithm used, "e" or "10".

Therefore, the competent environmental engineer must first clearly understand the units applicable to the data being supplied and/or collected on which engineering calculations will be performed or judgments rendered. Secondly, the units of the data supplied must be converted to those applicable to standard formula and/or reduced to a common basis. The appendix provides an extensive list of units commonly used in environmental engineering specialties and the conversion of each to units employed in engineering calculations.

Units have been compiled in the form of equalities and organized according to dimensional fundamentals. This organization makes application of the appendix to problem-solving straightforward. More attention has been given to conversions within the English system of units than routinely seen in standard texts. The Example 2-1 illustrates use of the appendix. Take particular note of Example 2-1 part b. which shows how to convert two different units to achieve the same result.

Example 2-1 — Applying the units in the appendix to problems of pressure conversions

a. Given a pressure of 5 atm, convert to ft of water

$$5 \text{ atm} \bullet \frac{33.9 \text{ ft H}_2\text{O}}{1 \text{ atm}} = 169.5 \text{ ft H}_2\text{O}$$

b. Given a pressure of 169.5 ft H_2O, convert to pounds per square inch, psi.

$$169.5 \text{ ft H}_2\text{O} \bullet \frac{14.696 \text{ psi}}{33.9 \text{ ft H}_2\text{O}} = 73.48 \text{ psi}$$

$$\text{Check: } 5 \text{ atm} \bullet \frac{14.696 \text{ psi}}{1 \text{ atm}} = 73.48 \text{ psi}$$

Concentrations of constituents are indigenous to all aspects of environmental engineering and are expressed as parts per million (ppm), milligrams per liter (mg/L), and grams per cubic meter (g/m³). PPM is very familiar, but many times the assumptions that apply are ignored. In *solutions*, ppm means the parts by weight of the reported material to one million weight units of the solution. For dilute solutions, ppm is approximately equal to mg/L with up to several thousand mg/L of dissolved material, only small errors are introduced with this assumption. In *air,* ppm generally refers to parts of material by volume to one million volumes of the gas. Obviously, ppm in air must be converted by the molecular weight of the material and the gas to reach units of weight/volume, mg/m³, for example.

Example 2-2 — Dimensional analysis

Sulfur dioxide has been measured as 35 ppm in a stack gas sample. Determine the mg/m³ concentration of SO_2.

$$\text{ppm} = \frac{V_s}{V_t} \qquad [2\text{-}1]$$

where: $V_s = n\,RT/P$ = volume of pollutant
$V_t = n_t\,RT/P$ = total volume
n_s = moles of material
    = weight (M_s)/molecular weight (MW_s)
n_t = moles total (assumed to be moles of gas because n_t is much greater than n_s)
R = universal gas constant = 0.08206 l-atm/gmol-°K.

or

$$\frac{M_s}{V_t} = \frac{(PV_s MW_s)}{(RTV_t)} \qquad [2\text{-}2]$$

With several steps, the relationship for P = 1 atm and T = 25°C can be developed

$$\text{Conc}\left(\frac{mg}{m^3}\right) = 4.09 \bullet 10^{-2}(\text{Conc}_{\text{ppm by volume}})MW_s \qquad [2\text{-}3]$$

Solution:

$$C_{mass} = 4.09 \bullet 10^{-2}(35)64 = 9.16\,\frac{mg}{m^3}$$

91.6

Dimensional analysis is the technique used to ensure that the units of the supplied data are manipulated through engineering formula and data conversions to produce the desired result. It entails conducting the appropriate algebraic operations on the dimensions alone. Proficiency in such analyses is essential for the competent environmental engineer.

A final complication is the mixed use of English units and Standard International (SI) or metric units. While there has been substantial pressure from government, academia, and other institutions to use SI units for more than 20 years, the profession has generally resisted this pressure. Ostensibly because of its extended his-

Chapter 2: Common Principles

TIP FOR THE P.E. EXAMINATION CANDIDATE

THERE IS NO BETTER WAY TO SUCCEED ON THE PROFESSIONAL ENGINEER'S EXAM THAN TO CHECK UNITS IN YOUR SOLUTIONS. The negative of this statement is that lack of attention to units will result in a wrong answer very quickly. As a hint, when working a problem, examine the units of the answer. If you are unsure of how to approach the problem, plot a path backwards from the answer's units, employing the provided data until all the units cancel, except those desired for the answer. This approach will also help locate simple errors that would otherwise be obscured deep in the solution.

tory, the profession continues to regularly communicate in a mix of English units and SI units.

2.4 Mass Balance

The mass balance is the foundation of many environmental engineering problem solutions. A competent environmental engineer is capable of applying this technique to many problems including reactor design, settler design, and any situation where there is a device that collects and treats a flow, be it water, air, or a solid. There are several ways to express the mass balance, but the simplest form is:

$$\text{Inputs} - \text{Outputs} \pm \text{Changes} = \text{Accumulation} \qquad [2\text{-}11]$$

Its application is illustrated in the following example.

Example 2-3 — Mass balance

Write the equation to describe the steady-state solids balance for a secondary clarifier of an activated sludge wastewater treatment plant (see diagram for variables). The recycle flow rate is $R = Q_r/Q_i$.

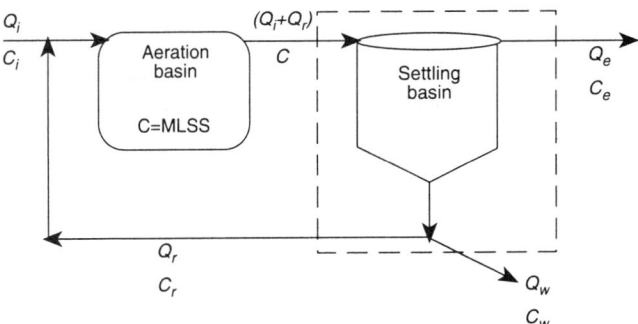

Solution: The dotted line in the above figure indicates the boundaries for the system to be modeled; all the flows entering are the *inputs* and the flows leaving are the *outputs*. *Changes* account for the reactions that occur inside the system. In the example, suspended solids are generally conserved inside a clarifier; i.e., there is no net growth or decay, and therefore it can be assumed that changes are zero. (If the example used a biological reactor instead of a clarifier, changes would be represented by the first order reaction rate equation, BOD_i (e^{-kt})). The change term can be developed from a fundamental understanding of the process being modeled. The last term, *accumulation*, accounts for the net gain or loss inside the system. In many environmental engineering applications, steady-state conditions exist or are assumed which means that there is no change in material inside the system, or accumulation is zero. From this discussion, our example can be solved as follows:

TIP FOR THE P.E. EXAMINATION CANDIDATE

There is no standardization of units on the examination or even in the same problem. In fact, there is a conscious decision to use the predominate units employed in practice for the subject media.

Inputs - Outputs ± Changes = Accumulation   [2-11]

$(Q_i + Q_r) C - Q_r C_r - Q_w C_w - Q_e C_e - 0 = 0$

or

$(Q_i + Q_r) C = Q_r C_r + Q_w C_w + Q_e C_e$

Note: $Q_r = R Q_i$, and from the diagram $C_w = C_r$

therefore:

$(1.0 + R) Q_i C = C_r(Q_r + Q_w) + Q_e C_e$

A final assumption often made recognizes that $C_r >> C_e$ which simplifies the equation by making $C_e = 0$

$(1.0 + R) Q_i C = C_r(Q_r + Q_w)$ [2-13]

A special mass balance case exists when a flow is introduced into a well-stirred tank. In this instance, the system is best modeled by assuming that the inflow is instantaneously and completely mixed with the contents in the tank. A complete-mix reactor can be described by the equation

$$\frac{dC}{dt}V = Q_i Ci - Q_i C + r_c V$$ [2-14]

where: C = concentration in the tank
r_c = reaction (change) rate
V = volume of tank
other variables are as shown on the figure

The steady-state solution to this equation assumes k, the first order rate constant, is applied when Equation 2-14 is integrated for change in rate with time, resulting in:

$$C = \frac{Ci}{1 + k(V/Q_i)} \qquad [2\text{-}15]$$

Metcalf and Eddy (1991) provides additional information on reactor mass balances and solutions for unsteady-state conditions and plug-flow reactor systems.

2.5 Engineering Economics

Engineering economics is an essential element of engineering practice. Examples of areas involving engineering economics include selecting the best option for air particulate collectors, choosing the best technology for a wastewater treatment application, or evaluating waste minimization alternatives in hazardous waste management. This section covers two principal forms of economic analysis, *Present Worth* and *Annual Cash Flow*.

TIP FOR THE P.E. EXAMINATION CANDIDATE

Engineering economics will not be a separate problem on the examination. In all cases, engineering economics may be a PART of the analysis required in any of the specialty specific questions. The requirement for the candidate is to understand the fundamental economic analyses and to apply these correctly in any environmental engineering scenario.

In general, economic analysis entails taking costs and benefits that occur at different times and, through application of the time value of money, bringing all to a common period of time. The time value of money may be represented by an inflation rate, an interest rate, or a depreciation rate, depending on the specific application, but all relate to the change in value on a time (usually annual) basis. Table 2-1 summarizes how to translate value, costs, or benefits, from future time to the present; how to annualize lump sum values; and how to extend present costs to future benefits.

Solving engineering economic problems can be simplified by translating the word description into the cash flow diagrams illustrated in Table 2-1 showing each of the given factors and then selecting the analysis approach (present worth, annualized cost, or others) that best fits the data as shown in the diagrams.

Many references have tables that compile the interest rate factors for a large range of interest rates (i) and time periods (n). Table 2-2 is an example for one interest rate.

Costs and benefits currently spent or earned, respectively, represent present values. However, most projects include initial costs, annual operating costs, and possibly some residual value at the end of its useful life. The project may also produce

Table 2-1
Summary of economic analyses

Name	Formula	Line Diagram	Symbols
Compound amount	$F = P(1+i)^n$		F = Future value P = Present value n = Time periods i = Interest rate
Present worth	$P = F(1+i)^{-n}$		
Sinking fund	$A = F\left[\dfrac{i}{(1+i)^n - 1}\right]$		F = Future value P = Present value n = Time periods i = Interest rate A = Periodic value
Series compound amount	$F = A\left[\dfrac{(1+i)^n - 1}{i}\right]$		
Capital recovery	$A = P\left[\dfrac{i(1+i)^n}{(1+i)^n - 1}\right]$		same as above
Series present worth	$P = A\left[\dfrac{(1+i)^n - 1}{i(1+i)^n}\right]$		
Gradient uniform series	$A = G\left[\dfrac{1}{i} - \dfrac{n}{(1+i)^n - 1}\right]$		G = Periodic gradient value
Gradient present worth	$P = \dfrac{G}{i}\left[\dfrac{(1+i)^n - 1}{i} - n\right]\left[\dfrac{1}{(1+i)^n}\right]$		same as above

Adapted from: Newnan, 1976.

Table 2-2
Compound interest factors at 6%

	Single payment		Uniform payment series				Gradient series	
	Compound Amount Factor	Present Worth Factor	Sinking Fund Factor	Capital Recovery Factor	Compound Amount Factor	Present Worth Factor	Gradient Uniform Series	Gradient Present Worth
	Find F given P	Find P given F	Find A given F	Find A given P	Find F given A	Find P given A	Find A given G	Find P given G
n	F/P	P/F	A/F	A/P	F/A	P/A	A/G	P/G
1	1.060	.9434	1.0000	1.0600	1.000	.943	0	0
2	1.124	.8900	.4854	.5454	2.060	1.833	.485	.890
3	1.191	.8396	.3141	.3741	3.184	2.673	.961	2.569
4	1.262	.7921	.2286	.22886	4.375	3.465	1.427	4.946
5	1.338	.7473	.1774	.2374	5.637	4.212	1.884	7.935
6	1.419	.7050	.1434	.2034	6.975	4.917	2.330	11.459
7	1.504	.6651	.1191	.1791	8.394	5.582	2.768	15.450
8	1.594	.6274	.1010	.1610	9.897	6.210	3.195	19.842
9	1.689	.5919	.0870	.1470	11.491	6.802	3.613	24.577
10	1.791	.5584	.0759	.1359	13.181	7.360	4.022	29.602
11	1.898	.5268	.0668	.1268	14.972	7.887	4.421	34.870
12	2.012	.4970	.0593	.1193	16.870	8.384	4.811	40.337
13	2.133	.4688	.0530	.1130	18.882	8.853	5.192	45.963
14	2.261	.4423	.0476	.1076	21.015	9.295	5.564	51.713
15	2.397	.4173	.0430	.1030	23.276	9.712	5.926	57.555
16	2.540	.3936	.0390	.0990	25.673	10.106	6.279	63.459
17	2.693	.3714	.0354	.0954	28.213	10.477	6.624	69.401
18	2.854	.3503	.0324	.0924	30.906	10.828	6.960	75.357
19	3.026	.3305	.0296	.0896	33.760	11.158	7.287	81.306
20	3.207	.3118	.0272	.0872	36.786	11.470	7.605	87.230
21	3.400	.2942	.0250	.0850	39.993	11.764	7.915	93.114
22	3.604	.2775	.0230	.0830	43.392	12.042	8.217	98.941
23	3.820	.2618	.0213	.0813	46.996	12.303	8.510	104.701
24	4.049	.2470	.0197	.0797	50.816	12.550	8.795	110.381
25	4.292	.2330	.0182	.0782	54.865	12.783	9.072	115.973
26	4.549	.2198	.0169	.0769	59.156	13.003	9.341	121.468
27	4.822	.2074	.0157	.0757	63.706	13.211	9.603	126.860
28	5.112	.1956	.0146	.0746	68.528	13.406	9.857	132.142
29	5.418	.1846	.0136	.0736	73.640	13.591	10.103	137.310
30	5.743	.1741	.0126	.0726	79.058	13.765	10.342	142.359
31	6.088	.1643	.0118	.0718	84.802	13.929	10.547	147.286
32	6.453	.1550	.0110	.0710	90.890	14.084	10.799	152.090
33	6.841	.1462	.0103	.0703	97.343	14.230	11.017	156.768
34	7.251	.1379	.0096	.0696	104.184	14.368	11.228	161.319
35	7.686	.1301	.0090	.0690	111.435	14.498	11.432	165.743
40	10.286	.0972	.0065	.0665	154.762	15.046	12.359	185.957
45	13.765	.0727	.0047	.0647	212.744	15.456	13.141	203.110
50	18.420	.0543	.0034	.0634	290.336	15.762	13.796	217.457
55	24.650	.0406	.0025	.0625	394.172	15.991	14.341	229.322
60	32.988	.0303	.0019	.0619	533.128	16.161	14.791	239.043
65	44.145	.0227	.0014	.0614	719.083	16.289	15.160	246.945
70	59.076	.0169	.0010	.0610	967.932	16.385	15.461	253.327
75	79.057	.0126	.0008	.0608	1300.949	16.456	15.706	258.453
80	105.796	.0095	.0006	.0606	1746.600	16.509	15.903	262.549
85	141.579	.0071	.0004	.0604	2342.982	16.549	16.062	265.810
90	189.465	.0053	.0003	.0603	3141.075	16.579	16.189	268.395
95	253.546	.0039	.0002	.0602	4209.104	16.601	16.290	270.437
100	339.302	.0029	.0002	.0602	5638.368	16.618	16.371	272.047

Source: Newnan, 1976.

a profit or a savings on a recurring basis which may be a uniform series payment or a uniform gradient payment. The *present worth* analysis brings all costs and benefits to a present value so that they can be summed to determine the life cycle cost (total cost) of a project for comparison with other alternatives to select a preferred option.

Table 2-1 provides the methods which take the interest rate (i) and time period (n) to determine the present worth of any cost or benefit. The simplest form is where there is some cost or benefit that will accrue n years in the future for which it is necessary to know its value in present dollars. That future amount (F) times the present worth factor, $(1 + i)^{-n}$, determines the present worth of that expense. Costs or benefits also accrue on a repeating basis over an extended period of time. Annual operating costs are an example. With a series present worth analysis, the compilation of recurring annual costs are converted to a total present worth value. Gradient series values are probably the least common and, therefore, the least understood, of interest calculations. A gradient present worth series converts the value of a series that increases by a fixed amount (G) over a period of years into a total value in today's dollars. The one limitation to using present worth analysis to compare alternatives is that all alternatives must have the life expectancy or service period. Present worth analysis can be more easily understood by using the methods and concepts in Table 2-1 and applying them to the following engineering situation example.

TIP FOR THE P.E. EXAMINATION CANDIDATE

For the PE examination a situation will be given and a specific analysis procedure directed. Describing the various costs and benefits in diagrams of present, future, gradient, and series values simplifies making a correct analysis.

TIP FOR THE P.E. EXAMINATION CANDIDATE

Interest Factor Tables can save a lot of time in the examination. If you use a set of these tables, be sure they are correctly applied. Annotating your Interest Factor Tables with the diagrams from Table 2-1 so you can keep track of the *Fs, Ps, As, Gs*, or any symbols your tables employ is recommended.

Example 2-4 — Present worth analysis

Use a present worth analysis to select the most cost-effective method from two options for a particulate control device on a boiler, each depreciating at a rate of 6%.

Option 1 — Electrostatic Precipitator

    Initial cost — $2.25 M

    Annual operating costs — $175,000 for the first year and increasing by $25,000 each year

Salvage value — $200,000

Life expectancy — 15 yr

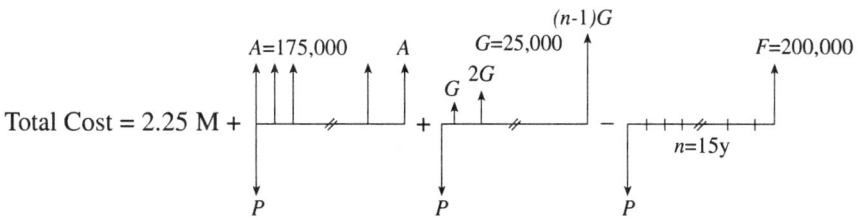

Step 1. Annual operating costs
uniform series $P = A$ (Series present worth
@ $i = 0.06, n = 15$ yr)
$P = \$175,000\ (9.7122) = \$1,699,635$
uniform gradient series present worth for $25,000
$P = G$ (gradient present worth @ $i = 0.06, n-1 = 14$ yr)
$P = \$25,000\ (51.73) = \$1,293,250$

Step 2. Salvage value
$P = F$ (present worth at @ $= 0.06, n = 15$ yr)
$P = \$200,000\ (0.4173) = \$83,460$

Present worth cost $= \$2,250,000 + \$1,699,635 + \$1,293,250 - \$83,460$

$= 5,159,425$

(Note: the sign of salvage is minus to represent return value)

Option 2 — Baghouse

Initial cost — $1.5 M

Annual operating costs — $150,000/yr

Bag replacement — $125,000 every three years

Salvage value — $ 50,000

Life expectancy — 15 yr

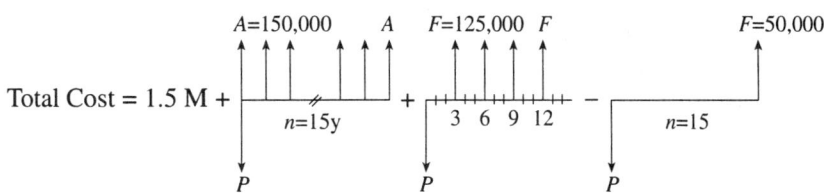

Step 1. Annual operating costs $P = \$150,000$ (Series present worth @ $i = 0.06$, $n = 15$ yr) $P = \$150,000 \, (9.7122) = \$1,456,830$

Step 2. Salvage value $P = F$ (present worth at @ $i = 0.06$, $n = 15$ yr) $P = \$50,000 \, (0.4173) = \$20,865$

Step 3. Bag replacement every three years, must spend $125,000 at the 3, 6, 9, and 12 yr periods.

$P = F($ present worth @ $i = 0.06$, $n = 3$ yr$)$
@3 yr $P = \$125,000 \, (0.8396)$ = $117,450
@6 yr $P = \$125,000 \, (0.705)$ = $88,125
@9 yr $P = \$125,000 \, (0.5919)$ = $73,988
@12 yr $P = \$125,000 \, (0.4970)$ = $62,125
                                                 TOTAL = $341,688

Total present worth cost = $1,500,000 + $1,456,830 + $341,688 - $20,865
= $3,277,653

Option 2, the Baghouse, is the most economical option; it saves $2,037,000 compared to electrostatic precipitatior.

Annual cash flow analysis converts all costs and benefits into an annual value. The total annual cost reveals the option which costs the least or provides the largest benefit. To calculate annualized costs requires the appropriate use of the capital recovery, sinking fund, and uniform and gradient series factors from Table 2-1. Example 2-5 demonstrates an annual cash flow analysis using the same data used in Example 2-4. Annual cash flow may be used for alternatives with unequal service lifes.

Example 2-5 — Annual cost analysis

Determine the annual cash flow cost for each particulate collector option—Electrostatic Precipitator and Baghouse—and select the most economical alternative.

Option 1 — Electrostatic Precipitator

        Initial cost $2.25 M
        Annual operating costs — $175,000 for the first year and increasing by $25,000 each year
        Salvage value — $200,000
        Life expectancy — 15 yr

Chapter 2: Common Principles

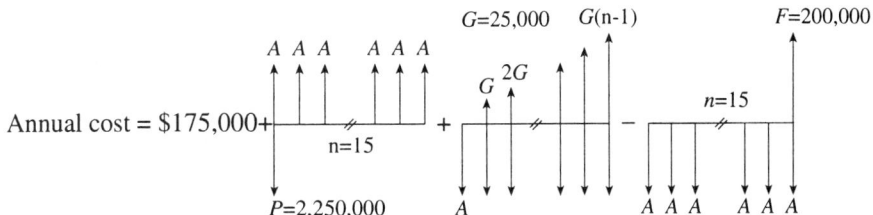

Step 1. The annual cost of $2,250,000
 $A = P$ (capital recovery factor @ $n = 15$, $i = 0.06$)
 $A = \$2,250,000 \, (0.103) = \$231,750$

Step 2. The uniform annual cost of $175,000/yr + $25,000/yr increase
 $A = \$175,000 + \$25,000$
 (gradient uniform series factor @ $n = 15$, $i = 0.06$)
 $A = \$175,000 + \$25,000 \, (5.926) = \$175,000 + \$148,150 = \$323,150$

Step 3. The annualized benefit of the salvage value $A = F$
 (sinking fund factor @ $n = 15$, $i = 0.06$
 $A = \$200,000(0.0430) = \$8,600$

Step 4. Sum annualized costs
 Annual cost of option 1 = $231,750 + $323,150 - $8,600 = $546,300

Option 2 — Baghouse

Initial costs — $1.5 M
Annual operating costs — $150,000 yr
Bag replacement — $125,000 every three years
Salvage value — $50,000

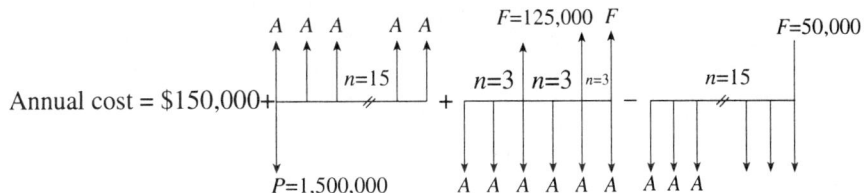

Step 1. The annual cost of $1,500,000
 $A = P$ (capital recovery factor @ $n = 15$, $i = 0.06$)
 $A = \$1,500,000(0.103) = \$154,5000$

Step 2. The uniform annual operating cost is $150,000/yr

Step 3. Annual cost of replacing bags every three years is $125,000
$A = F$ (sinking fund factor @ $n = 3$, $i = 0.06$)
$A = \$125{,}000 \ (0.3141) = \$39{,}263$

Note: Since this expense will occur again every three years, it will an annual cost for the life of the project. Therefore, no further calculations are required for years 6, 9, and 12.

Step 4. The annualized benefit of the salvage value
$A = F$ (sinking fund factor @ $n = 15$, $i = 0.06$)
$A = \$50{,}000 \ (0.0430) = \$2{,}150$

Step 5. Sum annualized costs
Annual cost of option 1 = $\$154{,}5000 + \$150{,}000 + \$39{,}263 - \$2{,}150$
= $\$341{,}613$

Again, Option 2, the Baghouse, is the most economical with annual costs $204,687 less than electrostatic precipitation.

There is a lot more to engineering economics than presented here, but most problems can be solved by correctly applying the terms presented in Table 2-1. Two other economic analyses, *capitalized costs* and *rate of return*, are worthy of mention. Capitalized cost is a special case of present worth analysis where all calculations are done for an infinite time period. This is accomplished by solving the interest equations setting n = a large number, 500 yr for example, and applying the interest factors in exactly the same way as in a standard present worth analysis. This special form of analysis is common in analyzing government and public works projects where services such as water supply must be continuous.

TIP FOR THE P.E. EXAMINATION CANDIDATE

Either annual cash flow or present worth analyses will enable making the correct economic decision, i.e., the most cost-effective, but the absolute dollar amount is different. Accordingly, ensure that the result provided as an answer is the one requested in the question.

Rate of return analysis uses the same factors presented in Table 2-1, but are solved to determine an interest rate accrued from a proposed project. This method is normally used when a project has initial costs that will then produce recurring cost savings or profits into the future and it is desired to know the interest rate the investment (initial costs) produced from the savings or income produced. This approach is common in business applications where investment dollars must justify an acceptable rate of return before the funds will be risked on a project.

2.5 Standard Notations

Symbol	Definition
A	periodic value
atm	atmosphere
BOD	biochemical oxygen demand
C_{mass}	standard international
F	future value
ft	feet
G	uniform gradient period
g	grams
gmol	gram-mole
i	interest rate for period n
η	moles
L	liter
lb	pounds
m	meters
mg	milligrams
M_s	mass
MW_s	molecular weight
n	interest period
P	pressure
ppm	parts per million
psi	pounds per square inch
V	volume

2.6 References

Davis, Mackenzie, and Cornwell, David. 1991. *Introduction to Environmental Engineering.* New York: McGraw-Hill.

National Council of Examiners for Engineering and Surveying (NCEES). 1993. *Fundamentals of Engineering Reference Handbook,* 1st Ed. Clemson, South Carolina: NCEES.

Newnan, Donald. 1976. *Engineering Economic Analysis.* Engineering Press.

Metcalf and Eddy, Inc. 1991. *Wastewater Engineering Treatment, Disposal, and Reuse*, 3rd ed. New York: McGraw-Hill.

Chapter 3

ESSENTIAL CHEMISTRY

by W. Christopher King, P.E., DEE

3.1 Introduction

Chemistry is fundamental to and the common denominator for all environmental media. The basic chemistry concepts apply to air, water, solid waste, and industrial hygiene. Therefore, these concepts have been consolidated in one chapter to provide a foundation for the remainder of the text. However, this chapter is not intended as a complete treatise on basic or environmental chemistry. The complexities of chemistry for environmental engineers transcend the basics provided. The references provided at the end of this chapter are just a few of several very good environmental chemistry texts in print at this time.

3.2 Definitions

Understanding the chemistry required for environmental engineering requires comprehension of nine essential terms. These terms and their definitions follow.

Molecular Weight — The sum of the atomic weights of the atoms that make up a molecule. (The atomic weights of the elements are provided in Table 3-1).

Example 3-1 — Molecular weight calulation

Molecular weight of sulfuric acid (H_2SO_4) = 2(1) + 32 + 4(16)

= 98 atomic mass units

TIP FOR THE P.E. EXAMINATION CANDIDATE

The text by Sawyer, McCarty, and Parkin (1994) contains a good background section that reviews basic chemistry and should bring practicing engineers up to the level necessary for the examination.

Mole — The weight of one Avogadro's number ($6.023 \cdot 10^{23}$) of molecules of a chemical. Moles can be in any mass units such as pound-moles or gram-moles. In Example 3-1, there are 98 grams in 1 gram-mole of sulfuric acid. Moles are used to measure the reactants and products from a chemical reaction.

Valence — The number of electrons an atom or molecule accepts or donates in reaction with another molecule. This number depends on the number of electrons in the outer electron shell of an atom. Atoms with nearly complete outer shells tend to take electrons to complete the shell (oxidizing agents), while atoms with almost empty outer shells readily give up electrons (reducing agents). However, for many common elements the valence changes depend on the molecule being formed. Table 3-2 provides valences for several elements common in environmental science.

Equivalent weight — The measure of how a molecule will combine with other atoms or molecules. The *Law of Multiple Proportions* states the amount of one element that combines with a fixed amount of another is always found to be in the ratio of small, whole numbers. This is the process of balancing equations. The equivalent weight of a molecule is calculated as:

equivalent weight = molecular weight / valence [3-1]

For the many elements with multiple valences, this process can be confusing. To the extent possible, valences for the common environmental chemicals have been provided in Table 3-2 or will be provided in this book where needed.

Table 3-1
Atomic weights of elements

	Symbol	Atomic number	Atomic weight		Symbol	Atomic number	Atomic weight
Actinium	Ac	89	[227]	Neodymium	Nd	60	144.24
Aluminum	Al	13	26.9815	Neon	Ne	10	20.180
Americium	Am	95	[243]	Neptunium	Np	93	[237]
Antimony	Sb	51	121.75	Nickel	Ni	28	58.69
Argon	Ar	18	39.948	Niobium	Nb	41	92.906
Arsenic	As	33	74.9216	Nitrogen	N	7	14.0067
Astatine	At	85	[210]	Nobelium	No	102	[259]
Barium	Ba	56	137.33	Osmium	Os	76	190.2
Berkelium	Bk	97	[247]	Oxygen	O	8	15.9994
Beryllium	Be	4	9.0122	Palladium	Pd	46	106.4
Bismuth	Bi	83	208.980	Phosphorus	P	15	30.9738
Boron	B	5	10.811	Platinum	Pt	78	195.08
Bromine	Br	35	79.904	Plutonium	Pu	94	[244]
Cadmium	Cd	48	112.41	Polonium	Po	84	[209]
Calcium	Ca	20	40.078	Potassium	K	19	39.098
Californium	Cf	98	[251]	Praseodymium	Pr	59	140.9077
Carbon	C	6	12.011	Promethium	Pm	61	[145]
Cerium	Ce	58	140.12	Protactinium	Pa	91	[231]
Cesium	Cs	55	132.905	Radium	Ra	88	[226]
Chlorine	Cl	17	35.453	Radon	Rn	86	[222]
Chromium	Cr	24	51.996	Rhenium	Re	75	186.2
Cobalt	Co	27	58.9332	Rhodium	Rh	45	102.905
Copper	Cu	29	63.546	Rubidium	Rb	37	85.4678
Curium	Cm	96	[247]	Ruthenium	Ru	44	101.07
Dysprosium	Dy	66	162.50	Samarium	Sm	62	150.4
Einsteinium	Es	99	[252]	Scandium	Sc	21	44.956
Erbium	Er	68	167.26	Selenium	Se	34	78.96
Europium	Eu	63	151.96	Silicon	Si	14	28.086
Fermium	Fm	100	[257]	Silver	Ag	47	107.868
Fluorine	F	9	18.9984	Sodium	Na	11	22.9898
Francium	Fr	87	[223]	Strontium	Sr	38	87.62
Gadolinium	Gd	64	157.25	Sulfur	S	16	32.066
Gallium	Ga	31	69.72	Tantalum	Ta	73	180.948
Germanium	Ge	32	72.61	Technetium	Tc	43	[98]
Gold	Au	79	196.967	Tellurium	Te	52	127.60
Hafnium	Hf	72	178.49	Terbium	Tb	65	158.925
Helium	He	2	4.0026	Thallium	Tl	81	204.38
Holmium	Ho	67	164.930	Thorium	Th	90	232.038
Hydrogen	H	1	1.0079	Thulium	Tm	69	168.934
Indium	In	49	114.82	Tin	Sn	50	118.71
Iodine	I	53	126.9045	Titanium	Ti	22	47.88
Iridium	Ir	77	192.22	Tungsten	W	74	183.85
Iron	Fe	26	55.847	Unnihlexium	Unh	106	[263]
Krypton	Kr	36	83.80	Unnilpentium	Unp	105	[262]
Lanthanum	La	57	138.91	Unnilquadium	Unq	104	[261]
Lawrencium	Lr	103	[260]	Unnilseptium	Uns	107	[262]
Lead	Pb	82	207.2	Uranium	U	92	238.03
Lithium	Li	3	6.941	Vanadium	V	23	50.94
Lutetium	Lu	71	174.97	Xenon	Xe	54	131.29
Magnesium	Mg	12	24.305	Ytterbium	Yb	70	173.04
Manganese	Mn	25	54.9380	Yttrium	Y	39	88.9059
Mendelevium	Md	101	[258]	Zinc	Zn	30	65.39
Mercury	Hg	80	200.59	Zirconium	Zr	40	91.22
Molybdenum	Mo	42	95.94				

(^{12}C = 12.0000 amu)
A value given in brackets denotes the mass number of the longest-lived isotope.

Table 3-2
Valences for common environmental elements

Element	Valence(s)
Al	+4, +3
C	-4 → +4
Ca	+2
Cl	-1
Cr	+6, +3
Cu	+2, +1
Fe	+3, +2
H	+1
K	+1
Mg	+2
Mn	+7, +6, +4, +3, +2
Na	+1
O	-2
P	+1
S	+6, +4, +2, -2
Zn	-2

Source: *University Chemistry*, by Bailar, Moeller, & Kleinberg

Oxidation — An atom or molecule is oxidized when it loses electrons. If one substance gives up electrons, another must receive them. The substance that receives electrons is called the oxidizing agent.

Reduction — The gain of electrons by a substance. The substance that gives up the electrons is called the reducing agent.

3.3 Equivalent Weight

An essential ingredient in solving environmental engineering problems involving chemical reactions is the calculation of equivalent weights of the involved chemicals and substances. Example 3-2 illustrates this calculation.

Example 3-2 — Equivalent weight calculation

Find the gram equivalent weight for sulfuric acid (H_2SO_4).

Hydrogen has only a +1 charge, therefore the positive charges are 2 • (+1) = +2. This must be balanced by the negative charge of the sulfate ion, SO_4. Sulfate always has a -2 valence. This can be deduced by summing the valences for the formation of sulfate from oxygen and sulfur. Oxygen always has a -2 valence, while sulfur can have any of three positive valences (see Table 3-2). Figure 3-1 depicts the outer electron shells in the formation of sulfate illustrating the valence for sulfate is -2.

3.4 The Basic Chemical Reaction

The stoichiometric or *balanced* chemical equation describes the equilibrium ratios of a chemical reaction. A balanced chemical equation equates the electron activities or the oxidation and reduction pairs. The electron activity is an individual chemical property of the elements primarily reflecting the completion of the outer electron shell. Example 3-3 illustrates a stoichiometric reaction.

Example 3-3 — A stoichiometric reaction for lime-soda ash water softening

Find the quantity of calcium hydroxide required to remove magnesium from water when the magnesium is in the form of magnesium sulfate. The magnesium sulfate concentration in the water is 1.0 mg/L.

[a] $MgSO_4 + Ca(OH)_2 \rightarrow Mg(OH)_2 \downarrow + CaSO_4$

[b] $\{24 + 32 + 4(16)\} + \{40 + 2(16 + 1)\} \rightarrow \{24 + 2(16 + 1)\} + \{40 + 32 + 4(16)\}$
$120 + 74 \rightarrow 58 + 136$

[c] $1\ mg/L + \dfrac{74}{120} \rightarrow \dfrac{58}{120} + \dfrac{136}{120}$

$1.0\ mg/L + 0.61\ mg/L \rightarrow 0.48\ mg/L + 1.13\ mg/L$

$1.61\ mg/L \rightarrow 1.61\ mg/L$

[d] $\dfrac{120}{58} + \dfrac{74}{58} \rightarrow 1 + \dfrac{136}{58}$

$2.07\ mg/L + 1.28\ mg/L \rightarrow 1\ mg/L + 2.35\ mg/L$

$3.35\ mg/L \rightarrow 3.35\ mg/L$

In the example, one mole of magnesium sulfate reacts with one mole of calcium hydroxide to yield 1 mole of magnesium hydroxide (a relatively insoluble precipitate, indicated by the down pointing-arrow [a]) and one mole of calcium sulfate. The equation is written with one (1) as the coefficient for each of the compounds to balance the valences for each element. Refering to Table 3-2, it is noted that calcium and magnesium always have valences of +2, which means there must be a total of -2 charge to balance these reactions. In Equation [a] the hydroxyl ion has a net charge of -1 because oxygen has a -2 valence and hydrogen has a +1, which then requires two hydroxyl ions to react with either calcium as the reactant or magnesium as the product.

Figure 3-1
Electron map of sulfate

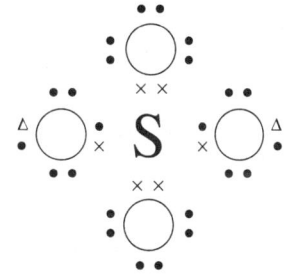

Δ — Electron acceptor site
$\times$ — Sulfur electrons
$\bullet$ — Oxygen electrons
The equivalent weight of sulfuric acid is then:
98/2 = 49 g/g-mole (or lb/lb-mole).

There are numerous ways to calculate the quantities of reactants and products in a stoichiometric reaction. Example 3-3 uses molecular weights to develop the mass relationships for reactants and products. Equation [b] provides the calculation of the molecular weights of each compound using the atomic weights in Table 3-1. Equation [c] calculates the ratios of the products and reactants based on the 1.0 mg/L concentration of magnesium sulfate. Equation [c] can be interpreted as follows: 1.0 mg/L of magnesium sulfate requires 0.61 mg/L of calcium hydroxide to complete the reaction and yields 0.48 mg/L of a magnesium hydroxide (precipitate) and 1.13 mg/L of calcium sulfate.

This approach can use any of the reactants or products as the basis for the calculation in Equation [c]. As an example, Equation [d] shows the procedure to calculate the quantities of reactants required to produce 1 mg/L of magnesium hydroxide precipitate. In this example, 2.07 mg/L of magnesium sulfate and 1.28 mg/L of calcium hydroxide are required as reactants to yield 1.0 mg/L of magnesium hydroxide, with 2.35 mg/L of calcium sulfate produced as a byproduct.

TIP FOR THE P.E. EXAMINATION CANDIDATE

Knowing how to calculate mass and concentrations in some form is essential for the P.E. exam candidate. It is probable that the examinations will require at least one of these calculations and possibly more

The example stoichiometric reaction represents the ideal. Putting the stoichiometric amounts of magnesium sulfate and calcium hydroxide in a reaction vessel would not instantly yield these products at the theoretical quantities. Environmental engineers must recognize that other factors affect these reactions. Time is the first consideration; the rate at which reactions occur depends, in turn, on many variables. Temperature, pressure, and concentration of reactants are other parameters that determine the rate and degree of completion for any reaction. Temperature and concentration are typically the control mechanisms in engineering chemical reactions. In Example 3-3, if the goal is to remove as much magnesium as possible, adding more than the stiochiometric quantity of calcium hydroxide will serve as a driving force to increase the rate of reaction and concomitantly the amount of precipitate produced. Chemical kinetics is not as critical for environmental engineers as it is for chemical engineers, but the topic must be understood by environmental engineers because it is a practical consideration in many environmental systems. Kinetics is discussed in more detail in Section 3.7.

Example 3-4 illustrates a practical application of the chemical principles discussed thus far for the environmental engineer.

Chapter 3: Essential Chemistry

Example 3-4 — Coagulation water treatment

Ferric chloride and lime are being used to enhance coagulation/flocculation in a water treatment plant. The stoichiometric reaction is as follows:

$$2FeCl_3 + 3Ca(OH)_2 \rightarrow 2Fe(OH)_3 \downarrow + 3CaCl_2$$

Based on this reaction and a flow of 5 Mgal/day, determine:

a) How many lb/day of ferric hydroxide sludge (on a dry weight basis) are generated assuming 15 mg/L of ferric chloride is added?

Step 1. $2FeCl_3 + 3Ca(OH)_2 \rightarrow 2Fe(OH)_3 \downarrow + 3CaCl_2$

Step 2. $2(56 + 3 \bullet 35.5) + 3(40 + 2\{16 + 1\})$
$\rightarrow 2(56 + 3\{16 + 1\}) + 3(40 + 2 \bullet 35.5)$

Step 3. $325 + 222 \rightarrow 214 + 333$

Step 4. $1 + \dfrac{222}{325} \rightarrow \dfrac{214}{325} + \dfrac{333}{325}$

Step 5. $1 + 0.68 \rightarrow 0.66 + 1.02$

15 mg/L of ferric chloride yields (0.66 • 15) or 9.9 mg/L of ferric hydroxide precipitate; converting to lb/day:

9.9 mg/L • 8.34 lb/gal • 5 Mgal/day
= 412.8 lb/day of ferric hydroxide sludge

b) How many lb/day of lime as calcium hydroxide are required per day (using the same ferric chloride feed rate as in a)?

As part of solving part a), the reaction ratio of ferric chloride and hydrated lime was found to be 1 to 0.68, which is applied as follows:

TIP FOR THE P.E. EXAMINATION CANDIDATE

The application of **8.34** is something every candidate **must know**. Understanding its application will be helpful in passing the examination. There are 8.34 pounds in a gallon of water. The second application of 8.34 is key to stoichiometric calculations as follows:

lb/day = (concentration, mg/L)(flow rate, Mgal/day)[8.34 lb/Mgal (mg/L)]

Step 1. 15.0 mg/L • 0.68 = 10.2 mg/L of calcium hydroxide (same as hydrated lime)

Step 2 10.2 mg/L • 8.34 lb/gal • 5 Mgal/day = 425.5 lb/day

3.5 Stoichiometric Calculations

While the preceding section presented one procedure for mass balancing reactants and products, this section addresses two other common approaches that are useful in certain environmental engineering calculations. The **molar** approach is illustrated using the following reaction:

$$2\,FeCl_3 + 3\,Ca(OH)_2 \rightarrow 2\,Fe(OH)_3 \downarrow + 3\,CaCl_2$$

Based on moles, this balanced equation states that 2 moles of ferric chloride react with 3 moles of calcium hydroxide to form 2 moles of ferric hydroxide and 3 moles of calcium chloride. The molar approach is preferred by many, particularly those with chemistry and chemical engineering backgrounds, because it is based on the reactivity of the compounds. It is important to recognize that 1 lb-mole of ferric hydroxide ($Fe(OH)_3$=56+51) weighs 107 lb or 1 g-mole of ferric chloride ($FeCl_3$=56+105.5) weighs 162 g, or more generally, the combined atomic weight of the elements comprising a compound may be applied in any mass unit for each mole.

Another approach to stoichiometric calculations is based on **valence**. Most chemical reactions depend on the loss, gain, or sharing of electrons between atoms. If all elements had a single valence or oxidation state, writing chemical reactions would be relatively simple. However, as shown in Table 3-2, many elements have numerous valences, depending on the compounds available to complete their outer electron shells, since the goal of all elements and compounds is to reach charge neutrality. Using the ferric chloride reaction below, calculation of the products and reactants based on oxidation is illustrated:

$$2\,FeCl_3 + 3\,Ca(OH)_2 \rightarrow 2\,Fe(OH)_3 \downarrow + 3\,CaCl_2$$

Using the valence approach begins by analyzing the reaction based on the elemental valences selected from Table 3-2. Start with the elements that have fixed valences: Ca = 2, O = -2, H = 1. OH is the hydroxyl ion, which also has a fixed valence of -1. This leaves Fe, which can have a +2 or +3 valence, and chlorine, which can have several valences. The valences of the variable atoms are derived based on their reactions with fixed valence atoms. Chlorine must have a -1 valence when two atoms react with one calcium to form calcium chloride. Therefore iron must be +3 because it reacts with three chlorines which have a -1 valence and 3 hydroxyl ions, each also -1 valence. The equivalent weights of each reactant and product can be calculated using Equation 3-1.

$$\text{molecular weight/valence} = \text{equivalent weight} \qquad [3\text{-}1]$$

$$FeCl_3 : \frac{6 + 3(35.5)}{3} = 54.1$$

$$Ca(OH)_2 : \frac{40 + 2(16+1)}{2} = 37$$

$$Fe(OH)_3 : \frac{56 + 3(16+1)}{3} = 35.6$$

$$CaCl_2 : \frac{40 + 2(35.5)}{2} = 55.5$$

Substituting these data into the equation expresses the reaction in terms of one equivalence of each reactant and product.

$$2FeCl_3 + 3Ca(OH)_2 \rightarrow 2Fe(OH)_3 \downarrow + 3CaCl_2$$
$$54.1 + 37 \rightarrow 35.6 + 55.5$$
$$91.1 \rightarrow 91.1$$

or putting these into a form based on one unit of ferric chloride:

$$\frac{54.1}{1} + \frac{37}{54.1} \rightarrow \frac{35.6}{54.1} + \frac{55.5}{54.1}$$
$$1 + 0.68 \rightarrow 0.66 + 1.02$$
$$1.68 \rightarrow 1.68$$

Reexamine Example 3-4; notice that the answer is the same as that achieved using the molecular weight and balanced equation technique. The valence approach is used when a balanced equation is not provided. Table 3-3 can be used to look up the equivalent weights for common environmental chemicals. When tabular values do not exist, the equivalence approach enables calculations based on only the molecular weight and charge of each constituent.

One of the most common applications of the equivalence approach in environmental engineering is in calculating chemical dosages for water softening and other water treatment processes which are discussed in Chapter 5.

Table 3-3
Chemicals in environmental engineering

Name	Formula	Molecular weight	Equivalent weight
Aluminum hydroxide	$Al(OH)_3$	78.0	26.0
Aluminum sulfate	$Al_2(SO_4)_3 \cdot 14.3H_2O$	600	100
Ammonia	NH_3	17.0	n.a.
Ammonium	NH_4^+	18.0	18.0
Ammonium sulfate	$(NH_4)_2SO_4$	132	66.1
Bicarbonate	HCO_3^-	61.0	61.0
Calcium bicarbonate	$Ca(HCO_3)_2$	162	81.0
Calcium	Ca	40.0	20.0
Calcium carbonate	$CaCO_3$	100	50.0
Calcium hydroxide	$Ca(OH)_2$	74.1	37.0
Calcium hypochlorite	$Ca(ClO)_2 \cdot 2H_2O$	179	n.a.
Calcium oxide	CaO	56.1	28.0
Calcium sulfate	$CaSO_4$	136	68.0
Carbon dioxide	CO_2	44.0	22.0
Carbonate	CO_3^{2-}	60.0	30.0
Chlorine	Cl_2	71.0	n.a.
Chlorine dioxide	ClO_2	67.0	n.a.
Copper sulfate	$CuSO_4$	160	79.8
Ferric chloride	$FeCl_3$	162	54.1
Ferric hydroxide	$Fe(OH)_3$	107	35.6
Ferric sulfate	$Fe_2(SO_4)_3$	400	66.7
Ferrous sulfate	$FeSO_4 \cdot 7H_2O$	278	139
Fluosilicic acid	H_2SiF_6	144	n.a.
Hydrochloric acid	HCl	36.5	36.5
Hydroxyl	OH^-	17.0	17.0
Hypochlorite	OCl^-	51.5	51.5
Magnesium carbonate	$MgCO_3$	84.3	42.1
Magnesium hydroxide	$Mg(OH)_2$	58.3	29.2
Magnesium	Mg	24.4	12.2
Magnesium sulfate	$MgSO_4$	120	60.1
Nitrate	NO_3^-	62.0	62.0
Orthophosphate	PO_4^{3-}	95.0	31.7
Oxygen	O_2	32.0	16.0
Potassium permanganate	$KMnO_4$	158	n.a.
Sodium aluminate	$NaAlO_2$	82.0	n.a.
Sodium bicarbonate	$NaHCO_3$	84.0	84.0
Sodium carbonate	Na_2CO_3	106	53.0
Sodium chloride	$NaCl$	58.4	58.4
Sodium flouride	NaF	42.0	n.a.
Sodium fluosilicate	Na_2SiF_6	188	n.a.
Sodium hexametaphosphate	$(NaPO_3)_n$	n.a.	n.a.
Sodium hydroxide	$NaOH$	40.0	40.0
Sodium hypochlorite	$NaClO$	74.4	n.a.
Sodium silicate	Na_4SiO_4	184	n.a.
Sodium sulfate	Na_2SO_4	142	71.0
Sodium thiosulfate	$Na_2S_2O_3$	158	n.a.
Sodium	Na	23	23
Sulfate	SO_4^{2-}	96.0	48.0
Sulfur dioxide	SO_2	64.1	n.a.
Sulfuric acid	H_2SO_4	98.1	49.0

3.6 Reaction Equilibrium

In any chemical reaction such as one expressed as:

$$aA + bB \rightarrow cC \qquad [3\text{-}2]$$

there are numerous assumptions inherent in expressing the reaction in this form. First, nearly all reactions are reversible to some extent. Equation 3-2 depicts a reaction where, with time, the reactants and products will reach an equilibrium and the rate of formation of products will equal the reverse reactions. The reversibility of a reaction depends on properties of the products. For example, in an aqueous reaction, if the product formed is a gas with low solubility, then the gas would transport to the gaseous phase and the reaction could not be reversed because of the loss of a product. Reactions that form insoluble salts are another example of this concept.

A second assumption included in Equation 3-2 is that all intermediate reactions can be summed into a single simplified reaction. In fact, the reaction proceeds in sequence as:

$$A + B \rightarrow AB \qquad [3\text{-}3]$$

$$AB \rightarrow C \qquad [3\text{-}4]$$

Summing these two reactions yields the reaction as written in Equation 3-2 and illustrates that the rate of the reaction and degree of completion may be affected by the properties of the intermediate, AB, in this example.

TIP FOR THE P.E. EXAMINATION CANDIDATE

To demonstrate competence as an environmental engineer, solutions to P.E. examination problems must account for equilibrium and reaction rates, not just basic equilibrium equations. This requires an understanding of the basis of reaction kinetics and the factors that affect equilibrium.

In reality, all the reactants, intermediates, and products in a chemical reaction either exist in an equilibrium or are progressing at some rate toward equilibrium. Chemical reactions can occur at varying rates; some may occur almost instantaneously, while others require years to reach a steady state. For engineered reactions, the latter rate is unacceptably slow.

Generally, equilibrium is described on the basis of the concentrations of the reactants and products. For the reaction

$$aA + bB \rightarrow cC \qquad [3\text{-}2]$$

it can be stated that

$$\frac{[C]^c}{[A]^a[B]^b} = K \qquad [3\text{-}5]$$

where the terms in brackets represent the molar concentrations of each chemical and K is the equilibrium constant for the reaction. If the goal is to remove [B] from an aqueous solution by forming [C] as a precipitate, Equation 3-5 indicates that the larger [A] becomes, the more of [C] is formed. In combination, Equations 3-2 and 3-5 demonstrate that increasing the reactants drives the reaction to the right, while increasing the products will move reaction to the left.

An extension of the foregoing concept describes the relationship between a solid compound and its component ions dissolved in a solution (usually water in environmental applications). Compounds are often treated as soluble or insoluble, but this is an oversimplification that can cause significant errors. Compounds treated as insoluble are really compounds of low solubility where it is possible in practice to ignore the dissolved fraction. In dealing with extremely toxic substances and environmental exposures, this assumption may not be valid. As an example, consider water chemistry's most popular "insoluble" compound, calcium carbonate. The solubility relation can be represented as:

$$\frac{[Ca^{+2}][CO_3^{-2}]}{[CaCO_3]} = K_{sp} = 5 \bullet 10^{-9} \, @ \, 25°C$$

This equation indicates that as long as the product of the molar concentrations of calcium and carbonate are below this very small number, the $CaCO_3$ remains in solution. Above this number, a precipitate will be formed and calcium can be removed.

3.7 Reaction Kinetics

The rate at which A plus B combine to yield C in Equation 3-2 is of major concern in all environmental engineering applications that use chemical and biological reactions.

First, the time required for a reaction determines the detention time, which dictates the size of the reactor and ancillary equipment. In engineering applications this detention time requires money for equipment and construction costs. Understanding how to optimize reaction rates becomes a primary engineering consideration in both the design and operating parameters for reactors of all types. The following discussion focuses on how to determine reaction rates.

Suppose one purpose is to remove B from a liquid waste stream by adding A to form a solid C as in Equation 3-2. This equation and its intermediates, Equations

3-3 and 3-4, are simplifications of a more complex system of reactions. The rate at which the overall reaction occurs is a function of the intermediate steps. The normal approach is to examine the reaction rate in relation to the concentration of the reactants. A reaction that occurs at a rate unrelated to the concentration of any of the products or reactants is referred to as a *zero-order reaction* and can be expressed as:

$$\frac{dC}{dt} = k \qquad [3\text{-}6]$$

where: dC = change in concentration
 dt = time between concentration measurements
 k = kinetic rate constant (minus sign would reflect decrease in concentration)

Integrating this expression yields:

$$Co - C_t = kt \qquad [3\text{-}7]$$

where: Co = initial concentration
 t = time
 C_t = concentration at time t

The form of kinetics most often seen in environmental engineering, for both chemical and biological reactions, is a *first-order reaction*. First order reactions are controlled by the concentration of one of the reactants (referred to as the rate limiting concentration), usually represented as:

$$\frac{dC}{dt} = kC \qquad [3\text{-}8]$$

TIP FOR THE P.E. EXAMINATION CANDIDATE

When working problems with given kinetic rate constants, be sure to note the base of the logarithm given in the problem statement.

Integrating this expression yields:

$$\ln\left(\frac{Co}{C_t}\right) = kt \qquad [3\text{-}9]$$

or

$$\log\left(\frac{Co}{C_t}\right) = \frac{kt}{2.30} \qquad [3\text{-}10]$$

Equation 3-9 or 3-10 can be used to determine the change in concentration of a reactant or a microorganism population for some time t, or to determine k with Co measured at two separate times.

The following references also have complete chapters on the subject: Stumm and Morgan (1996), Snoeyink and Jenkins (1980), and Morel and Hering (1993).

3.8 Acids and Bases

The essential chemical formulae in working with acids and bases are:

$$pH = -\log \{H^+\} \qquad [3-11]$$

and

$$pOH = -\log (OH^-) \qquad [3-12]$$

Equation 3-11 indicates that for pure water, where the concentration of H^+ equals $1 \cdot 10^{-7}$ moles/L, the pH is 7.00. A solution with a hydronium ion concentration above $1 \cdot 10^{-7}$ moles/L is classified as acidic. Equation 3-12 indicates that for pure water where the concentration of OH^- equals $1 \cdot 10^{-7}$ moles/L the pH is also 7.00. A solution with a hydroxide ion concentration above this level is basic.

Acids are compounds that readily release protons in reaction. Bases are compounds that easily accept protons. Acids and bases are classified as strong or weak, determined by the rate at which they react and degree to which reactions are completed. Acids are characterized as strong when they completely release their protons in water; weak acids only partially dissociate. Similarly, strong bases readily accept protons. Strong acids used in environmental engineering include sulfuric acid, nitric acid, hydrochloric acid, and hydrofluoric acid. Bases commonly used for environmental applications include sodium hydroxide, potassium hydroxide and calcium hydroxide. Weak acids and bases with their dissociation constants are shown in Table 3-4. The use of the dissociation constant is as follows:

for a weak acid represented as $HB \longrightarrow H^+ + B^-$

$$K_x = [H^+][B^-]/[HB] \qquad [3-13]$$

$$pK_x = -\log K_x \qquad [3-14]$$

Table 3-4
Dissociation constants for weak acids, bases, and salts at 25°C

Substance	Equilibrium equation	K_X	pK_X	Significance in environmental engineering
Acids				
Acetic	$CH_3COOH \Leftrightarrow H^+ + CH_3COO^-$	$1.8 \cdot 10^{-5}$	4.74	Organic wastes
Ammonium	$NH_4^+ \Leftrightarrow H^+ + NH_3$	$5.56 \cdot 10^{-10}$	9.26	Nitrification
Boric	$H_3BO_3 \Leftrightarrow H^+ + H_2BO_3^-$	$5.8 \cdot 10^{-10} (K_{A1})$	9.24	Nitrogen analysis
Carbonic	$H_2CO_3 \Leftrightarrow H^+ + HCO_3^-$	$4.3 \cdot 10^{-7} (K_{A1})$	6.37	Many applications
	$HCO_3^- \Leftrightarrow H^+ + CO_3^{2-}$	$4.7 \cdot 10^{-11} (K_{A2})$	10.33	
Hydrocyanic	$HCN \Leftrightarrow H^+ + CN^-$	$4.8 \cdot 10^{-10}$	9.32	Toxicity
Hydrogen sulfide	$H_2S \Leftrightarrow H^+ + HS^-$	$9.1 \cdot 10^{-8} (K_{A1})$	7.04	Odors, corrosion
	$HS^- \Leftrightarrow H^+ + S^{2-}$	$1.3 \cdot 10^{-13} (K_{A1})$	12.89	
Hypochlorous	$HOCl \Leftrightarrow H^+ + OCl^-$	$2.9 \cdot 10^{-8}$	7.54	Disinfection
Phenol	$C_6H_5OH \Leftrightarrow H^+ + C_6H_5O^-$	$1.2 \cdot 10^{-10}$	9.92	Tastes, industrial waste
Phosphoric	$H_3PO_4 \Leftrightarrow H^+ + H_2PO_4^-$	$7.5 \cdot 10^{-3} (K_{A1})$	2.12	Analytical buffer, plant nutrient
	$H_2PO_4^- \Leftrightarrow H^+ + HPO_4^{2-}$	$6.2 \cdot 10^{-8} (K_{A2})$	7.21	
	$HPO_4^{2-} \Leftrightarrow H^+ + PO_4^{3-}$	$4.8 \cdot 10^{-13} (K_{A3})$	12.32	
Propionic	$CH_3CH_2COOH \Leftrightarrow H^+ + CH_3CH_2COO^-$	$1.3 \cdot 10^{-5}$	4.89	Organic wastes, anaerobic digestion
Bases and Salts				
Acetate	$CH_3COO^- + H_2O \Leftrightarrow CH_2COOH + OH^-$	$5.56 \cdot 10^{-10}$	9.26	Organic wastes
Ammonia	$NH_3 + H_2O \Leftrightarrow NH_4^+ + OH^-$	$1.8 \cdot 10^{-5}$	4.74	Disinfection, nutrient
Borate	$H_3BO_3^- + H_2O \Leftrightarrow H_3BO_3 + OH^-$	$1.72 \cdot 10^{-5}$	4.76	Nitrogen analysis
Carbonate	$CO_3^{2-} + H_2O \Leftrightarrow HCO_3^- + OH^-$	$2.13 \cdot 10^{-4} (K_{B2})$	3.67	Many applications
	$HCO_3^- + H_2O \Leftrightarrow H_2CO_3 + OH^-$	$2.33 \cdot 10^{-8} (K_{B1})$	7.63	
Calcium hydroxide	$CaOH^+ \Leftrightarrow Ca^{2+} + OH^-$	$3.5 \cdot 10^{-2} (K_{B2})$	1.46	Softening
Magnesium hydroxide	$MgOH^+ \Leftrightarrow Mg^{2+} + OH^-$	$2.6 \cdot 10^{-3} (K_{B2})$	2.59	Softening

Source: Sawyer, McCarty, and Parkin, 1994.

Example 3-5 — Acid-base reaction

Find the pH of a solution of distilled water and nitric acid having a concentration of 0.01 moles/L.

Solution:

The molecular weight for HNO_3 can be calculated from Table 3-1 as

$(H = 1) + (N = 14) + (O = 3 \bullet 16) = 63$ g/g-mole

Note that the molecular weight for hydronium is 1 g/g-mole. Following Equation 3-13, this means a 0.01 molar concentration of nitric acid has a H^+ of 0.01 g/L. Since nitric is a strong acid, all of the H^+ dissociates. Therefore:

$pH = -\log \{H^+\} = -\log [1 \bullet 10^{-2}] = 2.0$

Acid-base reactions are an important part of environmental chemistry. The text by Snoeyink and Jenkins (1980) devotes an entire chapter to the subject because of the importance of these reactions and their complexity when weak acids and bases are involved.

3.9 Standard Chemistry Notations (Chemical symbols are listed in Table 3-1)

Symbol	Definition
C	concentration
g	gram
k	kinetic rate constant
K_x	dissociation constant
L	liter
lb	pound
mg	milligram
Mgal/day	million gallons per day
PK_x	log of K_x
t	time
[A]	molar concentration of A chemical

3.10 References

Manahan, Stanley E. 1991. *Environmental Chemistry*. 5th ed. Boca Raton, Florida: Lewis Publishers.

Morel, Francois M.M., and Janet G. Hering. 1993. *Principles and Applications of Aquatic Chemistry*. New York: John Wiley & Sons.

Sawyer, Clair N., Perry L. McCarty and Gene F. Parkin. 1994. *Chemistry for Environmental Engineers*. 4th ed. New York: McGraw-Hill, Inc.

Snoeyink, Vernon L., and David Jenkins. 1980. *Water Chemistry*. New York: John Wiley & Sons.

Stumm, Werner, and James J. Morgan. 1996. *Aquatic Chemistry*. 3rd ed. New York: John Wiley & Sons.

Chapter 4

FLUID MECHANICS

by Richard Dominguez, Ph.D., P.E.

4.1 Introduction

This chapter provides a review of fundamental concepts that apply to water, an *incompressible* fluid, and air, a *compressible* fluid, which are the media with which environmental engineers work. Environmental problems associated with these media can find solution through the application of fluid mechanics principles.

Only concepts considered essential to an environmental engineer's basic understanding are developed here. Other material is provided in abbreviated form or simply stated without proof. For some, it may be necessary or desirable to review basic fluid mechanics in greater detail. For this purpose, a list of suggested references is provided at the end of this chapter.

4.2 Fluid and Flow Properties

Fluid properties such as density, viscosity, specific weight, vapor pressure, and surface energy depend on the molecular structure of the fluid. *Flow properties* like velocity and pressure depend on both the fluid's properties and the dynamic conditions to which the fluid is subjected.

Fluids, both liquids and gases, are distinguished from solids by the inability to support shear stress while at rest. By contrast, in a solid the *shear stress* τ is proportional to the static shear deformation defined by the *shear strain* γ_s and the material's *shear modulus of elasticity G*. The relationship is given by Hook's Law:

$$\tau = G\gamma_s \qquad [4\text{-}1]$$

where: τ = shear stress
        G = shear modulus of elasticity
        γ_s = shear strain

Figure 4-1
Shear stress in a fluid

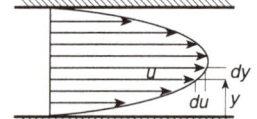
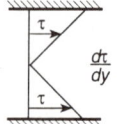

For shear stress to occur in a fluid, the fluid must be continuously deformed, as illustrated in Figure 4-1. This requires a variation in the velocity u with position y defining a local velocity gradient du/dy. This is expressed by the relationship:

$$\tau = \mu\, du/dy \qquad [4\text{-}2]$$

where μ is the constant of proportionality relating τ to du/dy, known as the *dynamic viscosity*. It is a property of the fluid.

Fluids that obey Equation 4-2 are known as *Newtonian fluids* and plot as straight lines whose slope is the *dynamic viscosity* μ as shown in Figure 4-2. Water and air are common Newtonian fluids. Fluids that do not conform to Equation 4-2 are called *non-Newtonian fluids* and are the subject of rheology.

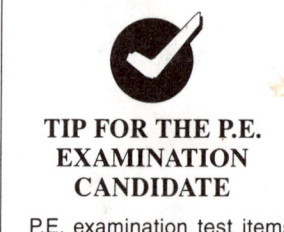

TIP FOR THE P.E. EXAMINATION CANDIDATE

P.E. examination test items only involve Newtonian fluids.

The dynamic viscosity μ must have fundamental units of F/L^2 to satisfy Equation 4-2. Fluid viscosity may also be expressed in units which are free of either force or mass units by means of the *kinematic viscosity* ν. The relationship between dynamic and kinematic viscosity is given by:

$$\mu = \rho\nu \qquad [4\text{-}3]$$

where ρ is the *mass density* of the fluid, defined as the mass per unit volume.

The *specific weight* γ of a fluid, defined as the weight per unit volume, is related to mass density in the following way:

$$\gamma = \rho g \qquad [4\text{-}4]$$

where *g* is the *gravitational constant* commonly assumed to have the value of 32.2 ft/sec² or 9.81 m/sec². The reciprocal of the specific weight is the *specific volume* $v = 1/\gamma$.

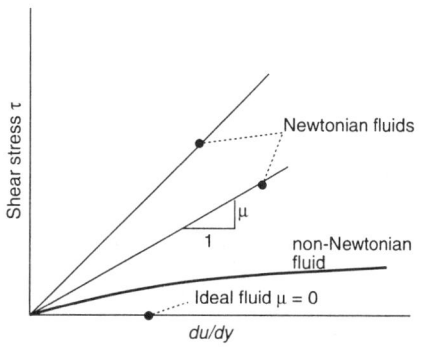

Figure 4-2
Dynamic viscosity defined by shear stress and rate of fluid deformation

Basic to the understanding of all flow problems is the relationship between the forces acting on the fluid and the fluid's motion. In theory, if all forces that act on the fluid can be correctly established, then its motion can be predicted. In some instances, it may be the motion that is known and the forces associated with it are sought.

The forces that act on a fluid particle are either *body forces*, as is the case when the fluid's weight must be considered, or *surface forces* composed of shear forces, normal forces, or both. A *shear force* F_t is calculated as $F_t = \tau A$, where *A* is the surface area that the shear stress acts on tangentially.

A *normal force* F_p is calculated as the product of the normal stress (which in fluid mechanics is the pressure *p*) and the surface area *A*, which is normal to *p*. Thus $F_p = pA$.

4.3 Hydrostatics

When the forces acting on a fluid particle are such that the vector sum of the forces is 0, then by Newton's Second Law of Motion, the flow is non-accelerating since $\Sigma F = ma = 0$. Since *m* is not equal to 0, *a* must be 0.

When this is the case for all particles that form the flow field, it means either that the velocity is 0 everywhere, which defines a hydrostatic condition, or that a non-accelerating, constant-velocity condition exists.

In such a case, the pressure variation from point-to-point is given by:

$$dp/dh = -\rho g \qquad [4\text{-}5]$$

Here h denotes the elevation of the point at which the pressure is computed. To compute the pressure variation, the relationship between ρ and either p or h must be known. If the variation in ρ is small, as is the case with water, the fluid may be considered *incompressible* and ρ treated as a constant, independent of either pressure or elevation. In this situation, Equation 4-5 simplifies to:

$$p/\rho + gh = \text{constant}$$

or

$$p = -\rho gh = -\gamma h \qquad [4\text{-}6]$$

This is the hydrostatic pressure equation.

4.4 Conservation Principles

4.4.1 Conservation of Mass The principle of conservation of mass is defined by considering a streamtube of varying cross-section as shown in Figure 4-3.

Figure 4-3
Streamline and streamtube definition

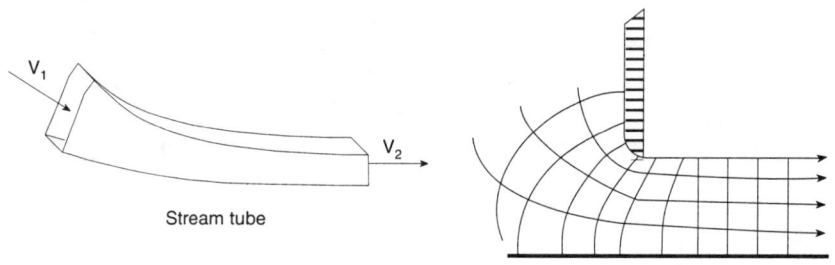

Stream tube

The mass passing through any differential stream tube area is $dQ_m = \rho v dA$. The total mass entering and leaving any section of the streamtube must be equal, as mass is neither created, destroyed, or stored. Thus, the total mass flow Q_m passing any section is:

$$Q_m = \int \rho v dA = \rho A_1 V_1 = \rho A_2 V_2 = \text{constant} \qquad [4\text{-}7]$$

where V_1 and V_2 are the average velocities at sections 1 and 2.

Chapter 4: Fluid Mechanics

For incompressible flow i.e., water, $\rho_1 = \rho_2 = \rho$ and Equation 4-7 reduces to

$$Q = A_1 V_1 = A_2 V_2 = \text{constant} \qquad [4\text{-}8]$$

where Q is the volumetric discharge. This is a form of the *continuity equation* applicable to one-dimensional, steady flow.

4.4.2 Momentum
Newton's second law of motion:

$$\Sigma F = m\mathbf{a} = m\, dv/dt \qquad [4\text{-}9]$$

states that the sum of all forces acting on a body equals the product of the body's mass m and the resultant acceleration $\mathbf{a}$ that the forces cause. Since $\mathbf{a}$ by definition is dv/dt, the time rate of change of velocity (either magnitude or direction), Equation 4-9 may be regarded as expressing the principle that the sum of all forces acting on a body must equal the change in its linear momentum mv. Accordingly, Equation 4-9 may be written as:

$$\Sigma F\, dt = m\, dv$$
sum of impulse forces = change in linear momentum $\qquad [4\text{-}10]$

In this form, the *impulse — momentum principle* is applicable to a solid. By considering a control volume as shown in Figure 4-4 which all fluid passes into and out of, Equation 4-10 can be extended to fluid behavior. Only the forces acting on the boundaries need be considered in calculating the change of momentum, as all internal fluid particle interactions cancel (Newton's Third law — for every action there is an equal and opposite reaction).

Figure 4-4
Control volume for momentum balance

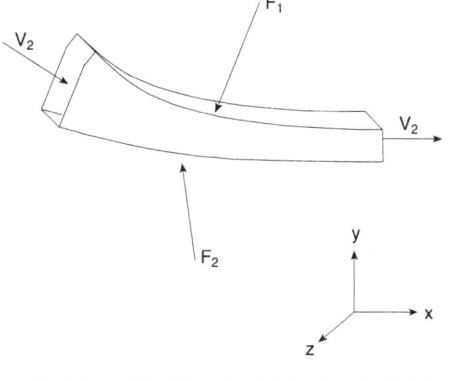

Since a continuous stream of flow is involved, the impulses take the form of continuously applied external forces. The mass passing through any section of the control section in a time interval dt is $\rho Q dt$. Using the average velocity in the x, y, and z directions, Equation 4-10 may be written in a convenient component form for application, which can be applied in the x, y, z directions

$$\Sigma F_x = \rho Q(V_2 - V_1)_x$$

$$\Sigma F_y = \rho Q(V_2 - V_1)_y \qquad [4\text{-}11]$$

$$\Sigma F_z = \rho Q(V_2 - V_1)_z$$

The ΣF_x, ΣF_y, and ΣF_z, are the forces acting on the fluid boundaries, such as pressure and shear force, as well as body forces, such as fluid weight. In applying Equation 4-11 it must be remembered that the momentum equation is a vector equation and the force and velocity terms must be treated as vectors.

4.4.3 Conservation of Energy

Energy, which is defined as the ability to do work, may take several forms in fluids. These include mechanical, thermal, and chemical energy.

The principle of conservation of energy requires that:

$$E_k + E_p + E_z + E_M + E_T = \text{constant} \qquad [4\text{-}12]$$

where:
- E_k = the kinetic energy
- E_p = the pressure energy
- E_z = the potential energy
- E_M = the mechanical energy added or removed from the fluid
- E_T = the thermal energy

The first three terms, E_k, E_p, and E_z, are associated with the energy the fluid has at a point on a streamline forming the flow pattern. The last two terms, E_M and E_T, account for the addition or loss of mechanical energy from the streamline by boundary resistance (fluid friction), pumps, turbines, or heat exchangers.

Fluid mechanics is concerned with the mechanical forms of energy and with mechanical energy loss through transformation to thermal energy. Newton's Second Law ($\Sigma F = ma = 0$) can be used along with the principle of work to derive the *energy equation* for an incompressible fluid.

Referring to Figure 4-5, Equation 4-12 may be written for any two points, 1 and 2, along a streamline as:

$$V_1^2/2g + p_1/\gamma + z_1 + E_{M1} + E_{T1} = V_2^2/2g + p_2/\gamma + z_2 + E_{M2} + E_{T2} = \text{constant} \qquad [4\text{-}13]$$

If there are no energy losses or additions between points 1 and 2, then Equation 4-13 becomes:

$$V_1^2/2g + p_1/\gamma + z_1 = V_2^2/2g + p_2/\gamma + z_2 = \text{constant} \qquad [4\text{-}14]$$

Figure 4-5
Energy change along a streamline

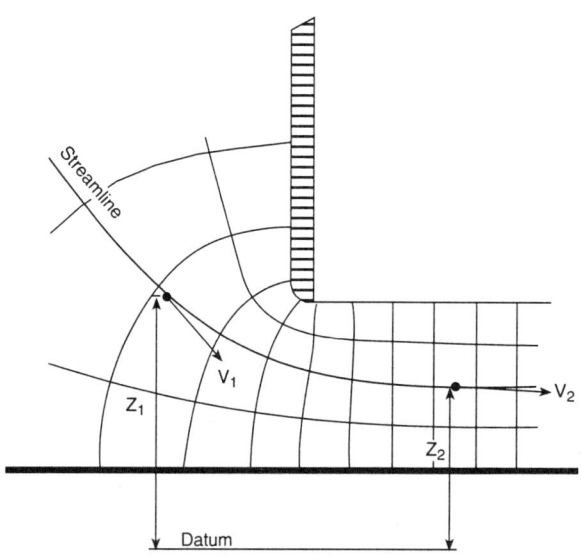

Each term in Equation 4-14 has the basic units of energy per unit weight. Expressed in any set of consistent units, this is FL/F or units of length L. This is the reason that the term *head* has found usage in hydraulics. It expresses the energy as the equivalent height above a datum that a unit weight of fluid must be placed to have a given energy level.

Thus, the constant in Equation 4-14 may be viewed as the total head H_T at a point. Therefore,

$$V_1^2/2g \;\; + \;\; p_1/\gamma \;\; + \;\; z_1 \;\; = \;\; H_T$$
$$\text{velocity head} \quad \text{pressure head} \quad \text{elevation head} \quad \text{total head}$$

[4-15]

Equation 4-15 is the *energy* or *Bernoulli equation* applied along a streamline or streamtube. It may be applied across streamlines only when the flow can be reasonably treated as being *irrotational*. When this is not the case, the application of Equation 4-15 across streamlines will lead to inexact results.

The development of the energy equation in the above form is based on a number of important assumptions: the flow is steady (no variation of velocity at a point with time); it is incompressible (γ is a constant); the energy is constant along the streamline; and the flow can be treated as either one-dimensional or irrotational.

When energy losses or additions occur between two points, Equation 4-13 may be modified to reflect this by adding these terms to the left and right hand sides of the equation in the following manner:

$$V_1^2/2g + p_1/g + z_1 + E_{M1 \to 2} + E_{T1 \to 2} = V_2^2/2g + p_2/g + z_2 + \Sigma H_L \quad [4\text{-}16]$$

in this equation ΣH_L is the sum of all energy loss terms associated with boundary resistance and flow transition.

4.5 Conduit Flow

Closed conduit flow (also called *pipe flow*), refers to the motion of a fluid within a conduit caused by pressure forces under conditions when the conduit is *full* at all times. If the conduit flows partially full (a special use applicable to water only) and therefore has a free surface, the principal force controlling the motion is gravity, and fluid motion must be analyzed as open-channel flow. This type of flow will be covered in Chapter 5 on Hydraulics and Hydrology.

TIP FOR THE P.E. EXAMINATION CANDIDATE

Bernoulli's equation is one of the most important, if not the most, in environmental engineering. It can be used to solve most any problem involving the movement of water or air.

Flow within a conduit may be *laminar, fully turbulent*, or *transitional*, depending on the velocity, conduit diameter, conduit wall roughness, and the viscosity of the fluid. Which of the three flow regimes occurs can be predicted by the *Reynolds number* N_R, using the conduit diameter as the characteristic length.

$$N_R = \rho V d / \mu = V d / \nu \quad [4\text{-}17]$$

where: d = inside conduit diameter
V = average velocity

The Reynolds number is a dimensionless quantity that represents the *ratio of inertial forces to viscous forces*. Thus, the smaller the value of the N_R, the larger the viscous forces are with respect to inertial forces associated with the fluid motion. The nominal value at which laminar flow structure changes to turbulent is considered to be $N_R \approx 2000$. Figure 4-6 illustrates the development of the velocity profile from the point at which flow enters the conduit until it becomes fully developed and the profile no longer changes. Fully developed laminar and turbulent velocity profiles are plotted and shown in Figure 4-6, which may be compared with the average velocity V, which is also shown for comparison.

4.5.1 Major Energy Losses Energy loss that occurs as the fluid passes through the system is paramount in the design of conduit systems. It depends on the *regime* under which the system operates.

Figure 4-6
Fully developed laminar and turbulent profiles having the same average velocity

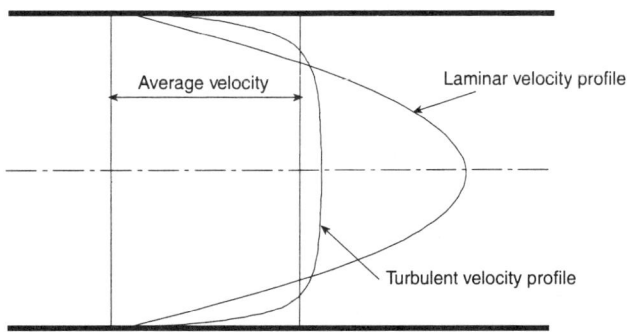

Flow in closed conduits has been the subject of investigation since man began to transfer water from one location to another. Modern design practice has its roots in the experimental work of Nikuradse; the subsequent development of equations that fit his data (the Colebrook-White equations for application to commercial pipes); and the graphical solution of these equations in the form of the Moody diagram shown in Figure 4-7.

The energy lost between two points in a conduit under fully developed, uniform, steady-flow conditions can be evaluated by:

$$h_f = (fL/d)V^2/2g \qquad [4\text{-}18]$$

where h_f is head loss, expressed as some multiple of the velocity head $V^2/2g$ by the numerical value of the term fL/d; L is the length of the pipe over which the loss is calculated; d is the inside pipe diameter; and f is the Darcey-Weisbach resistance coefficient, which is a function of the N_R and the relative pipe roughness ε/d. The value of f is obtained from the Moody diagram or by direct solution of the Colebrook-White equations. Figure 4-8 can be used to determine the ε/d for common pipe materials.

The *Moody diagram* shows the complex dependence the resistance coefficient has on the Reynolds number and pipe roughness. For low Reynolds numbers (less than 2,000), the flow is laminar and the resistance is *independent of pipe roughness*. Conversely, at high Reynolds numbers (>4,000), the flow is turbulent and, if the pipe is hydraulically *rough, f* will be independent of the Reynolds number and

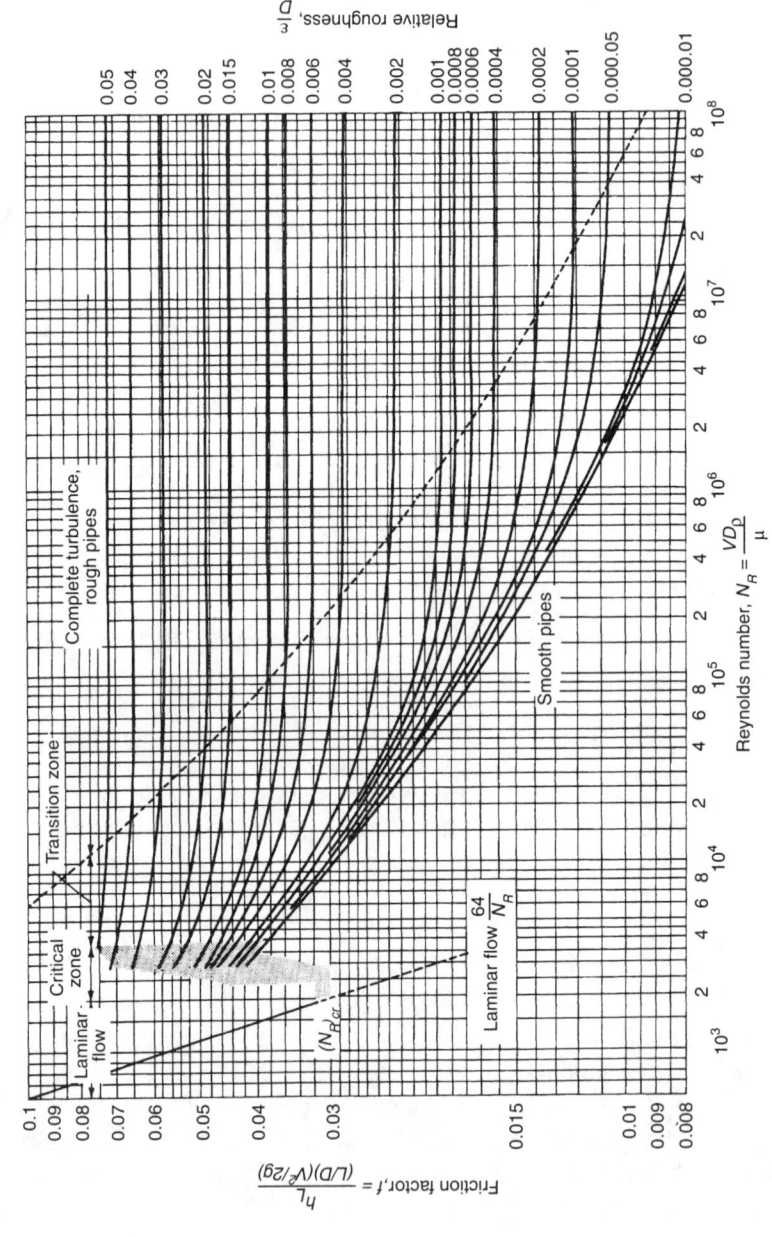

Figure 4-7
Moody diagram for determination of the Darcy-Weisbach friction factor in pipes

Figure 4-8
Moody diagram for relative roughness as a function of diameter for pipes constructed of various materials

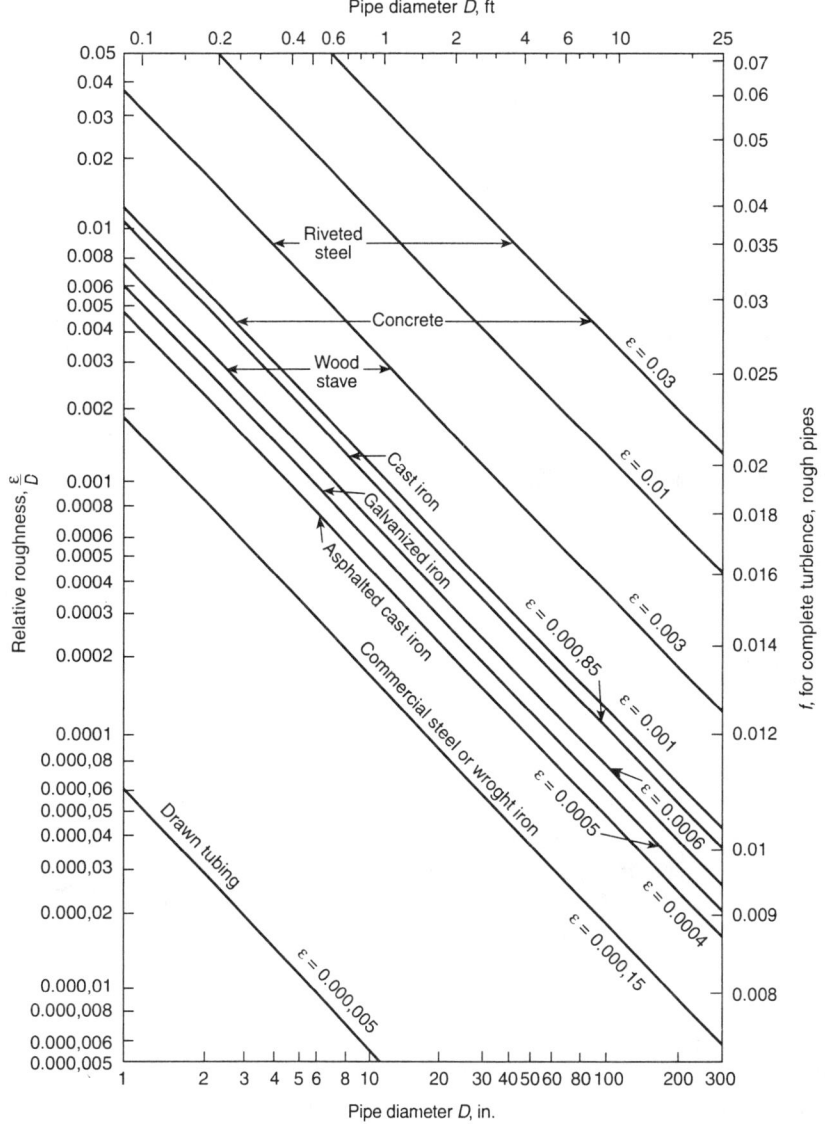

only dependent on ε/d. However, if the pipe is hydraulically *smooth*, then like laminar flow, f, depends only on the Reynolds number and can be determined by:

$$f = 0.316 / N_R^{1/4} \qquad [4\text{-}19]$$

An understanding of this seemingly paradoxical behavior rests on understanding the growth and stability of the boundary layer that forms at the pipe wall and extends to the centerline of the pipe. At the pipe wall, the velocity for a real fluid is zero and rapidly increases in the direction normal to the pipe wall, reaching its maximum value at the center of the pipe. If the pipe is hydraulically smooth, then a region of laminar flow will form close to the wall (the laminar sublayer). At increasing distance from the wall, the velocity is great enough that a laminar structure cannot be sustained and the flow structure begins to break down into intermittent zones of turbulence. At some further distance from the wall, the velocity is so large that a laminar flow structure cannot be sustained even momentarily. This defines where fully turbulent flow begins. The flow structure regions are illustrated in Figure 4-9.

Figure 4-9
Three flow structure regions

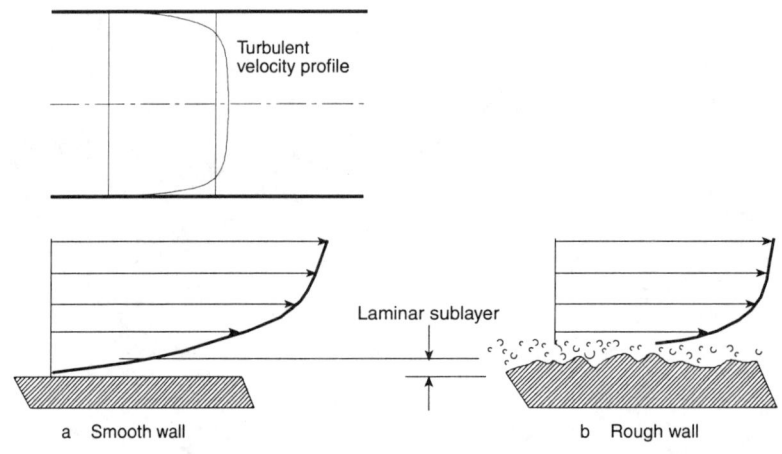

If the wall roughness elements are large in comparison to the thickness of a potential laminar sub layer, the local flow next to the wall will be disturbed to the extent that a laminar flow structure cannot form, in which case the flow structure is turbulent across the entire pipe as shown in Figure 4-9b.

Between smooth and rough turbulent flow falls the *zone of transition* on the Moody diagram. Pipe flow resistance in the zone depends on both the Reynolds number and relative roughness.

4.5.2 Minor Energy Losses
Energy losses calculated by Equation 4-18 are called *major losses* and account for the energy lost over pipe sections where the flow is uniform. Additionally, other important losses occur as a result of non-uniform flow conditions. These occur at entrance and exit points, valves, pipe diameter transitions *bends*, and other places where significant boundary changes exist. Such losses are termed *minor losses*, although in fact they may contribute to a large part of the total energy loss in the pipeline system. Minor losses are calculated by:

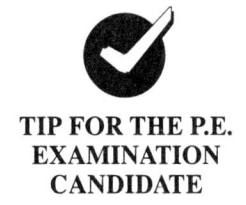

TIP FOR THE P.E. EXAMINATION CANDIDATE

Do not neglect minor hydraulic losses in P.E. examination test items unless directed to do so in the problem statement.

$$h_L = C_L V^2 / 2g \qquad [4\text{-}20]$$

where h_L is the head loss and C_L is an experimentally-determined coefficient accounting for the loss. A limited list of loss coefficients is given in Table 4-1. An extensive list of minor loss coefficients can be found in Davis (1952).

Table 4-1
Loss coefficents for commercial pipe fittings

Fitting	C_L
Globe valve, fully open	10
Angle valve, fully open	5
Swing check valve, fully open	2.5
Closed return bend	2.2
Tee, through side outlet	1.8
Short radius elbow	0.9
Medium radius elbow	0.8
Long radius elbow	0.6
45-degree elbow	0.4
Gate valve, fully open	0.2
Gate valve, 3/4 open	1
Gate valve, 1/2 open	5.6
Gate valve, 1/4 open	24

4.5.3 Energy and Hydraulic Grade Lines
The *energy grade line* (EGL) defines the total energy the fluid has at any point along the hydraulic system. The slope of the EGL is a measure of the rate at which energy is lost or gained by the fluid at that point in the system. The EGL is the sum of all terms in the energy equation Equation 4-16. The *hydraulic grade line* (HGL) defines the *piezometric head* at any point in the system and is the sum of the pressure and elevation heads. It reflects the height that a water column would reach in a manometer if connected to the system at a point. Figure 4-10 illustrates the variation in hydraulic and energy grade lines and the effect on energy loss through boundary transitions that produce non-uniform flow.

Piping analysis and design problems normally fall in one of three groupings:

Case 1. The energy loss in the system is sought knowing the velocity or discharge and both the pipe and fluid properties. This is a direct solution of Equation 4-18 using the Moody diagram to obtain f.

Figure 4-10
EGL and HGL for a closed conduit system

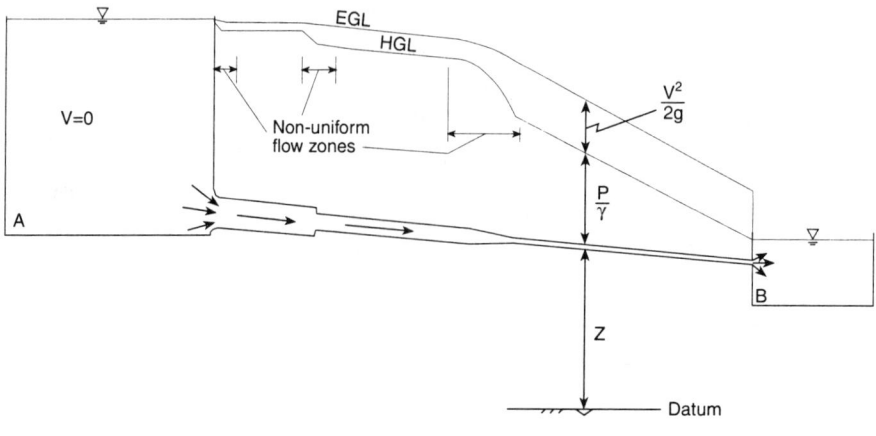

Case 2. The *velocity* or *discharge* is sought knowing the energy loss, the pipe and the fluid properties. This takes the form of a trial-and-error solution where an estimate of velocity or discharge is made to calculate a Reynolds number, which is then used to get an estimate of the friction coefficient f. The discharge is then checked by:

$$Q = [\pi^2 g H_L d^5 / 8fL]^{1/2} \quad [4\text{-}21]$$

until reasonable agreement between the values of Q are obtained.

Case 3. The *pipe diameter* is sought knowing either the velocity or discharge, the energy loss, pipe length, roughness and fluid properties. This also is a trial-and-error solution process. A trial d is estimated, which permits the calculation of the Reynolds number and relative roughness. The Moody diagram is then used to obtain a value for f, which is then used to calculate the energy loss, which is compared with the known value. If agreement is found, d is correct. If not, a new estimate is made and the process repeated.

TIP FOR THE P.E. EXAMINATION CANDIDATE

A sketch of a hydraulic problem statement like that shown in figure 4-10 can assist in correctly solving such problems.

The preceding discussion is illustrated by means of examples.

Example 4-1

Referring to the figure below, two large detention basins, A and B, are connected by a cast-iron pipeline with ε = 0.001 that has a gate valve to control flow. The pipeline is constructed as a 12 in., 120 ft section connected to a 4 in., 100 ft section by means of a transition.

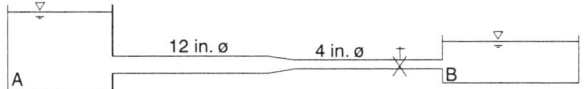

Determine the maximum flow possible when the elevation between the two basins is 30 ft.

Solution:

At maximum flow, the gate valve will be fully open and the energy loss that occurs in the pipeline between A and B is equal to the elevation difference, thus

Step 1. $\Sigma h_f + \Sigma h_L = \Sigma H_L = 30 \text{ ft}$

Step 2. $\varepsilon_1 / d_1 = 0.001 / 1 = 0.001$

$\varepsilon_2 / d_2 = 0.001 / 0.333 = 0.003$

Assume rough turbulent flow. From the Moody diagram

$f_1 = 0.0195$
$f_2 = 0.026$

Step 3. For the 12 in. pipe:

$$h_{f1} = (f_1 L_1 / d_1) V_1^2 / 2g$$
$$= (0.0195)(120) V_1^2 / 2g$$
$$= 2.34 V_1^2 / 2g$$

For the 4 in. pipe:

$$h_{f2} = (f_2 L_2 / d_2) V_2^2 / 2g$$
$$= (0.026)(100 / 0.333) V_2^2 / 2g$$
$$= 7.81 V_2^2 / 2g$$

Step 4. Minor losses: $h_L = C_L V^2 / 2g$

Item	C_L
Square entrance	0.5
Gate valve full open	0.2
Transition $A_2/A_1 = 0.11$	0.43
Square exit	1.0

Step 5. Substituting into the step 1 relationship for energy loss gives:

$$(2.34 + 0.5 + 0.2)V_1^2 / 2g + (7.81 + 0.43 + 1.0)V_2^2 / 2g = 30$$

Step 6. V_2 may be eliminated by means of the conservation-of-mass requirement (Equation 4-8)

$$Q = A_1 V_1 = A_2 V_2 \quad A_1 = 0.785 \,\text{ft}^2 \quad A_2 = 0.0873 \,\text{ft}^2$$
$$V_2 = (0.785 / 0.0873)V_1 = 9V_1 \text{ thus,}$$
$$759 V_1^2 / 2g = 30 \quad \text{with } g = 32.2 \,\text{ft/sec}^2$$
$$V_1 = 1.59 \,\text{ft/sec} \text{ and } V_2 = 14.31 \,\text{ft/sec}$$

Step 7. Check assumed pipe friction coefficients:

$$N_{R1} = V_1 d_1 / v = 1.59(1) / 1.217 \bullet 10^{-5} = 1.3 \bullet 10^5$$
$$N_{R2} = V_2 d_2 / v = 14.31(0.33) / 1.217 \bullet 10^{-5} = 3.88 \bullet 10^5$$

With the above Reynolds numbers, obtain revised values of f from Moody diagram.

$$f_1 = 0.0218 > 0.0195 \text{ and } f_2 = 0.26 \text{ as assumed}$$

If the loss associated with the determination of f_1 were significant, the calculations would be repeated using the new estimate of f_1. However in this case, since the losses associated with the 12-in. pipe are less than one half of 1% of the total loss, further calculation is not warranted.

Step 8. $Q = A_1 V_1 = 0.785(1.59) = 1.25 \,\text{ft}^3 / \text{sec}$

or

$= 1.25(449) = 560 \,\text{gal/min}$

Example 4-2

It is proposed to replace all piping between the detention basins with a new, extremely smooth plastic pipe and a gate valve to control flow. What is the minimum diameter pipe that can be used to achieve the same maximum flow as in the original system?

Solution:

This is a trial-and-error solution (Case 3) and since the flow is hydraulically "smooth" the friction coefficient can be obtained from either the Moody diagram or directly by Equation 4-19. The head across the pipe system is 30 ft, thus a pipe diameter is required that creates a flow that will result in a 30 ft energy loss across the pipe.

$$\Sigma h_f + \Sigma h_L = \Sigma H_L = 30 \text{ ft}$$

A convenient means for solving the problem is in a tabular format using the following relations to compute needed values:

$$f = 0.316 / N_R^{1/4}$$
$$N_R = Vd / v$$
$$A = 0.785 d^2$$
$$V = Q / A$$
$$h_L = C_L V^2 / 2g$$
$$h_f = fL / d V^2 / 2g$$

with

$$L = 220 \text{ ft}$$
$$v = 1.217 \bullet 10^{-5} \text{ ft}^2 / \text{sec}$$
$$Q = 1.25 \text{ ft}^3 / \text{sec}$$
$$\Sigma C_L = 0.5 + 0.2 + 1.0 = 1.7$$

Trial d in.	A ft²	V ft/sec	N_R 1•10⁵ ft²/sec	f ft	V²/2g ft	Σh_L ft	Σh_f ft	Head Loss ft
6.0	0.196	6.37	2.61	0.014	0.63	1.07	3.87	4.94
5.0	0.136	9.17	3.14	0.013	1.31	2.22	9.21	11.43
4.0	0.087	14.33	3.93	0.013	3.19	5.42	26.57	31.99

With a 4-in. pipe, the estimated head loss is 31.99 ft, which is slightly greater than the actual head loss, which must be 30 ft. We can now determine the flow rate that a 4-in. pipe would permit by using Equation 4-21.

$$Q = [\pi^2 g H_L d^5 / 8fL]^{1/2}$$
$$= [(3.14)^2 (32.2)(30)(4/12)^5 / 8(0.013)(220)]^{1/2}$$
$$= 1.31 \text{ ft}^3/\text{sec} > 1.25 \text{ ft}^3/\text{sec, therefore select 4-in. diameter pipe.}$$

4.5.6 Graphical solution When the required pipe diameter d is sought, obtaining the friction factor f from the Moody diagram requires both the relative roughness and Reynolds number, both of which involve the unknown pipe diameter. This situation, is the same as illustrated in Example 4-2 in that it involves a trial and error approach. A direct solution for the pipe diameter is possible using the *Li diagram* shown in Figure 4-11. The Li diagram is a modification of the Moody diagram (Figure 4-7) arranged so that the pipe diameter d appears in only one of the three coordinates. The use of the Li diagram is illustrated in the following example.

Figure 4-11
Li diagram for graphical solution the pipe diameter d

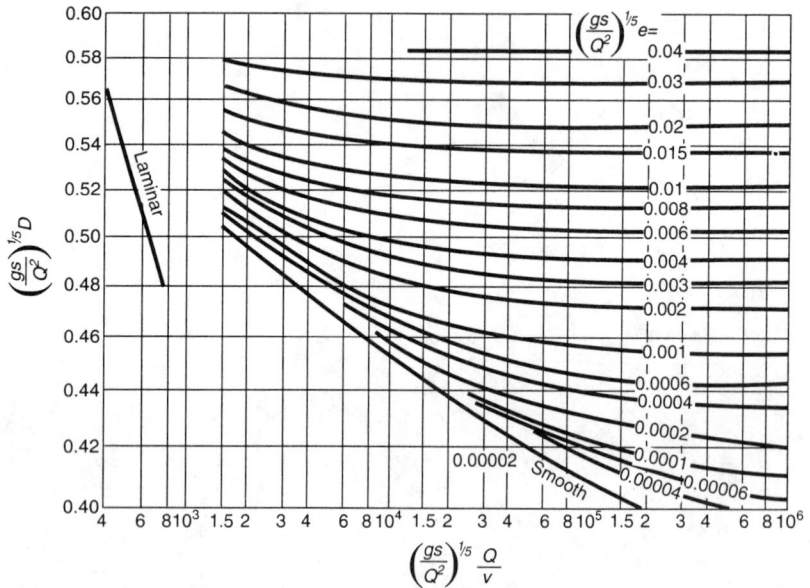

Example 4-3

What diameter steel pipe is required to convey 3 ft³/sec of water at 50°F if the maximum pressure drop over a 2,000-ft horizontal length is 45 psi?

Solution:

Step 1 $v = 1.41 \bullet 10^{-5} \text{ ft}^2 / \text{sec}$

$\gamma = 62.4 \text{ lb} / \text{ft}^3$

$\varepsilon = 0.0002 \text{ ft}$

$h_f = (45)(144) / 62.4$
$= 103.8 \text{ ft}$

Step 2. Calculate the value of the plotting term $[gS_f / Q_2]^{1/5}$ used with d and Q/v. The slope of the energy grade line is:

$$S_f = 103.8 / 2000 = 0.0542$$

$$[gS_f / Q^2]^{1/5} = [(32.2)(0.0542) / (3)^2]^{1/5} = 0.720$$

$$[gS_f / Q^2]^{1/5} Q / v = (0.720)(3) / 1.41 \bullet 10^{-5} = 1.53 \bullet 10^5$$

$$[gS_f / Q^2]^{1/5} \varepsilon = (0.720)(0.0002) = 1.44 \bullet 10^{-4}$$

Step 3. Enter the Li diagram with $1.53 \bullet 10^5$ and $1.44 \bullet 10^{-4}$ and read the value of

$[gS_f / Q^2]^{1/5} d = 0.41$ thus,

$d = 0.41 / 0.720 = 0.569 \text{ ft}$ or 6.83 in.

Select the nearest commercially available size larger than 6.83 in. Use 8 in.

Step 4. *Check Solution* by Equation 4-18 by first calculating

$N_R = Qd / Av = (3)(0.667) / (0.349)(1.41 \bullet 10^{-5}) = 4 \bullet 10^5$

$\varepsilon / d = 0.0002 / 0.667 = 0.0003$

From the Moody diagram $f = 0.0165$

$h_f = (fL / d)V^2 / 2g = (0.0165)(2000 / 0.667)(8.60)^2 / (2)(32.2)$
$= 56.8 \text{ ft}$ or 25 psi, which is less than 45 psi.

4.7 Standard Fluids Notations

Symbol	Definition	Units
a	Acceleration	L/T^2
A	Area	L^2
cfs	cubic feet per second	
C_L		
d	Pipe Diameter	L
E	Energy	none
EGL	Energy grade line	
E_k	Kinetic Energy	FL
E_p	Pressure Energy	FL
E_z	Potential Energy	FL
E_M	Mechanical Energy	FL
E_T	Thermal Energy	FL
f	Friction Factor	none
F	Force = ma[1]	F or ML/T^2
g	Gravitational Constant	L/T^2
G	Modulus of Elasticity (shear modulus)	none
h	Elevation	
HGL	Hydraulic grade line	
H_L	Head Loss	L
h_f	Head Loss	L
L	Length	
m	mass	M or FT^2/L
n	Manning Head Loss Friction Factor	none
N_R	Reynolds Number	none
p	Pressure	F/L^2
q	—	—
Q	Volumetric Flow Rate	L^3/T
Q_{gpm}	Volumetric Flow Rate in gpm	L^3/T
Q_m	Mass Flow Rate	M/T
t	Time	T
v	Velocity Vector	L/T
u	Velocity Component in x — direction	L/T
v	Velocity Component in y — direction	L/T
V	Average Velocity for One-dimensional Flow	L/T
y	Distance from the wall	L

---

[1] This relationship may be used to convert any set of units from force to mass or mass into force.

z	Elevation head	L
Greek Symbols		
γ_s	Shear strain	none
γ	Specific Weight	F/L^3
ν	Kinematic Viscosity & specific volume	L^2/T
μ	Dynamic Viscosity	FT/L^2
π	3.14159	none
ρ	Mass Density	M/L^2
τ	Shear Stress	F/L^2

4.8 References

Albertson, M.L., J.R. Barton and D.B. Simons. 1960. *Fluid Mechanics for Engineers*. Upper Saddle River, New Jersey: Prentice Hall.

Bird, R.B., W.E. Stewart and E.N. Lightfoot. 1960 *Transport Phenomena*. New York: Wiley & Sons.

Chow, Ven Te, ed. 1964. *Handbook of Applied Hydrology*. New York: McGraw-Hill.

Chow, Ven Te, D.R. Maidment and L.W.Mays. 1988. *Applied Hydrology*. New York: McGraw-Hill.

Davis, C.V. 1952. *Handbook of Applied Hydraulics*. New York: McGraw-Hill.

King, H.W., and E.F. Brater. 1963. *Handbook of Hydraulics*. 5th ed. New York: McGraw-Hill.

Roberson, J.A., and C.T. Crowe. 1985. *Engineering Fluid Mechanics*. 3rd ed. Boston: Houghton-Mifflin.

Shames, I.V. 1982. *Mechanics of Fluids*. 2nd ed. New York: McGraw-Hill.

Simon, A.L. 1981. *Practical Hydraulics*. 2nd ed. New York: John Wiley & Sons.

Chapter 5

HYDRAULICS & HYDROLOGY

by Richard Dominguez, PhD, P.E.

5.1 Introduction

The work of many environmental engineers involves supplying water for domestic and industrial consumption and/the management of rainfall runoff and contaminated water after its primary use. Accordingly, water is, for environmental engineers, a fluid of special interest. *Hydraulics* is the branch of engineering mechanics that deal with the behavior of water. The flow of water in closed and open conduits and the selection of pumps,are discussed in this chapter.

This chapter also discusses essential aspects of *hydrology*, the science of water, its properties and distribution phenomena in both surface and underground regimes.

5.2 Hazen-Williams pipe flow analysis

The preceding chapter describes the strongest scientific basis for predicting water flow in pipes flowing full and provides the most accurate basis for calculations.

However, there are several empirical equations that have historically been used to predict pipe flow. The most widely used, still in use today, is the Hazen-Williams formula:

$$Q = 0.285 C d^{2.63} S_f^{0.54}$$ [5-1]

where:
- Q = the discharge in gallons per minute
- d = the pipe diameter in inches
- S_f = the slope of the energy grade line

Pipe resistance is accounted for by the Hazen-Williams friction factor C in the equation. A list of friction coefficients for use in the Hazen-Williams equation can be found in Table 5-1.

TIP FOR THE P.E. EXAMINATION CANDIDATE

The Hazen-Williams formula is most commonly used for full pipe flow calculations as in waster supply situations.

Table 5-1
Hazen-Williams C Values

Pipe description	C
Asbestos cement	140
Brass	130-140
Cast iron	
New, unlined	130
Old, unlined	40-120
Cement lined	130-150
Tar-coated	115-135
Concrete or concrete-lined	
Steel-formed	140
Wooden-formed	120
Copper	130-140
Fire hose (rubber-lined)	135
Galvanized iron	120
Glass	140
Lead	130-140
Plastic	140-150
Steel	
Coal-tar enamel lined	145-150
New, unlined	140-150
Riveted	110
Clay	100-140
Old pipes in bad condition	60-80

Source: Simon, 1981.

It should be noted that the Hazen-Williams equation makes no provision for fluid properties and, therefore, can only be used for water, within the temperature range associated with municipal cold water distribution systems. Also, pipeline velocities must be less than 10 ft/sec and pipe diameters greater than 2 in. Application of the Hazen-Williams formula is illustrated in Example 5-1.

Example 5-1

What is the value of C that would predict a 3 ft³/sec discharge in an 8-in. diameter pipeline that is 2,000 ft long and has a headloss of 56.8 ft.

Solution:

Step 1. Convert problem data for use in the Hazen-Williams formula

$$Q = 3\,\text{ft}^3/\sec = (3)(449) = 1347\,\text{gal}/\min$$
$$S_f = 56.8/2000 = 0.0284\,\text{ft}/\text{ft}$$

Step 2. The calculated values are then substituted into the Hazen-Williams equation and C solved for:

$$Q = 0.285 C d^{2.63} S_f^{0.54}$$
$$1347 = 0.285 C (8)^{2.63} (0.0284)^{0.54}$$
$$C = 136$$

5.3 Pipe networks

Analysis of the flow in a system of two or more pipes joined at a common junction is based on determining the flow distribution that will satisfy the following conditions:

1. The net flow in and out of the system and at each pipe junction *must be zero* in order to satisfy the conservation-of-mass requirement.

2. Flow is always in the direction of a negative pressure gradient $-dp/dL$, meaning the flow is always from a point of high pressure to a point of lower pressure.

TIP FOR THE P.E. EXAMINATION CANDIDATE

Analysis of pipe networks require labor-intense calculations or computer-based techniques. Accordingly, such problems are typically included in the P.E. examination. However, the candidate should be prepared to analyze a four-pipe network as illustrated in Figure 5-1.

3. The pressure at a pipe junction *must be the same* for all pipes joined at that point. This implies that the flow distribution within the pipe network must be such that the velocity or discharge in each pipe will result in an energy dissipation rate that will lead to the same pressure at pipe junctions. Thus, the hydraulic grade lines for each pipe leading into a junction must slope so that they have a common value at the junction as illustrated in Figure 5-1.

Figure 5-1
HGL for a system of pipes

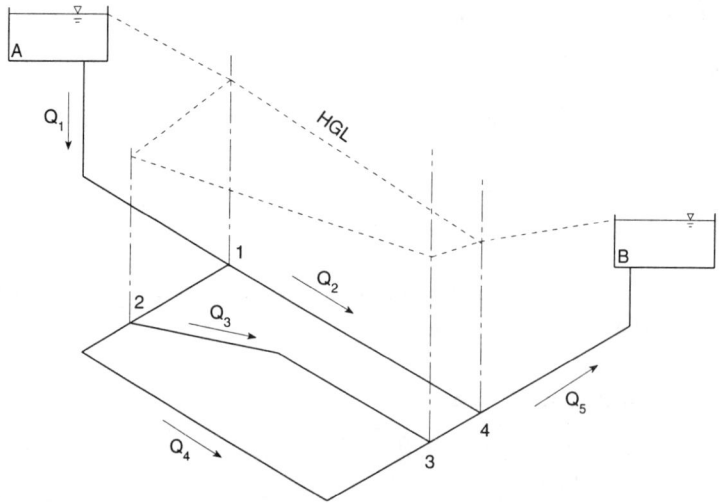

Pipe networks are is commonly encountered in water distribution systems. Analysis and design of a network of pipes requires an iterative solution process based on imposing the above hydraulic conditions on the physical network. Common solution algorithms are based on numerical or successive approximation methods such as the *Hardy Cross Method*.

5.4 Pumps

5.4.1 Classification and Characteristics   Pumps are devices that convert mechanical energy into fluid energy. Pumps are classified according to the mechanical principle used to accomplish this energy transfer. Common types of pumps are centrifugal, piston, gear, and screw pumps. Centrifugal pumps fall under the classification of *turbo machinery*, as their principle of operation is the same as for turbines, torque converters and other similar machines. Piston, gear and screw pumps are classified as *positive displacement* pumps.

The most common types of pumps encountered in environmental engineering applications are *centrifugal pumps* because of their high flow capacity. A centrifugal pump has two primary elements: an *impeller* and a *casing* that encloses the impeller. The impeller is rotated by means of a motor that supplies a torque force to the impeller. The impeller in turn, accelerates angularly until the counter torque, caused by the reaction of the fluid within the casing, equals that of the motors-supplied torque. At this point, no further acceleration occurs and the pump impeller turns at a constant *angular velocity* ω.

The velocity at any point on the impeller v_t is a function of the angular of velocity vector ω, and the radial distance r from the center of rotation, expressed as the vector product $\omega \cdot r$. The fluid enters the casing in the vicinity of the shaft and moves outward along the vane with a velocity v_r. The absolute velocity v that the fluid acquires is the vector addition of v_t and v_r (Figure 5-2).

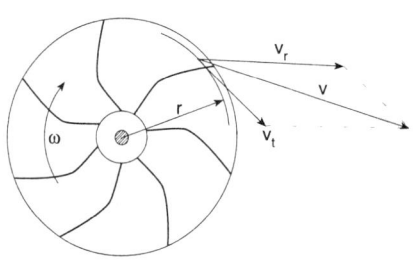

Figure 5-2
Velocity vector relationship for a rotating impeller

$$v = v_t + v_r \qquad [5\text{-}1]$$

The Power, P, which is the rate at which the pump does work, can be expressed in terms of torque as:

$$P = T\omega \qquad [5\text{-}2]$$

where: T = torque applied to the pump impeller
                ω = the impeller angular velocity

Power can also be expressed in terms of the fluid, as a function of the flow rate at a given head:

$$P = Q\gamma H \qquad [5\text{-}3]$$

where: Q = the volumetric discharge
                γ = the specific weight of the fluid
                H = is the pressure head at which the discharge Q takes place.

If power is expressed in units of ft-lbs/sec then Equation 5-3 can be expressed to yield power in units of horse power:

$$P_{hp} = Q\gamma H / 550 \qquad [5\text{-}4]$$

Pump *Efficiency* is defined as the ratio of energy out of the system to energy into the system.

$$e = \text{Energy Out} / \text{Energy In} \qquad [5\text{-}5]$$

Thus, a pump that imparts a discharge at a head condition equivalent to 8 horsepower and requires 10 horsepower from the motor to turn its impeller would have an efficiency of 0.8 or 80%. The *combined efficiency*, which accounts for both the pump and motor can be expressed as:

$$e = e_m e_p \qquad [5\text{-}6]$$

where: e_p = the pump efficiency
e_m = the motor efficiency.

A common problem in selecting and sizing pumps is how to scale up or down from a known set of pump operation conditions where ω and γ are constant. Discharge Q, head H and power P will scale up as exponential functions of the diameter of the impeller d_I. The relationship where ω and γ are constant is:

$$Q_1 / Q_2 = d_{I1} / d_{I2}$$
$$H_1 / H_2 = [d_{I1} / d_{I2}]^2 \qquad [5\text{-}7 \text{ a,b,c}]$$
$$P_1 / P_2 = [d_{I1} / d_{I2}]^3$$

Alternately, where the impeller diameter d_I and the specific weight γ are constant, the relationship between discharge, head and power is:

$$Q_1 / Q_2 = \omega_1 / \omega_2$$
$$H_1 / H_2 = [\omega_1 / \omega_2]^2 \qquad [5\text{-}8 \text{ a,b,c}]$$
$$P_1 / P_2 = [\omega_1 / \omega_2]^3$$

5.4.2 Specific Speed Comparison of Equations 5-7 and 5-8 shows that Q, H, and P each bear a different relationship to the angular velocity and pump impeller diameter. This complicates pump selection to meet discharge and head requirements and to relate it to power requirements that will result in efficient operation and power utilization. A dimensionless parameter useful for pump evaluation is the specific speed N_s which is defined as:

$$N_s = \omega (Q_{gpm})^{1/2} / H^{3/4} \qquad [5\text{-}9]$$

where: ω = the pump's angular velocity in rpm
Q_{gpm} = the discharge in gal/min
H = the pump operating head in ft

TIP FOR THE P.E. EXAMINATION CANDIDATE

The candidate is expected to be knowledgeable in applying a pump curve.

Pumps are tested by their manufacturer to establish a performance rating curve. A typical pump curve is shown in Figure 5-3. The specific speed of the pump is normally calculated at the pump's maximum efficiency and is provided as part of the pump specifications. Knowing the specific speed, corresponding to maximum efficiency, it is possible to select a pump that has a combination of discharge, pressure head and operating speed that will yield maximum efficiency.

5.4.3 Pressure and NPSH Considerations The location of a pump is an important consideration. A pump should be located so that all points within the pump are at a pressure that exceeds the vapor pressure of the fluid under all anticipated operating conditions. If the pressure falls below the vapor pressure, *cavitation* will take place, resulting in reduced flow, loss of efficiency, noise, and damage

Figure 5-3
Typical centrifugal pump performance curves

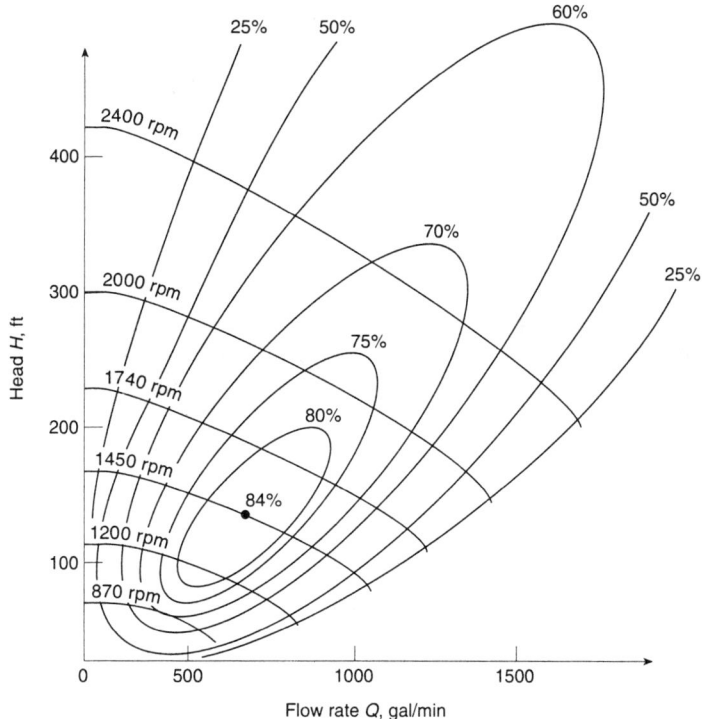

Source: Simon, 1981.

to the pump surfaces in the vicinity of the imploding cavitation bubbles. The zones of lowest pressure in the pump correspond to the locations where the fluid reaches its highest velocity, such as in the vicinity of the impeller tips. The pressure distribution in the pump is a function of the pump's geometrical configuration, operating speed, and fluid flowing through it.

To aid in selection and installation, pump manufacturers provide a value known as the *net positive suction head* (NPSH). The NPSH is the maximum elevation at which the pump may be located above a supply surface assumed to be at atmospheric pressure. By knowing the NPSH and evaluating all head losses leading up to the pump intake, the pressure at the intake can be calculated and compared to the NPSH of the pump. If the pressure is less than the required NPSH, the pump installation must be modified by either lowering the pump or changing other system characteristics to raise the pressure at the intake.

TIP FOR THE P.E. EXAMINATION CANDIDATE

In problems involving pumps, it is important to ensure that a pump's NPSH is exceeded.

5.4.4 Series and Parallel Pump Installation All pumps have a limited range of discharge, head, and rotational velocities within which they can operate efficiently. When either discharge or head vary over a wide range, the most practical solution may be the installation of several pumps, either in parallel or in series. Normally, it is also more economical to control discharge rates from a pump by using a valve on the discharge side of the pump, which introduces additional resistance and produces an increase in the head against which the pump must operate. The variable resistance will range from zero with the valve fully open to an infinite resistance with the valve fully closed.

A series installation is illustrated in Figure 5-4 and in the example problem 5-2 where the choice is between using one large pump or a series of smaller pumps to provide a desired flow rate and overcome the elevation difference and resistance of the pipe system. The nature of a series pump installation is that discharge is essentially unchanged but the *head is increased*.

Figure 5-4
Single or series pump installation

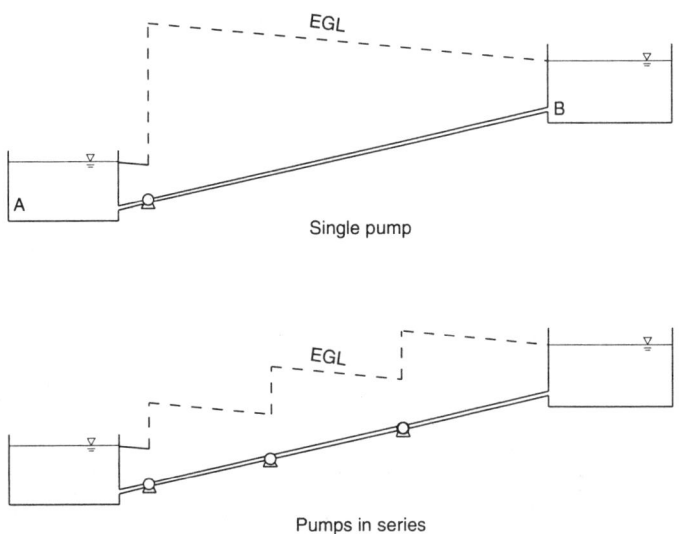

Example 5-2

Water is pumped from the lower to the upper reservoir at 500 gal/min by a pump using 7.46 kw of power, operating at 1200 rpm. It is desired to increase the pumping capacity to 1350 gpm. The pipe is 5-in., old cast-iron, with an e = 0.002 ft. Minor losses in the system are 2.4 velocity heads.

    a) At what speed must the pump be operated?

    b) What is the present combined efficiency of the pump-motor?

    c) What is the power requirement at 1350 gal/min assuming the efficiency is the same?

Solution:

Step 1. Determine all energy losses:

$A = 0.136 \text{ ft}^2$ $\varepsilon / d = 0.002 / (5/12) = 0.0048$

500 gal/min   **1350 gal/min**

$V = Q / A = (500 / 449) / 0.136 = 8.19 \text{ ft/sec}$

$N_R = Vd / v = (8.19)(5/12) / 1.217 \bullet 10^{-5} = 2.8 \bullet 10^{-5}$

$V = 22.1 \text{ ft/sec}$

$N_R = 7.56 \bullet 10^{-5}$

Step 2. From the Moody diagram $f = 0.031$ and $f = 0.030$ are obtained and applying Equations 4-16, 4-18, and 4-20

$h_f + h_L = \{(fL/d) + \Sigma C_L\}(V^2 / 2g)$

$= \{(0.031)(200 / 0.417) + 2.4\}(8.19)^2 / (2)(32.2) = 15.5 + 2.5 = 18 \text{ ft}$

$= \{(0.031)(200 / 0.417) + 2.4\}(22.1)^2 / (2)(32.2) = 109 + 18 = 127 \text{ ft}$

Step 3. Total head which must be pumped against at 500 and 1350 gal/min is:

$H = 30 + 18 = 48 \text{ ft}$ and $30 + 127 = 145 \text{ ft}$

Step 4. Using Equation 5-8b:

$H_1 / H_2 = [\omega_1 / \omega_2]^2$

$48 / 145 = [1200 / \omega_2]^2$ $\omega_2 = 2085 \text{ rpm}$

Step 5. Power out of pump at 500 gpm is (Equation 5-4):

$P_{hp} = Q\gamma H / 550 = (500 / 449)(62.4)(48) / 550 = 6.06 \text{ hp}$

Power into pump is 7.46 kw = 10 hp. Thus, from Equation 5-5:

$e = 6.06 / 10 = 60.6\%$

Step 6. At 1350 gal/min:

$P_{hp} = Q\gamma H / 550 = (1350 / 449)(62.4)(145) / 550 = 49.46 \text{ hp}$

Power required $= 49.46 / e = 49.06/.606 = 80.9 \text{ hp} = 60.4 \text{ kw}$

Chapter 5: Hydraulics & Hydrology

Example 5-3

For Example Problem 5-2, what is the minimum number of pumps, placed in series, needed to obtain a 1350 gal/min flow rate between the reservoirs? At what spacing should they be placed? Sketch the EGL for the system.

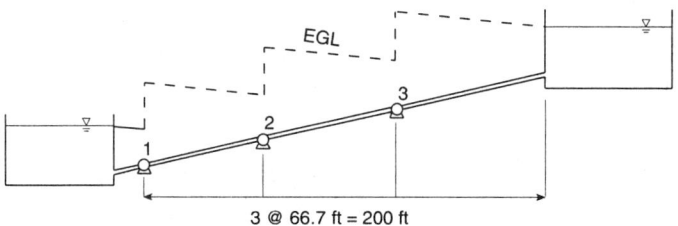

3 @ 66.7 ft = 200 ft

Solution:

Step 1. The total head necessary at 1350 gal/min is 145 ft.

Step 2. The number of pumps required is = 145/48 = 3.02 (use 3)

Step 3. Spacing = 200/3 = 66.7 ft.

TIP FOR THE P.E. EXAMINATION CANDIDATE

The ability to correctly design or analyze multiple pumps applications is essential.

When a very wide range of discharge is needed, the solution may be to connect pumps in para*llel* rather than operate a larger single pump over a wide discharge range. Such situations are common in the design of sanitary and storm sewers because of the flow variability in these systems. This is illustrated in Figure 5-5 which shows a typical lift station. In parallel pump installations, discharge is increased proportionally to the number of pumps but the *head remains constant*.

Figure 5-5
Sewer lift station example

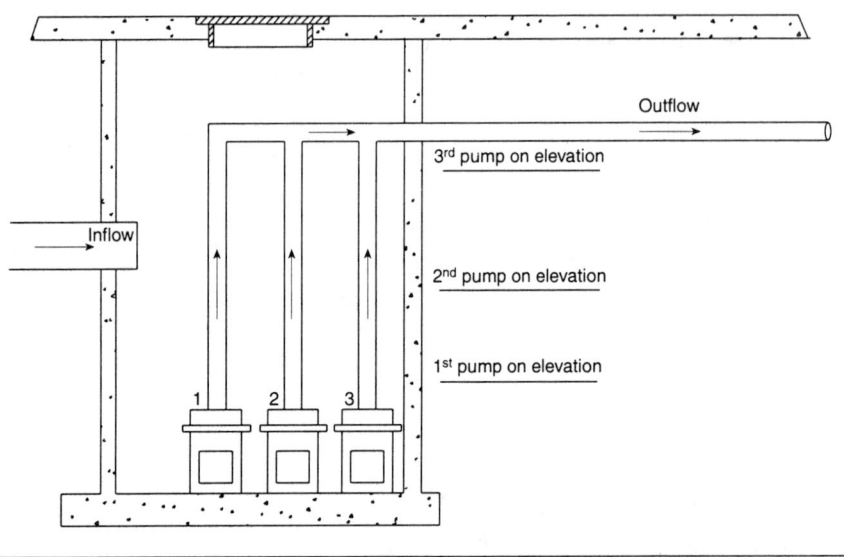

Example 5-4

A sanitary sewer lift station must pump 1,080 gal/min a distance of 24 ft (from pump to outlet). The major and minor losses in the piping total 10 ft. An 8 in. pump has the following discharge characteristics

Discharge (gal/min)	Head (ft)
0	42
175	40
260	38
363	34
440	28
530	20
630	7

a) What size pump of the same type and manufacture as the 8 in. pump described, would be required for a single pump installation?

Solution:

Assume the pump speed is the same and the head corresponds to the maximum value. Therefore the pump diameter will scale according to Equation 5-7a:

$Q_1 / Q_2 = d_{f1} / d_{f2}$

$363 / 1080 = 8 \text{ in.} / d_{f2}$

$d_{f2} = 23.8$ say 24 in.

b) How many 8 in. pumps with characteristics described would be needed for a parallel pump installation?

Solution:

Step 1. The total head that must be pumped against is determined and corresponds to the maximum water elevation difference between the sump and the outlet end, 24 ft. To this we add the sum of all major and minor losses between the pumps and the outlet, which are calculated to be 10 ft. Therefore, the total pump head is 24 + 10 = 34 ft.

Step 2. From the pump discharge curve, a discharge value (Q) will be 363 gal/min at a head of 34 ft.

Step 3. Since the pumps are connected in parallel and discharge to a common manifold in which the pressure can be considered to be essentially constant, the pump performance curve of the three pumps is a simple multiple of the discharge values at the same corresponding heads.

Thus, the number of pumps required is:

1080/363 = 2.97, therefore 3 pumps are needed.

The performance curves for 1, 2, and 3 pumps are shown in Figure.

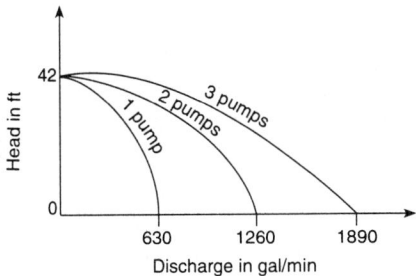

5.5 Free Surface Flow or Open Channel Flow

The behavior of natural rivers and streams, artificial channels, runoff from pavements, and the flow of liquids in partially full pipes are all examples of *free surface* or *open channel flow*. What differentiates free surface flow from closed conduit flow is that one of the boundaries *is not fixed*. A free surface implies that the surface is subject only to the pressure of another fluid, normally a gas. If the gas is air, the surface pressure is local atmospheric pressure. Analysis of free surface flows is complicated by the fact that the location of the free surface is not usually known. It becomes a function of the channel discharge and the rate at which energy is being transferred into or out of the flowing fluid.

In closed conduit flow, if the conduit's cross-sectional dimensions and the discharge are known, the average flow velocity can be determined using the continuity equation. By contrast, knowing an open channel's dimensions and the discharge is not sufficient to determine the average velocity of the flow unless we know the position of the free surface. Since energy loss is a function of velocity and boundary resistance, it is not possible to calculate the energy loss directly.

TIP FOR THE P.E. EXAMINATION CANDIDATE

Review the application of the conservation equation discussed in Chapter 4.

Free surface flows can be thought of as being *self adjusting*. For a constant discharge and uniform channel geometry, when the depth goes down the velocity must go up, as will the rate of energy loss. Conversely, when the depth goes up, the velocity will drop and the rate of energy loss will diminish. Free surface flow solutions are based on the application of the conservation equations discussed in Chapter 4, namely the continuity Equation 4-8, the momentum Equation 4-10 and the energy Equation 4-15.

In Figure 5-6 water enters a channel of constant cross-section, slope, and resistance from an upstream reservoir by the opening of a gate. When lifted, water flows out at the initial depth of the gate opening. As the water moves down stream, its depth changes. If, as shown, the depth decreases, both velocity and hydraulic resistance increase. At some point downstream the depth will remain constant, as long as the channel slope and boundary resistance do not change. At this point the flow has reached an energy balance between the energy being transferred to the fluid and the energy being dissipated by the fluid. This can be understood by applying the energy equation. At point 1, the total energy would be

$$V_1^2/2g + p_1/\gamma + z_1 = E_1$$

Figure 5-6
Steady flow velocity and depth change of a channel

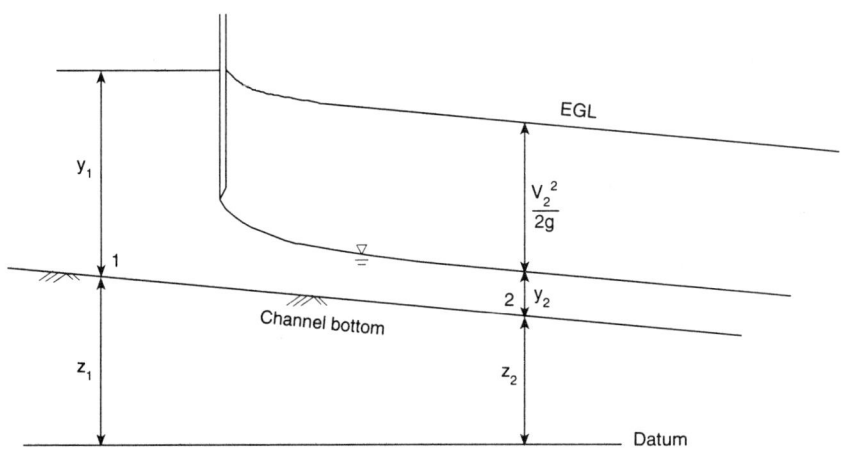

If the reservoir is sufficiently large that the velocity at point 1 is zero, then the energy reduces to the sum of the elevation head and the pressure head (which is equal to the depth of water, y_1).

$$0 + y_1 + z_1 = E_1$$

At point 2, the total energy would be:

$$V_2^2 / 2g + y_2 + z_2 = E_2$$

Between points 1 and 2, energy is lost because of boundary resistance and non-uniform flow conditions. The loss is equal to the difference in energy between the points, reflected by EGL. We can express this relationship as follows:

$$E_1 + E_2 = \Sigma H_L$$

or

$$\Sigma H_L = E_2 - E_1 = V_2^2 / 2g + (y_2 - y_1) + (z_2 - z_1)$$

5.5.1 Flow Classification There are several terms essential to a clear understanding of free surface flow and its analysis.

> *One-dimensional flow* refers to a flow situation in which either fluid or flow properties change in only one direction. A two-dimensional flow would have properties that can vary in two directions. Free surface flow is one-dimensional if the depth y is a function of the distance x measured along the channel normally positive in the direction of flow, the discharge Q, and the time t.

Steady flow exists when the depth at a fixed point does not change with time.

Figure 5-7
Various types of open-channel flow

Unsteady flow exists when the depth at a fixed point does change with time.

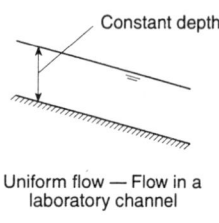

Uniform flow — Flow in a laboratory channel

If the depth is the same at all points along the channel, the flow is termed *uniform*.

When depth varies with position along the channel it is termed *varied*. It may be either *gradually-varied or rapidly-varied*, depending on the rate at which the depth changes and its impact on energy loss. This distinction will be taken up later in the discussion of the calculation of energy losses.

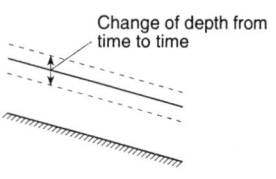

Unsteady uniform flow — Rare

Spatially varied flow exists when depth changes as a result of the addition or subtraction of flow. This occurs where streams branch together or divide, and where flow enters or leaves a channel at points along it's length, as with direct runoff or seeping.

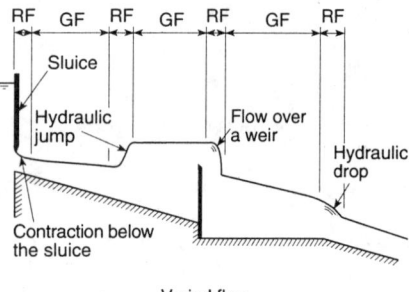

Varied flow

Figure 5-7 illustrates the SE classifications.

5.5.2 Channel Properties There are some terms that need to be defined as they apply to open channel flow. Referring to Figure 5-7, the *cross-sectional area* of the channel, A, is normal to the direction of flow. For channels of small slope, this direction is assumed to be vertical. For channels of large slope, the cross-sectional area is measured perpendicular to the free surface and the channel bottom. The *top width T* is the distance from bank to bank at the free surface.

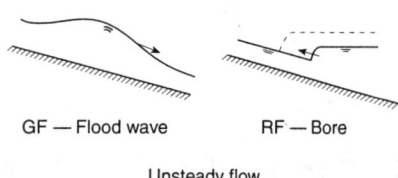

GF — Flood wave   RF — Bore

Unsteady flow

GF = gradually varied flow
RF = rapidly varied flow
Source: Chow, 1959.

The *hydraulic depth D* is a computed value and is defined as:

$$D = A/T \qquad [5\text{-}10]$$

For a rectangular section, the hydraulic depth is the actual depth y. For other cross-sections, it is an average depth based on the use of the top width dimension to define the cross-sectional area.

Figure 5-8
Hydraulically most efficient channel cross-sections

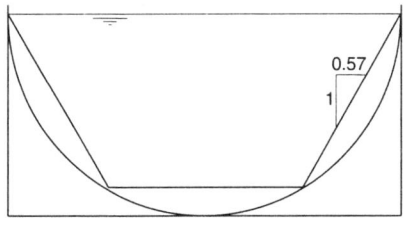

The *hydraulic radius R* for a channel is the cross-sectional area divided by the wetted perimeter W_p. The *hydraulically most efficient channel* will be one that has the smallest wetted perimeter for a given area. Figure 5-8 shows the most efficient dimensions for common channel cross-sections.

Another useful computed geometric channel property is that of the section factor defined as:

$$z = AR^{2/3} \qquad [5\text{-}11]$$

Geometric properties, for common channel shapes can be found in Table 5-2. A straight channel whose cross-section, roughness and slope are constant is termed *prismatic*.

TIP FOR THE P.E. EXAMINATION CANDIDATE

The candidate should be prepared to employ any of the common channel shapes in P.E. examination problems.

Example 5-5

The depth of flow is measured to be 1.75 ft in a trapezoidal channel that has a bottom width of 10 ft and a side slope of 1:2. Determine the cross-sectional area, hydraulic radius, hydraulic depth and section factor for the channel.

Solution:

1. Cross-sectional area $A = (b + zy)y = [10 + (2)(1.75)]1.75 = 23.63 \text{ ft}^2$
2. Wetted perimeter
   $W_p = b + 2y[1 + z^2]^{1/2} = 10 + (2)(1.75)[1 + 2^2]^{1/2} = 17.83 \text{ ft}$
3. Hydraulic Radius $R = A / W_p = 23.63 / 17.83 = 1.32 \text{ ft}$
4. Top width $T = b + 2zy = 10 + (2)(2)(1.75) = 17 \text{ ft}$
5. Hydraulic depth $D = A / T = 23.63 / 17 = 1.38 \text{ ft}$
6. Sector factor $z = AR^{2/3} = (23.63)(1.32)^{2/3} = 28.43$

Table 5-2
Geometric elements of channel sections

Section	Area (A)	Wetted perimeter (P)
Rectangle	by	$b + 2y$
Trapezoid	$(b + zy)y$	$b + 2y\sqrt{1+z^2}$
Triangle	zy^2	$2y\sqrt{1+z^2}$
Circle	$\frac{1}{8}(\theta - \sin\theta)d_0^2$	$\frac{1}{2}\theta d_0$
Parabola	$\frac{2}{3}Ty$	$T + \frac{8y^2}{3T}$ *
Round-cornered rectangle ($y > r$)	$(\frac{\pi}{2} - 2)r^2 + (b + 2r)y$	$(\pi - 2)r + b + 2y$
Round-bottomed triangle	$\frac{T^2}{4z} - \frac{r^2}{z}(1 - z\cot^{-1}z)$	$\frac{T}{z}\sqrt{1+z^2} - \frac{2r}{z}(1 - z\cot^{-1}z)$

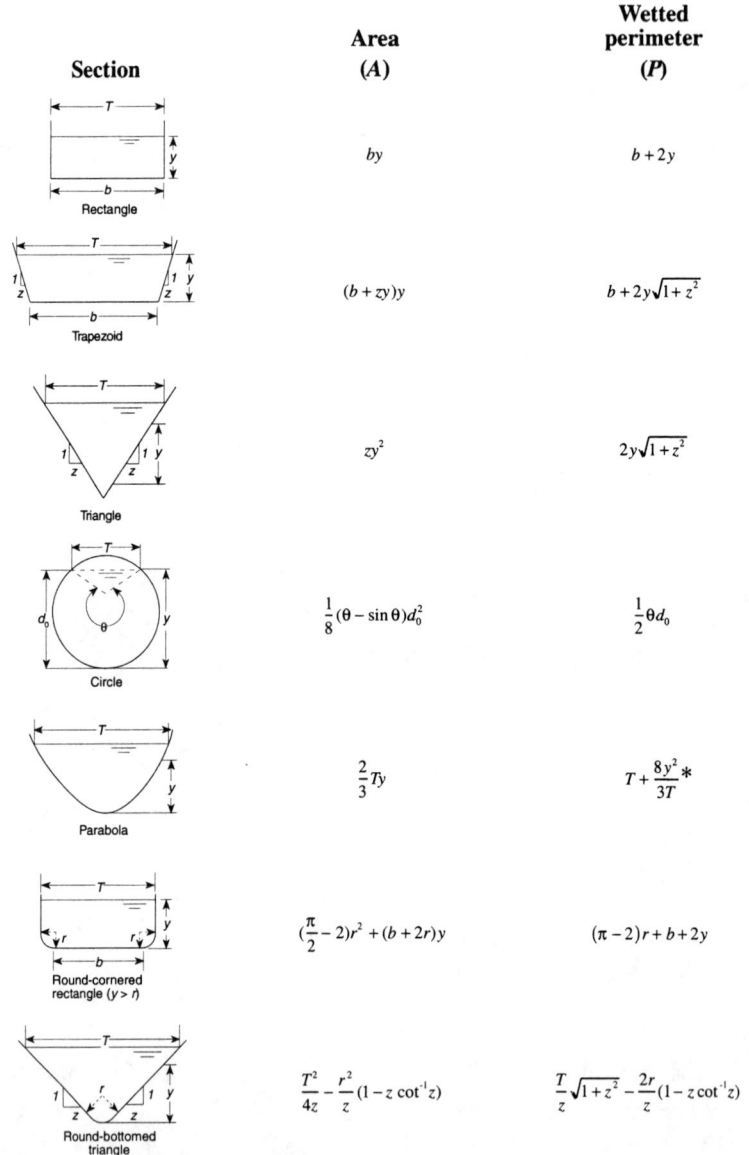

*Satisfactory approximation for the interval $0 < x \leq 1$, where $x = 4y/T$.
When $x > 1$, use the exact expression $P = (T/2)[\sqrt{1+x^2} + 1/x \ln(x + \sqrt{1+x^2})]$
Source: Chow, 1959.

Hydraulic radius (R)	Top width (T)	Hydraulic depth (D)	Section factor (Z)
$\dfrac{by}{b+2y}$	b	y	$by^{1.5}$
$\dfrac{(b+zy)y}{b+2y\sqrt{1+z^2}}$	$b+2zy$	$\dfrac{(b+zy)y}{b+2zy}$	$\dfrac{[(b+zy)y]^{1.5}}{\sqrt{b+2zy}}$
$\dfrac{zy}{2y\sqrt{1+z^2}}$	$2zy$	$\dfrac{1}{2}y$	$\dfrac{\sqrt{2}}{2}zy^{2.5}$
$\dfrac{1}{4}(1-\dfrac{\sin\theta}{\theta})d_0$	$(\sin\tfrac{1}{2}\theta)d_0$ or $2\sqrt{y(d_0-y)}$	$\dfrac{1}{8}(\dfrac{\theta-\sin\theta}{\sin\tfrac{1}{2}\theta})d_0$	$\dfrac{\sqrt{2}}{32}\dfrac{(\theta-\sin\theta)^{1.5}}{(\sin\tfrac{1}{2}\theta)^{0.5}}d_0^{2.5}$
$\dfrac{2T^2y}{3T^2+8y^2}$	$\dfrac{3A}{2y}$	$\dfrac{2}{3}y$	$\dfrac{2}{9}\sqrt{6}\,Ty^{1.5}$
$\dfrac{(\pi/2-2)r^2+(b+2r)y}{(\pi-2)r+b+2y}$	$b+2r$	$\dfrac{(\pi/2-2)r^2}{b+2r}+y$	$\dfrac{[(\pi/2-2)r^2+(b+2r)y]^{1.5}}{\sqrt{b+2r}}$
$\dfrac{A}{P}$	$2[z(y-r)+r\sqrt{1+z^2}\,]$	$\dfrac{A}{T}$	$A\sqrt{\dfrac{A}{T}}$

Circular sections flowing partially full are very common and deserve special consideration. Figure 5-9 provides a plot of area, hydraulic radius, wetted perimeter, top width, hydraulic depth and section factor for critical flow (discussed later), in a dimensionless format, obtained by normalizing all parameters by the section full values, denoted by the subscript 0. To use the figure, enter with the ratio of y/d_0 (actual depth/full depth) and read the normalized parameter value. Then multiply it by the section full value, which is easily computed.

5.5.3 Flow Regimes Flow in open channels is analogous to that in pipes. The flow can be l*aminar, transitional or turbulent.* The structure that occurs depends on the fluid's viscosity, the flow rate and the boundary roughness. If the boundaries are hydraulically smooth, a laminar sublayer can form in a turbulent flow. However, the distinctive difference between open channel flows and pipe flows is that methods of predicting energy losses are not as complete or accurate.

As noted in Chapter 4, the commonly accepted upper limit for laminar flow in pipes is at a Reynolds number of 2000, based on pipe diameter as the characteristic

Figure 5-9
Partially full circular channel section properties normalized

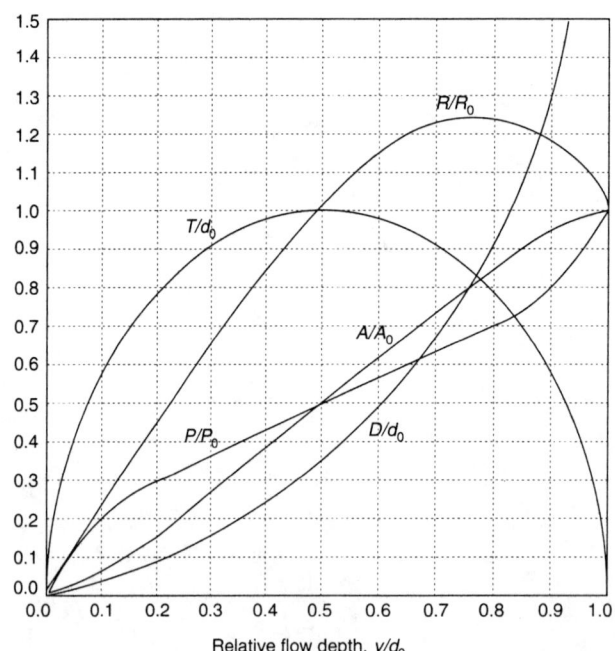

Source: Chow, 1959.

length. For open channels, the corresponding Reynolds number value is 500 when defined on the basis of the channel hydraulic radius.

$$N_R = VR / \nu \qquad [5\text{-}12]$$

The *Reynolds number* expresses the ratio of viscous forces to inertial forces. Thus, there are three regimes based on flow structure:

Laminar	$N_R < 500$
Transitional	$500 < N_R < 2{,}000$
Turbulent	$N_R > 2{,}000$

Free surface flows are a gravity-driven phenomena, in contrast to being pressure driven as occurs in closed conduits. Gravitational forces play the dominant role in free surface flow, however, viscous effects are present and may also be very important.

The dimensionless parameter that relates gravitational forces to inertial forces is the *Froude number* N_F defined as follows:

$$N_F = V / [gL]^{1/2} \qquad [5\text{-}13]$$

where: V = the average velocity;
g = the acceleration of gravity; and
L = a characteristic length.

For open channels, the velocity used is the average velocity based on dividing the discharge by the cross-sectional area. The characteristic length is normally the depth y, but D and R may be used also. The Froude number also expresses the ratio of two velocities, that of the flow velocity V to c the speed of propagation of gravity wave in shallow water. The propagation speed of a wave is also termed the wave celerity and is defined as:

$$c = [gy]^{1/2} \qquad [5\text{-}14]$$

Note that the speed a wave travels at is a function of depth y.

Critical Flow occurs when the Froude number equals 1, the velocity in both the numerator and denominator are equal, i.e., the flow velocity equals the velocity at which a gravity wave would travel in shallow water. This fact is very important to understanding how boundary effects are propagated either upstream or downstream and ultimately affect the flow depth and velocity in the channel. If the Froude number is less than 1, flow is termed *subcritical*. If the number is greater than 1, it is termed *supercritical*.

$N_F < 1$	subcritical
$N_F = 1$	critical
$N_F > 1$	supercritical

To fully characterize a free surface flow requires four specifications based on how the depth changes spatially, how it changes with time, the flow structure, and the flow velocity compared to wave celerity in a given depth of fluid. The classification matrix is given in Table 5-3.

Table 5-3
Free surface flow classification matrix

y as a function of x	y as a function of t	N_R	N_F
Uniform	Steady	Laminar	Subcritical
or	or	Transitional	Critical
Nonuniform	Unsteady	Turbulent	Supercritical

Based on the criteria listed in Table 5-3, it is possible to describe a particular surface flow as non-uniform, steady, turbulent, and subcritical.

5.5.4 Uniform Flow
Uniform flow in channels or conduits is commonly encountered in the practice of environmental engineering. The channel depth that corresponds to the depth at which uniform flow occurs is called the *normal depth*. This is illustrated in Figure 5-10.

Figure 5-10
Relationship of channel bottom, free surface, and EGL slopes in uniform flow

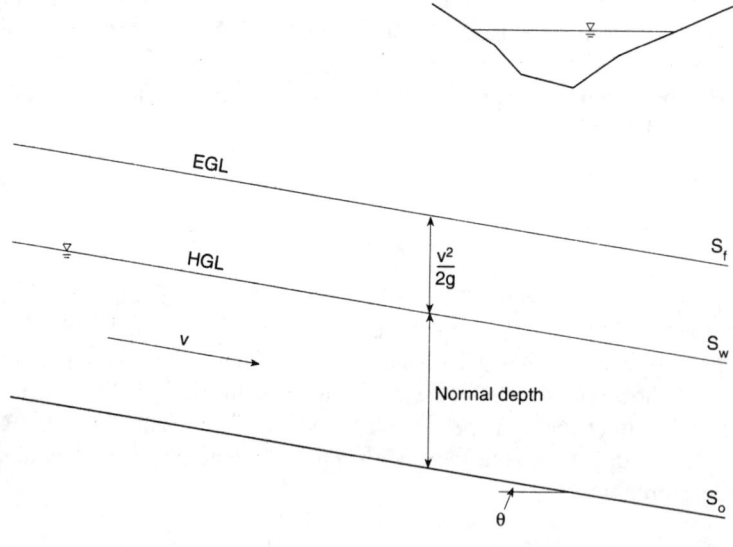

In this situation, depth does not change, which means the flow is neither accelerating nor decelerating. Thus, all the forces that act on the fluid in the direction of flow, namely the component of the weight in the flow direction Wx, the boundary resistance force t0 WpL, and the pressure forces P1 and P2, are in balance and all of the forces in the x direction sum to zero. This is illustrated in Figure 5-11. Also, the slope of the channel bottom S_0, slope of the free surface S_w, and the slope of the energy grade line S_f are the same. For open channel flow, the free surface is the hydraulic grade line (HGL).

Figure 5-11
Elemental segment of uniform flow

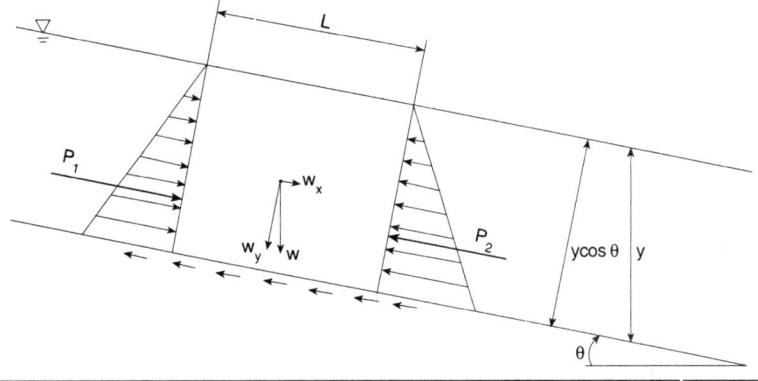

The unique hydraulic conditions described above, common in open channel flow, enable development of the Chezy Equation. Referring to Figure 5-10 again, the forces in the direction of flow can be equated to obtain:

$$\Sigma F_x = P_1 - P_2 - \tau_0 W_p L + \gamma L A \sin\theta = 0$$

For uniform flow $P_1 = P_2$, thus

$$\tau_0 W_p L = \gamma L A \sin\theta \quad [5\text{-}15]$$

If the boundary shear stress τ_0 is assumed to be constant or it represents an average value, and we then substitute R for $= A / W_p$ and note for small angles, $\sin\theta = S_0 = S_f = S_w$. The above equation reduces to:

$$\tau_0 = \gamma R S_f \quad [5\text{-}16]$$

Equation 5-16 makes it possible to determine the average shear stress, knowing the channel configuration and the depth of flow, when flow is uniform.

It is desirable to be able to relate τ_0 to the physical boundary conditions. To do this, apply the principles governing pipe flow, keeping in mind that resistance in open channel flow is analogous to that of pipe flow. Major distinctions, however, are that the relative roughness cannot be expressed as simply on the basis of a single physical dimension (like the pipe diameter which is used in pipe flow). It has been found that using the hydraulic radius in place of diameter to define relative roughness provides an acceptable way to account for both boundary resistance and channel configuration.

The Darcy-Weisbach friction factor f for pipes depends on both *Reynolds number* and *relative roughness*. When the hydraulic radius is used in place of pipe diameter to define both Reynolds number and relative roughness, open channel resistance based on the Darcy-Weisbach friction factor can be established. The boundary shear stress in a pipe is related to the Darcy-Weisbach friction factor by the following relation:

$$\tau_0 = f\rho V^2 / 8 \qquad [5\text{-}17]$$

Substituting Equation 5-17 into Equation 5-16 and solving for V the *Chezy equation* results:

$$V = C_c [RS_f]^{1/2} \qquad [5\text{-}18]$$

where C_c is the Chezy coefficient that accounts for boundary resistance and is related to f as follows:

$$C_c = [8\gamma / f\rho]^{1/2} = [8g / f]^{1/2} \qquad [5\text{-}19]$$

An empirical equation that has seen wide use since its introduction in 1889 is the *Manning's equation*.[8] Like the Chezy equation, it can be used to predict the relationship between either the velocity or discharge, geometric channel properties, and depth of flow under uniform, free-surface flow conditions.

$$V = (1.486 / n) R^{2/3} S_f^{1/2} \qquad [5\text{-}20]$$

where V = ft/sec
       R = ft

An alternate form of the Manning equation (Equation 5-20) may be written by using the continuity relationship:

$$Q = AV = (1.486 / n) AR^{2/3} S_f^{1/2} \qquad [5\text{-}21]$$

where Q = ft³/sec
       A = ft²

Table 5-4 presents values for n, Manning's coefficients.

Table 5-4
Values of the Manning coefficient, *n*

	Wetted perimeter	**n**
A.	Natural channels	
	clean and straight	0.030
	sluggish with deep pools	0.040
	major rivers	0.035
B.	Floodplains	
	pasture, farmland	0.035
	light brush	0.050
	heavy brush	0.075
	trees	0.15
C.	Excavated earth channels	
	clean	0.022
	gravelly	0.025
	weedy	0.030
	stony, cobbles	0.035
D.	Artifically lined channels	
	glass	0.010
	brass	0.011
	steel, smooth	0.012
	steel, painted	0.014
	steel, riveted	0.015
	cast iron	0.013
	concrete, finished	0.012
	concrete, unfinished	0.014
	planed wood	0.012
	clay tile	0.014
	brickwork	0.015
	asphalt	0.016
	corrugated metal	0.022
	rubble masonry	0.025

Source: Chow, 1959.

Flow resistance in the above equations is accounted for by the Manning coefficient *n*. A common problem in hydraulics is the need to find the depth of flow knowing the discharge, and the channel geometry. In Equation 5-21, this situation can be accomodated by solving for the *section factor* $AR^{2/3}$. Note that the only terms embodying the depth *y* are contained in the section factor. The exact form this expression will take depends on the shape of the channel. This is illustrated in the following example.

Note: Equations 5-20 and 5-21 *require* the specific units shown.

Example 5-6

For a trapezoidal channel with a bottom width of 10 ft and a side slope 1:2, flowing 1.75 ft deep, determine the discharge if the slope is 0.0008 and the Manning n is 0.015.

Solution:

The normal depth is known, therefore Mannings Equation 5-21 can be applied directly after calculating the section factor Z_s (see example 5-5).

$$Q = (1.486/n)AR^{2/3}S_f^{1/2} = (1.486/0.015)(28.43)(0.0008)^{1/2} = 79.7 \text{ ft}^3/\text{sec}$$

Example 5-7

A triangular cross-section sluice channel is constructed on a 2% grade with sides sloped at 45° and has an n value of 0.01. Determine the normal depth when the flow in the channel is 10 ft³/sec.

When the channel geometry is complex the simplest process is that of trial-and-error to find the normal depth. However, in this case the geometry is such that y can be obtained directly. Both procedures will be demonstrated.

Direct Solution:

    Step 1. Use Equation 5-21 to express the unknown terms involving y in the form of the section factor Z_s term:

$$AR^{2/3} = Qn/(1.486)S_f^{1/2}$$
$$= (10)(0.01)/(1.486)(0.02)^{1/2}$$
$$= 0.48$$

    Step 2. From Table 5-2 the expressions for A and R are obtained, noting that $z = 1$ for a 45° side slope:

$$(zy^2)[(zy/2(1+z^2)^{1/2}]^{2/3} = (y^2)[(y/2(2)^{1/2}]^{2/3} = y^{8/3}/1.999)$$

$$AR^{2/3} = 0.48 \text{ from Step 1}$$

$$\frac{y^{8/3}}{1.999} = 0.48$$

$$y = 0.98 \text{ ft}$$

Trial-and-Error Solution:

Step 1. Same as Step 1 for direct solution.

Step 2. Set up a table and pick a trial value for y, compute the section factor and compare it to the required value for $AR^{2/3}$.

trial y (ft)	$AR^{2/3}$
2.0	3.18
1.0	0.50
0.99	0.49
0.98	0.48 = 0.48 from Step 1 y = 0.98 ft

Therefore: The normal depth corresponding to 10 ft³/sec is 0.98 ft.

The Darcy-Weisbach equation, the Chezy equation and the Manning equation can be and are all used to analyze open channel flow. The relationship between the resistance factors used in each of the three equations is

$$C_c = [8g/f]^{1/2} = 1.486 R^{1/6} / n \qquad [5\text{-}22]$$

5.5.5 Specific Energy If we exclude the elevation term from the energy equation (Equation 4-14, Chapter 4), the result is termed the *specific energy* defined as:

$$E = y + V^2 / 2g \qquad [5\text{-}23]$$

This is the energy per pound of fluid measured with respect to the channel bottom. Examining how this term changes from point to point along a channel provides a useful means for predicting how depth will be altered when channel slope or flow resistance changes. By excluding the elevation term z, that accounts for the flow's potential energy, the specific energy measures the energy *contained within the fluid* in the form of pressure and kinetic energy. As the channel slope changes, so will the flow depth and velocity that define the value of E.

Using the continuity equation, Equation 5-23 becomes:

$$E = y + Q^2 / 2gA^2 \qquad [5\text{-}24]$$

Since the area A is a function of the depth y for a given discharge Q, the *specific energy is a function of the depth only*.

If depth is plotted against specific energy as in Figure 5-12, the result is called the *specific energy diagram*. For channels of small slope, the specific energy curve has an upper leg asymptote that is a line having a 45° slope and a lower leg asymptote that is a line having a 0° slope. For a specific energy value, there exist two possible depths called *alternate depths*, y_1 and y_2 shown in Figure 5-12. The upper

Figure 5-12
Specific energy diagram

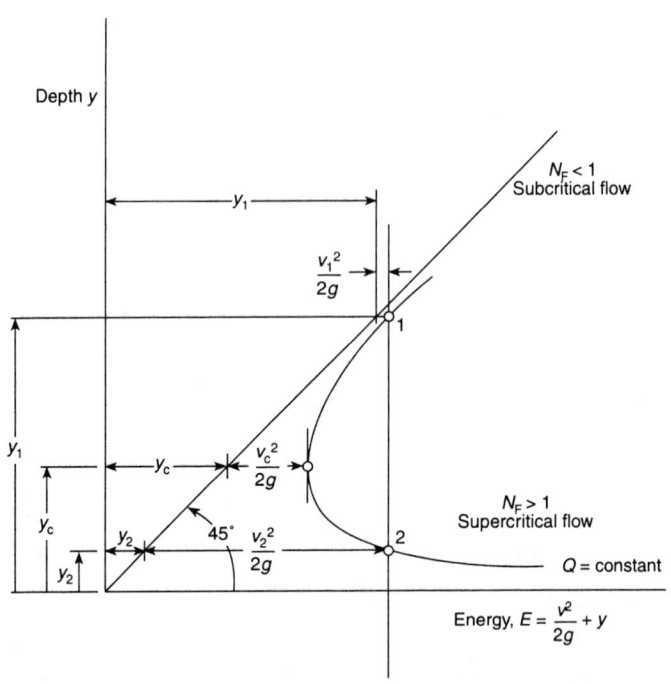

stage of the specific energy curve defines the range of *subcritical* depths; the lower stage defines *the supercritical* depths. The only exception to having alternate depths is at the point where the minimum value of specific energy occurs. Here, there is only one depth, the *critical* depth, y_c.

For a given discharge, there is only one critical depth, which corresponds to the state of minimum specific energy. This fact is often used to determine channel discharge. If the depth is known to be critical and can be measured, then the channel's discharge can be calculated. The form of the relationship between critical depth y_c and Q depends on the channel cross-section. For a rectangular channel the relation is:

$$y_c = [q^2 / g]^{1/3} \qquad [5\text{-}25]$$

where q is the discharge per unit width of channel. The total channel discharge Q would be qb where b is the width of the channel.

Critical flow is essential to the understanding and correct prediction of free surface flow behavior. The characteristics of critical flow follow:

- For a given discharge, the specific energy is a *minimum*.
- For a specific energy value, the flow is a *maximum*.
- $N_F = 1$
- $V = c$
- The hydraulic depth is one half the velocity head expressed as:

$D/2 = V^2/2g$

- The surface is unstable, allowing waves to form easily since a small change in specific energy causes a relatively large change in depth at or near critical depth.

Examination of the specific energy curve indicates that, when the flow is subcritical, if E is reduced the depth will go *down*. However, if the flow is supercritical and E is reduced, the depth will go *up*.

Example 5-8

A rectangular channel has three long sections whose bottom slopes are 0.00008, 0.001 and 0.000012. The channel has an $n = 0.015$ and a bottom width of 20 ft. When the discharge is 45 ft³/sec determine:

a) Normal depth in each section.

b) Classify the flow.

c) Determine the critical depth.

d) Plot the points shown on a specific energy diagram.

Solution:

Step 1. By use of Equations 5-13 and 5-21, normal depth, velocity, and the Froude number for each section is calculated.

Section	y_n (ft)	V (ft/sec)	N_F	Classification	$E = y + V^2/2g$
1	1.80	1.25	0.164	Subcritical	1.82
2	0.80	2.81	0.554	Subcritical	0.92
3	3.50	0.64	0.061	Subcritical	3.51

Step 2. $q = Q/b = 45/20 = 2.25$ ft³ / sec - ft
From Equation 5-25, $y_c = [q^2/g]^{1/3} = [(2.25)^2/32.2]^{1/3} = 0.54$ ft

Step 3. Points plot in the upper half of the specific energy diagram.

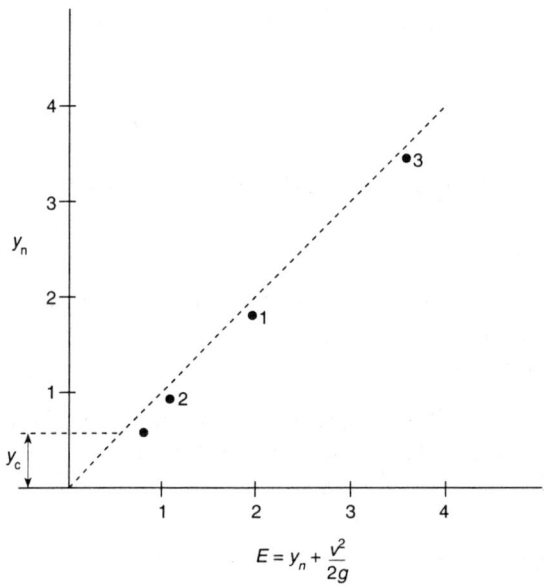

Example 5-9

Determine the alternate depths for channel section 3 in example 5-8.

Solution:

Step 1. From example 5-8, $y_n = 3.50$ ft, the flow is subcritical and $E = 3.51$.

Step 2. The alternate depth must be in the supercritical range and have the same specific energy. Thus,

$$E = y_n + Q^2/2gA^2 = y_n + (45)^2/(64.4)(20y_n)^2 = 3.51$$
$$y_n + 0.0786/y_n^2 = 3.51$$
$$y_n^3 - 3.51y_n^2 + 0.786 = 3.51 \; 0$$
$$y_n = 0.15 \text{ and } 3.50 \text{ ft}$$

Chapter 5: Hydraulics & Hydrology 119

5.5.6 Critical Flow Sections Knowing where critical flow occurs in an open channel system is important because at that point the relationship between discharge and depth is known. The computation of flow profiles throughout the system starts at a *control point*, where a known stage-discharge relationship exists, and then proceeds either upstream or downstream, depending on whether the flow is subcritical or supercritical. A point of critical flow is always a control point. Figure 5-13 shows common situations that provide control points.

Figure 5-13
Examples of free surface flow control points

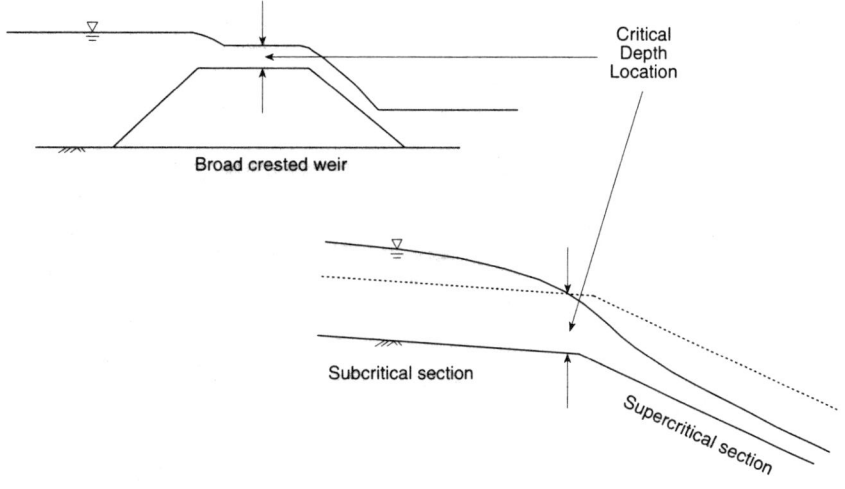

A means for computing critical depth is to use the section factor for critical flow computation, which is a function of channel geometry. It is defined as:

$$Z = A[D]^{1/2} \qquad [5\text{-}26]$$

The relationship Z has to discharge at critical flow, is:

$$Z = Q/[g]^{1/2} \qquad [5\text{-}27]$$

When the channel geometry and discharge are known, the procedure is to use Equation 5-26 and Equation 5-27 to solve for y_c. The exact form of Equation 5-26 will depend on the channel geometry that defines both A and D. Figure 5-14 can be used to determine Z for common channel shapes.

When supercritical flow is immediately followed by subcritical flow downstream of it, an abrupt change in the water surface will take place, called a hydraulic jump (illustrated in Figure 5-15).

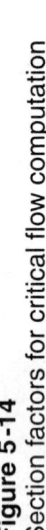

Figure 5-14
Section factors for critical flow computation

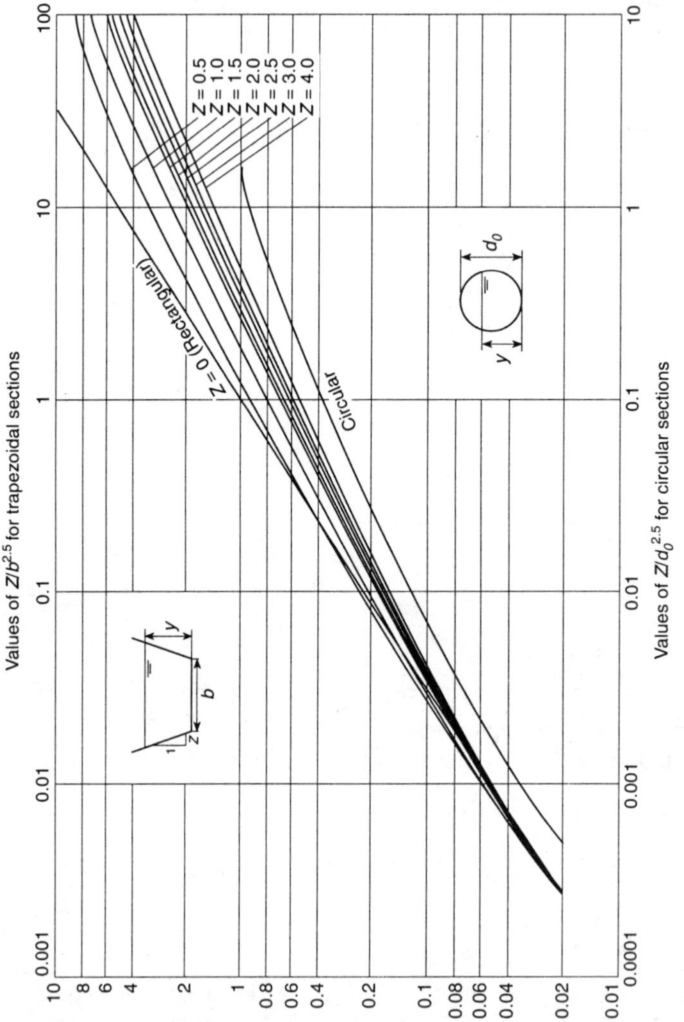

Figure 5-15
Supercritical to subcritical flow transition— the hydraulic jump

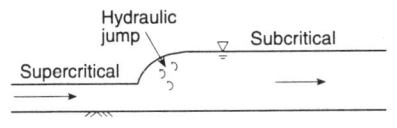

A hydraulic jump occurs because the depth in supercritical flow can not increase without a reduction in specific energy. When this occurs, the specific energy values move along the specific energy curve toward the minimum value at critical depth. Without a change in the channel slope to cause an increase in the specific energy, the depth cannot continue to increase once it has reached critical depth. Therefore, the subcritical depth cannot be achieved by gradual depth increase passing through critical depth. To accomodate the foregoing constraints, the water surface goes directly from a supercritical depth y_1 to the subcritical depth y_2 with an accompanying loss in energy ΔE. The upstream and downstream depths y_1 and y_2 that produce a hydraulic jump are called *conjugate depths* and are determined by momentum considerations. The energy loss intensity of the jump, the point where it will occur and its length all can be determined by a combination of theoretical and supporting experimental considerations. For a rectangular channel of small slope, the relationship between conjugate depths is:

$$y_1 / y_2 = 1/2[(1+8N_{F1}^2)^{1/2} - 1] \qquad [5\text{-}28]$$

where N_{F1} is the upstream Froude number defined as:

$$N_{F1} = V_1 / (gy_1)^{1/2} \qquad [5\text{-}29]$$

The energy loss across the jump ΔE is:

$$\Delta E = [y_2 - y_1]^3 / 4y_1 y_2 \qquad [5\text{-}30]$$

TIP FOR THE P.E. EXAMINATION CANDIDATE

Hydraulic jumps can be used to accomplish engineering purposes, such as mixing, in addition to being hydraulic phenomena requiring understanding.

Figure 5-16 shows a comparison of hydraulic jump profiles for different Froude numbers.

5.5.7 Flow Profiles There are many situations where the flow is not uniform and it is necessary to predict how the depth will change as differing slopes, resistance, channel geometry, and the effects of hydraulic structures are encountered. It may not always be necessary to have numerical values; rather, it may be sufficient simply to know how the depth will change. For example, is there a possibility of a hydraulic jump forming? If so, where? Will the depth increase or decrease? Where will the control point be? Will it be necessary to perform surface profile computations in the upstream or downstream directions?

Figure 5-16
Hydraulic jump profile and energy loss dependence on Froude number

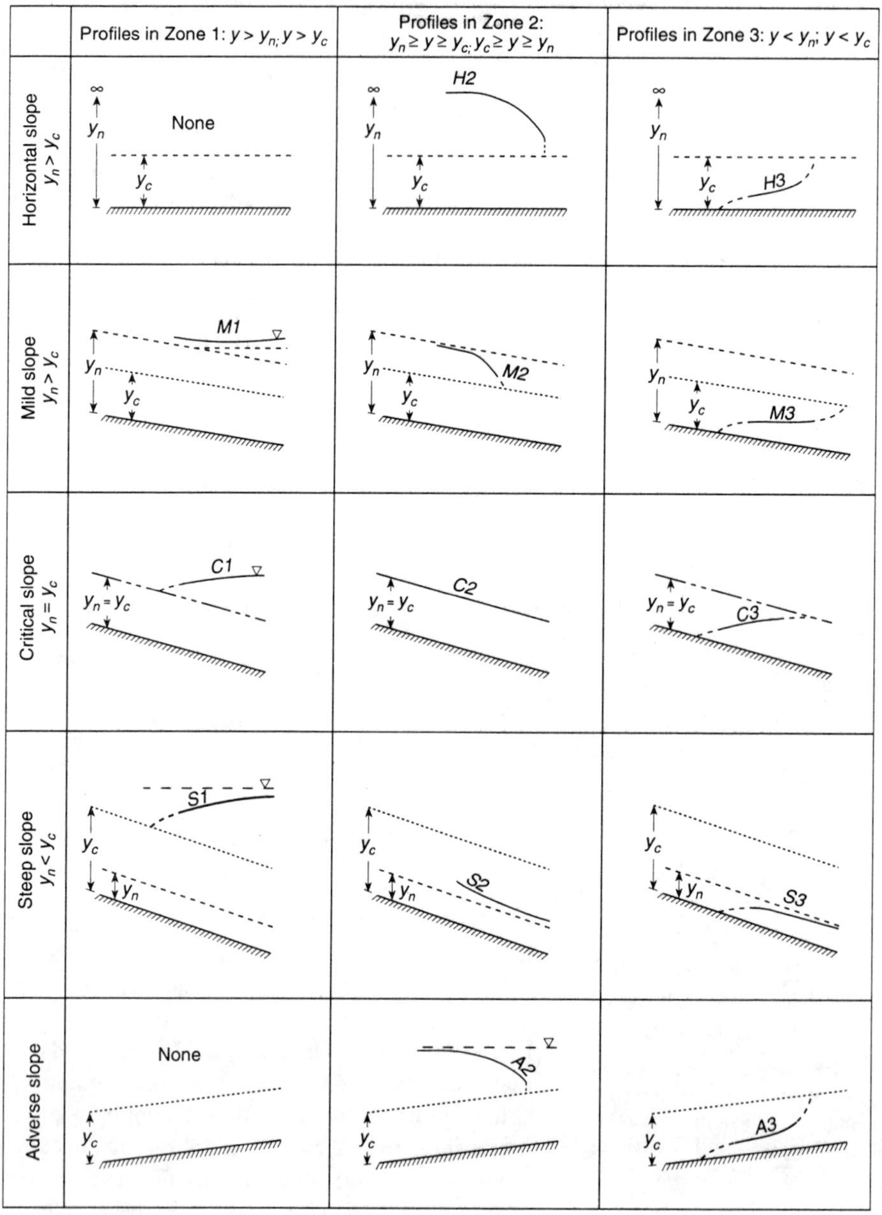

	Profiles in Zone 1: $y > y_n;\ y > y_c$	Profiles in Zone 2: $y_n \geq y \geq y_c;\ y_c \geq y \geq y_n$	Profiles in Zone 3: $y < y_n;\ y < y_c$
Horizontal slope $y_n > y_c$	None	H2	H3
Mild slope $y_n > y_c$	M1	M2	M3
Critical slope $y_n = y_c$	C1	C2	C3
Steep slope $y_n < y_c$	S1	S2	S3
Adverse slope	None	A2	A3

Source: Simon, 1981.

These are all questions that can be addressed by computing the normal and critical depths, along with the Froude number, then sketching the surface profile for the system, based on a classification that hydraulic engineers have developed utilizing the 13 basic profiles that can occur. These are shown in Figure 5-16. Each profile is given a letter reflecting the slope of the channel and a number subscript that defines the zone in which the profile will form. Zones are determined by the lines representing channel bottom, critical depth, and normal depth. A profile falling above the first encountered line is in zone 1; between the first and the second, zone 2; and the second line and the channel bottom, zone 3. The letter designations are as follows:

TIP FOR THE P.E. EXAMINATION CANDIDATE

Understanding the effects of non-uniform flom and their detemination is essential for environmental engineers.

$N_F < 1$	Subcritical or Mild	M
$N_F = 1$	Critical	C
$N_F > 1$	Supercritical or Steep	S
-	Horizontal	H
-	Adverse	A

Note: Horizontal or adverse slopes cannot sustain uniform flow, so a normal depth line is not present.

5.6 Hydrology

Hydrology deals with the distribution of water on and within the earth and its atmosphere. The field can be divided into surface hydrology, groundwater hydrology, and meteorology for the purpose of grouping the processes that control the distribution of water on the surface of the earth, below its surface, and in the earth's atmosphere. This section addresses the essential concepts of surface and groundwater occurrence and movement.

An objective of surface hydrology is to predict the expected quantity of water at a particular location at some future time. Given the uncertainty of rainfall and the complexity and interdependence of the processes that control surface runoff, it is not surprising that engineering hydrology in practice relies on theory, empirical measurement, and field observation, which draws heavily upon the sciences, engineering hydraulics, and statistics.

The total volume of water in the atmosphere and on and below the earth's surface is constant. This accounts for water in all phases—liquid, solid, and vapor. The energy that powers phase change and water distribution comes from the sun. The process that describes the movement of water through its phases is called the *hydrologic cycle,* depicted in Figure 5-17. The cycle entails evaporation from surface waters (major sources are earth's oceans and lakes), movement through the atmo-

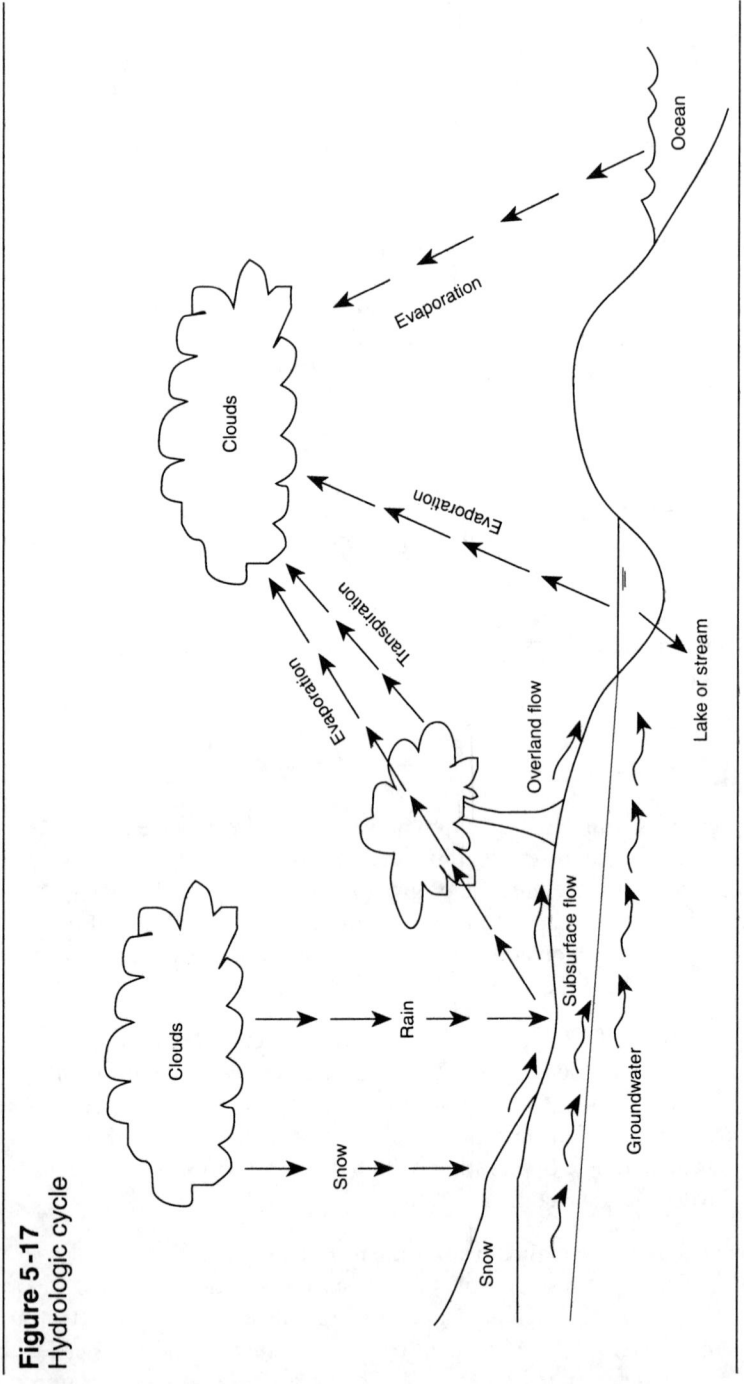

Figure 5-17
Hydrologic cycle

Chapter 5: Hydraulics & Hydrology

sphere, saturation leading to precipitation (rain or snow), infiltration to the ground, surface storage, runoff, groundwater movement, and eventually stream flow, which leads the water back to the lakes and oceans to complete the cycle.

5.6.1 Rainfall Characteristics When it is necessary to estimate the maximum possible rainstorm that might occur, and there is an absence of extensive rainfall records for the location of interest, a useful approach is to estimate the *probable maximum precipitation*, known as the PMP. This concept is based on an evelution of the world's rainstorm records, in which the maximum rainfall for a given duration has been measured.

An envelope of all storms is described by the equation:

$$R = 15.3 D^{0.486} \qquad [5\text{-}31]$$

where: R = the rainfall depth in in.
D = the storm duration in hrs

The PMP is based on the fact that for rain to continue to fall undiminished, it must be replaced at a rate equal to that at which it is precipitating from the atmosphere. Thus, maximum possible rainfall is based on the atmosheric conditions associated with a given locality and the availability of moisture to feed a storm. Figure 5-18 shows the world records for recorded storms.

Figure 5-18
Envelope of the world's record storms

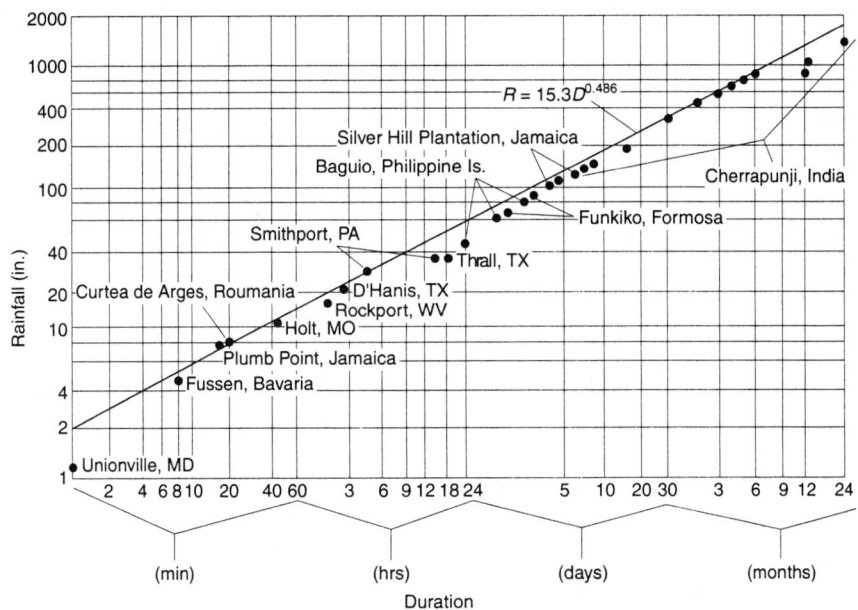

There are specific procedures for developing a PMP for a given location that the reader should refer to if needed. However, a conservative worst-case approach can be had by use of Equation 5-31.

Rainfall over a drainage basin is measured by rainfall gauges of two types. The simplest and most widely used is an *accumulative gauge*. This type of gauge simply records the total volume of water that enters the gauge and represents an equivalent depth of rainfall. A more sophisticated means of measuring rainfall is the use of a *continuous recording gauge*, which will give the total accumulation at any point during the rainfall. Regardless of which type of gauge is used, the gauge represents a measurement of the rainfall at a single point within the drainage basin.

How a single gauge reading can be extrapolated over the drainage basin is the subject of gauge weighting. There are three commonly used methods — *gauge averaging*, the *isohyetal method*, and the *Thiessen method*. The methods are illustrated in Figure 5-19.

Gauge averaging is simply an arithmetic average of all available gauges influencing the drainage area. It is the simplest but least accurate of the three methods.

The isohyetal method is based on creating isohyetal lines that represent lines of equal rainfall accumulation. These are constructed by interpolating between gauge points the same way contour lines are created using elevation points. The rainfall amount falling between any two isohyetal lines is the area between the two lines multiplied by the average of the accumulation on each of the lines.

The Thiessen method is based on creating areas defined by a polygon surrounding each of the rain gauges. The procedure is to connect the gauges with a straight line, subdivide them by a line drawn perpendicular at the midpoint, extended until it contacts a similar line drawn from the other gauges. This defines a polygon area surrounding each of the gauges. The total basin rainfall, then, is the polygon area times the gauge value, which are then summed for all gauges in the drainage basin.

5.6.2 Surface Hydrology The amount of surface runoff that will be produced by a rainfall event is of interest to environmental engineers. *Surface runoff* refers to rain that falls to the ground and is neither *intercepted* or *infiltrated* into the soil. It has two components: *overland flow*, the water that travels over the surface, and *channel flow*.

Water that is intercepted is stored in depressions that can range in size from a dimple on a leaf to a large lake. Depression storage eventually evaporates or infiltrates into the ground. Infiltration describes the process that accounts for the water that seeps into the soil and becomes part of the groundwater system. Surface runoff lost to infiltration will become *soil moisture, unsaturated flow* moving down through the soil, or *saturated groundwater flow*. Water retained in the root zone may later be drawn up by plants and trees and returned to the atmosphere by the

Chapter 5: Hydraulics & Hydrology 127

Figure 5-19
Classification of free surface flow profiles

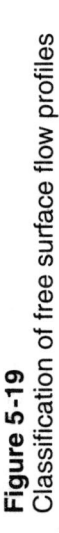

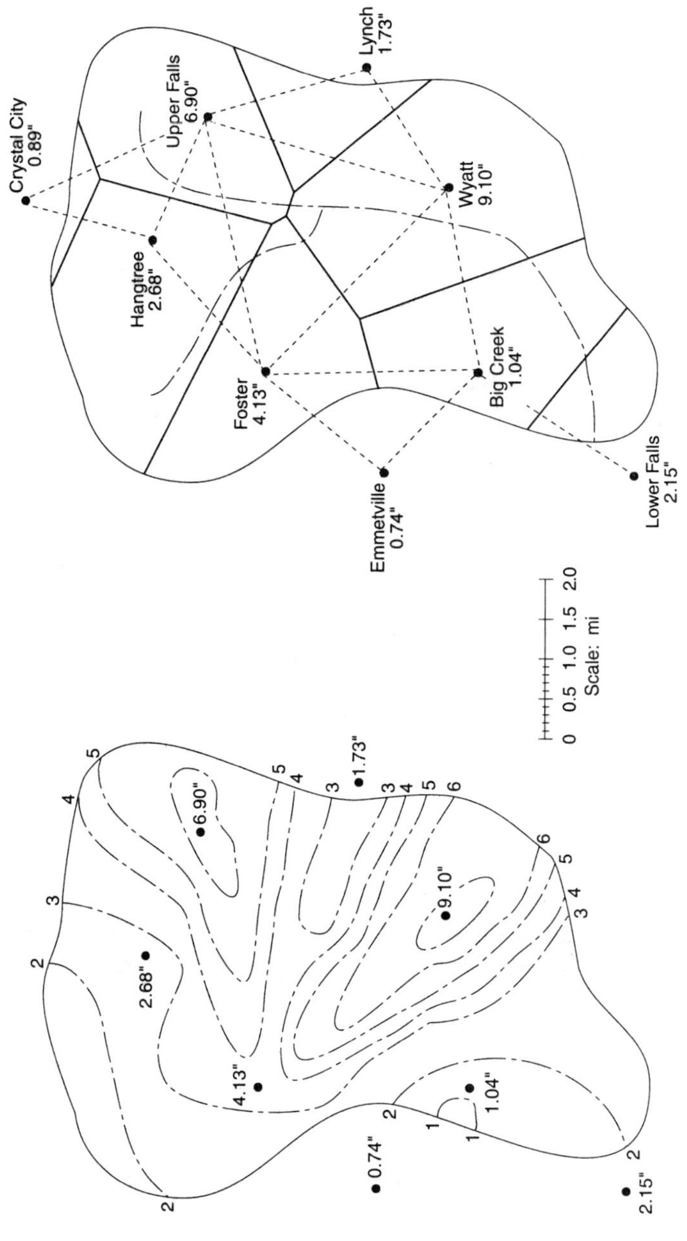

process of *transpiration*. Water can enter the soil until the point of saturation, termed the *field capacity* of the soil. After this point is reached, water must either migrate through the soil or become surface runoff.

One of the very important and widely used concepts in hydrology is the *linear storage equation*:

$$I - O = dS / dt \qquad [5\text{-}32]$$

where: I = inflow
          O = the outflow
          S = the volume of water stored in the system
          t = time

Equation 5-32 is a form of the conservation-of-mass principle, expressing that the sum of all inflow to a hydrologic system minus all outflow equals the change in storage of the system. The equation can be used to evaluate a total water budget of a drainage basin, to route hydrographs through channels, and to model the storage operational characteristics of a dam and reservoir system.

Hydrographs represent the output response of a drainage basin to a rainfall event input, presented in the form of a stream discharge history versus time. A rainfall history in the form of a plot of either *intensity* or rainfall *accumulation* is called a *hyetograph*. Figure 5-20 shows both a hyetograph and resulting runoff hydrograph that a rainfall event produces for a drainage basin.

Referring to Figure 5-20, it is noted that hydrographs have several common characteristics that depend on the size of the drainage area, hydraulic resistance associated with runoff, the slope of drainage area, soil and vegetation characteristics, and other important features that control the runoff process. A maximum discharge is termed the peak discharge. The time from the beginning of runoff to peak discharge is known as time *to peak*. The positively sloping part of the hydrograph is termed the *rising limb*. The negative sloping part is referred to as the *recession limb*. The area underneath the hydrograph represents a volume of water. For streams fed by groundwater, the discharge will not be zero either before the rainfall or following rainfall and all runoff. It is therefore necessary to be able to distinguish the volume of water that is attributable to runoff only. To do this, the stream flow caused by the groundwater contribution, which is known as the *baseflow*, must be separated.

The simplest means to estimate baseflow is by defining the point at which the hydrograph starts to respond to the runoff process and then draw a straight line to a point on the recession limb. Everything above the line will be the runoff hydrograph volume and everything below the line, the baseflow. The stream's naturally-occurring baseflow discharge will normally be greater than at the outset of the runoff. This is because, as the stream rises, water moves from the stream into what was

Figure 5-20
Typical hydrograph showing basin response to rainfall hyetograph

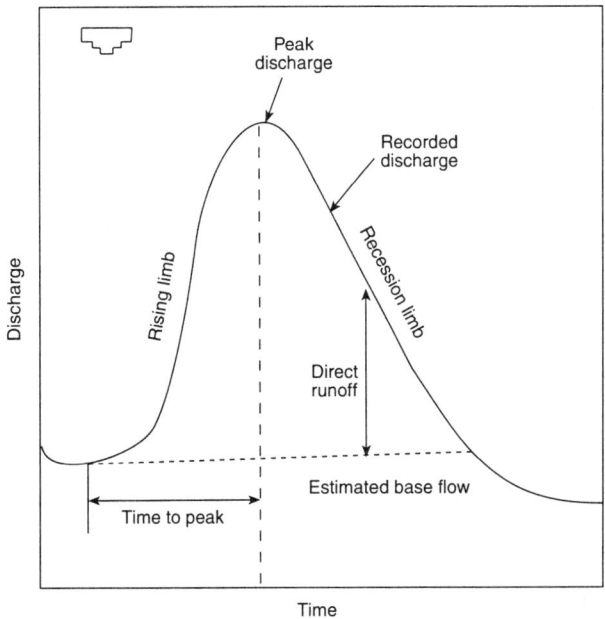

previously a dry river bank. This effect is called *bank storage*. When the storm ceases and the water levels go down, water in the riverbank flows out until the groundwater level is reduced to what it was originally.

The time that it takes a rain drop that falls at the most remote position in the watershed to reach the watershed outlet is known as the *time of concentration t_c*. Knowing the time of concentration for a given watershed is important in that maximum peak discharge will only be achieved if the duration of the rainfall event exceeds the time of concentration of the watershed basin.

Figure 5-21 represents the effect that different parameters would have on the hydrograph defining the response of the drainage basin. Important considerations are the effects of the parameter on the peak discharge, the total volume of runoff, and the time that it takes to complete the runoff process.

A unit hydrograph is defined as the hydrograph that results from one inch of rain uniformly distributed over the drainage basin.[11] There are several methods for developing a unit hydrograph. To

TIP FOR THE P.E. EXAMINATION CANDIDATE

Unit hydrographs are particularly helpful in analyzing surface runoff situations.

Figure 5-21
Factors which influence runoff hydrograph characteristics

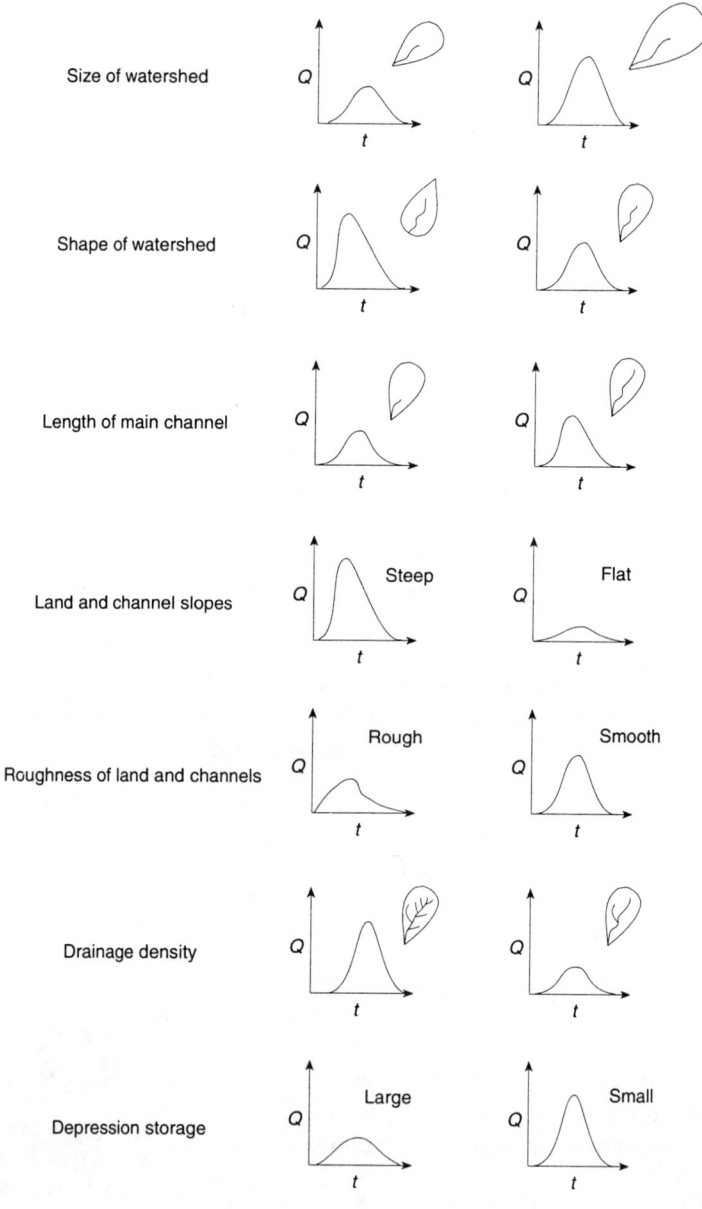

Source: Gumaji, M. FEMA *Flood Migration Course*. 1986.

Chapter 5: Hydraulics & Hydrology 131

illustrate unit hydrograph development one will be obtained from a stream record. Once a unit hydrograph for a particular drainage basin has been determined, it is possible to develop the runoff response for rainfalls other than one inch. The procedure assumes a linear response. For example, if we know what the peak discharge at the outlet of the drainage basin would be for a one-inch rainfall, then a ten-inch rainfall would produce exactly ten times the value of the one-inch rainfall. A unit hydrograph is defined for a specific duration, such as a three-hour unit hydrograph. The concept of lineararity can be directly applied as long as the storm event is three hours. By combining unit hydrographs and lagging their beginning times, it is possible to create a unit hydrograph for times other than that accounted for by the original data. This is illustrated in Example 5-10. It is assumed that the characteristics of the drainage basin are constant and that the hydrograph is a function of the basin characteristics.

Example 5-10

The stream record measured before, during, and after a rainfall that deposited rain over a period of 4 hrs is shown. The area of the drainage basin is 8.3 mi^2. Determine the 4 hr unit hydrograph for the drainage basin. What would be the peak discharge if 3 in. of rain fell on the drainage basin? Also determine the peak stream flow and when it will be reached after the estimated start of rain.

Solution:

The solution is most easily carried out in tabular form. Columns 1 and 2 contain the time and corresponding stream flow. The baseflow is determined by plotting the stream flow and noting the points where flow starts and stops responding to runoff—in this case, at hours 10 and 44 respectively. A linear variation is assumed between the points. Baseflow values are shown in Column 4, which is then subtracted from column 2 to give the runoff shown in column 5. The runoff volume is then calculated by averaging the flow rate at the beginning and end of each 2-hr interal and multiplying by the seconds in 2 hrs (3600). The total runoff is 7,264,800 ft^3/sec. We then divide this volume by the drainage area to obtain the equivalent depth of rain — 0.031 ft or 0.377 in. All runoff values in column 5 are then multiplied by 1/0.377 to obtain the unit hydrograph in column 7, which is then multiplied by 3 and the baseflow added back to give the expected stream flow for a 4-hr, 3-in. storm.

Hydrology — Unit Hydrograph from stream data and combined storm event

Start hour = 6 Rain duration (hr) = 4
Time interv (hr) = 2 Watershead area in mi^2 8.3

1	2	3	4	5	6	7	8
Time (hr)	Q ft^3/sec	Runoff start & end	Baseflow ft^3/sec	Runoff ft^3/sec	Vol. of runoff ft^3	Unit hydrograph ordinates ft^3/sec	4 hr storm 3 in.
0	44		44	0	0	0	44
2	43		43	0	0	0	43
6	44		44	0	0	0	44
8	44		44	0	0	0	44
10	46	<< Start	46	0	0	0	46
12	61		48	13	46,800	35	152
14	67		50	17	108,000	45	185
16	68		52	16	118,000	42	179
18	94		54	40	201,600	106	373
20	115		56	59	356,400	157	526
22	126		58	68	457,200	180	599
24	152	Linear	60	92	576,000	244	793
26	175		62	113	738,000	300	962
28	178	Variation	64	114	817,200	303	972
30	177		66	111	810,000	295	950
32	153	Assumed	68	85	705,600	226	745
34	160		70	90	630,000	239	787
36	141		72	69	572,400	183	621
38	133		74	59	460,800	157	544
40	119		76	43	367,200	114	418
42	98		78	20	226,800	53	237
44	80	<< End	80	0	72,000	0	80
46	75		75	0	0	0	75
48	71		71	0	0	0	71
50	67		67	0	0	0	67
52	66		66	0	0	0	66
54	65		65	0	0	0	65
56	65		65	0	0	0	65

Total = 7,264,800 ft^3/sec Rain depth (ft) = 0.031
Area (mi^2) = 8.3 Rain depth (in.) = 0.377
Area (ft^2) = 231,390,720 Unit hdro mult = 2.654

The peak discharge would be 972 ft^3/sec. If the rain is assumed to start at hour 8 of the stream record, then peak flow is reached 20 hrs later.

Example 5-11

Determine the storm hydrograph for the drainage basin in Example 5-10, where a rainfall hyetograph defining a storm sequence is:

time (hrs)	Rainfall (in.)
0-4	1.3
8-12	3.1
12-16	0.8

What is the maximum discharge and the time to peak?

Solution:

Step 1. Using the unit hydrograph developed in Example 5-10, the appropriate storm hydrograph can be obtained by lagging the 4-hr unit hydrographs.

1	2	3	4	5	6	7	8
	Unit hydrograph			Rainfall accumulation			
Time hr	ordinates ft^3/sec			1.3	3.1	0.8	Storm hydrograph
0	0			0			0
2	35	8 hr lag		45.5			45.5
4	45			58.5			58.5
6	42	___	12 hr lag	54.6			54.6
8	106	0		137.8	0		137.8
10	157	35	___	204.1	108.5		312.6
12	180	45	0	234	139.5	0	373.5
14	244	42	35	317.2	130.2	28	475.4
16	300	106	45	390	328.6	36	754.6
18	303	157	42	393.9	486.7	33.6	914.2
20	295	180	106	383.5	558	84.8	1026.3
22	226	244	157	293.8	756.4	125.6'	1175.8
24	239	300	180	310.7	930	144	1384.7
26	183	303	244	237.9	939.3	195.2	1372.4
28	157	295	300	204.1	914.5	240	1358.6
30	114	226	303	148.2	700.6	242.4	1091.2
32	53	239	295	68.9	740.9	236	1045.8
34	0	183	226	0	567.3	180.8	748.1
36		157	239	0	486.7	191.2	677.9
38		114	183	0	353.4	146.4	499.8
40		53	157	0	164.3	125.6	289.9
42		0	114	0	0	91.2	91.2
44			53	0	0	42.4	42.4
46			0	0	0	0	0

Step 2. The combined storm hydrograph (column 8) is obtained by the linear addition of each unit hydrograph multiplied by the appropriate rainfall accumulation (columns 5 to 7). The scaled unit hydrographs and storm hydrographs are plotted.

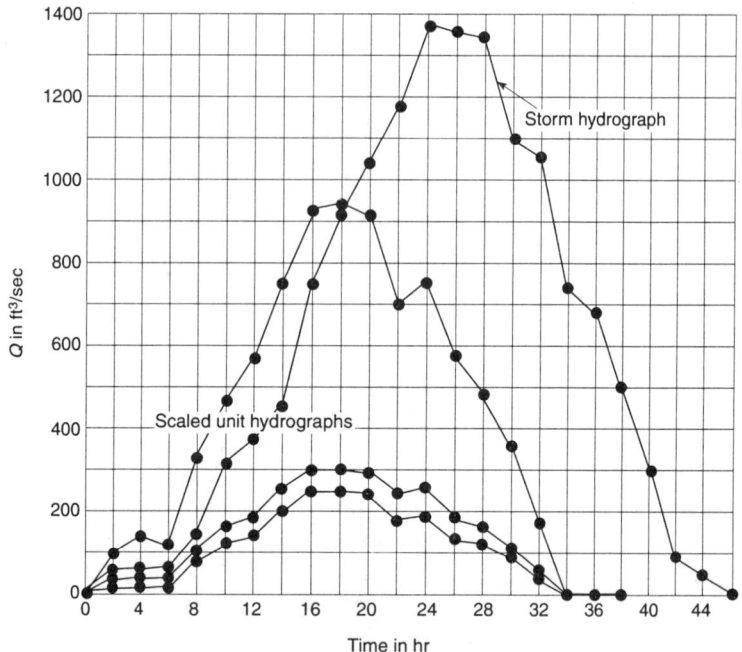

Step 3. The maximum discharge is 1384.7 ft^3/sec and occurs 24 hr after start of rain.

Development of a unit hydrograph from a stream record is a relatively straightforward process. This method requires sufficient data from a gaged drainage basin. When data is not available, there are other approaches including the construction of *synthetic hydrographs*. These methods are outside the scope of this text, but a detailed introduction can be found in Chow, et al. (1988).

5.6.3 Small Watershed Analysis There are often times where it is neither necessary nor warranted to develop a complete runoff hydrograph. A common example is the estimation of design flows for storm sewers. In this instance, a reasonable estimate of the maximum discharges that can be expected is needed so that the drainage system can be sized to accommodate maximum flows without flooding. A simple means of doing this if the use of the *rational equation*:

$Q = CiA$ [5-33]

where:
- A = the area of the drainage area in acres;
- i = the rainfall intensity in inches per hour;
- C = the runoff coefficient, which is a function of the physical characteristics of the drainage area; and
- Q = the peak discharge.

The shape of the assumed hydrograph is triangular. This equation has an empirical basis and can be used where the drainage basin is relatively small, 20 acres or less. This equation is a very simplified representation of the runoff process. In order to estimate the peak discharge that can be achieved in a given drainage area, the rainfall event must exceed the time of concentration. This will be the case if the drainage basin is kept relatively small and the intensity represents rainfalls of practical interest to design. Table 5-5 gives values of the runoff coefficient (e), which will vary from 0 to 1.0. A value of 1 would define an area in which all rainfall becomes runoff. A value of 0 would mean that all rainfall is intercepted or infiltrated.

5.7 Groundwater Hydrology

Groundwater hydrology encompasses the occurrence and movement of water below the earth's surface. Water from the ground is an important, and in many places the primary source of water for domestic, agricultural, or industrial needs. Increasingly, more attention is being paid to the recognition that waste disposal processes, runoff from agricultural uses, and other contributors that have contaminated and are continuing to contaminate groundwater resources. Effective prevention and remediation depends on understanding what controls the movement of water below the surface.

Water bearing geologic formations made up of soils or fractured rocks are known as *aquifers*. Aquifers may be *confined* or *unconfined*. Figure 5-22 illustrates each.

An unconfined aquifer will have a fluctuating upper surface known as the *phreatic line* that defines the water table. Below the phreatic surface, the soil is saturated, meaning that all of the voids are filled with water. Above the phreatic surface, water will be found as a result of capillary action, but not in a saturated state. A confined aquifer will be bounded by a less permeable layer of either rock or soil, which produces a channel of lesser hydraulic resistance within which the groundwater will move.

Table 5-5
Rational equation runoff coefficients

Type of area or development	C
Type of development	
Urban business	0.70 - 0.95
Commercial office	0.50 - 0.70
Residential development	
Single-family homes	0.30 - 0.50
Condominiums	0.40 - 0.60
Apartments	0.60 - 0.80
Suburban residential	0.25 - 0.40
Industrial development	
Light industry	0.50 - 0.80
Heavy industry	0.60 - 0.90
Parks, greenbelts, cemeteries	0.10 - 0.30
Railroad yards, playgrounds	0.20 - 0.40
Unimproved grassland or pasture	0.10 - 0.30
Type of surface areas	
Asphalt or concrete pavement	0.70 - 0.95
Brick paving	0.70 - 0.80
Roofs of buildings	0.80 - 0.95
Grass-covered sandy soil	
Slopes 2% or less	0.05 - 0.10
Slopes 2% or 8%	0.10 - 0.16
Slopes over 8%	0.16 - 0.20
Grass-covered clay soils	
Slopes 2% or less	0.10 - 0.16
Slopes 2% or 8%	0.17 - 0.25
Slopes over 8%	0.26 - 0.36

Source: Robertson, 1985.

5.7.1 Flow in Porus Media The volume of water that a soil can contain is a function of the amount of voids available to be filled. The means for expressing the soil's capacity to hold water is to define its *porosity* η, defined as:

$$\eta = \frac{V_{void}}{V_{total}} \qquad [5\text{-}34]$$

where: V_{void} = the volume of the voids; and
                V_{total} = the total volume of solids and voids, often referred to as the *bulk volume*.

When water is withdrawn from the water-bearing formation, either through pumping or gravitational drainage, some water will be retained within the voids due to

Figure 5-22
Aquifer types

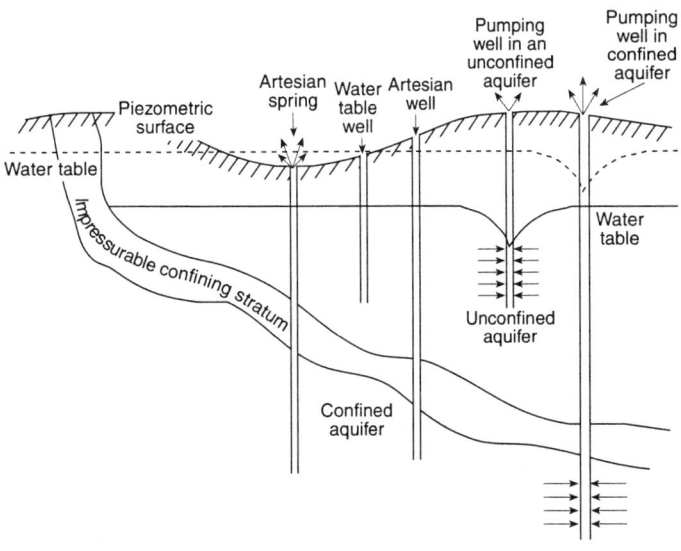

the coating of the surfaces and molecular attraction. That part of the water that can be removed is referred to as the *specific yield*. The remaining water that stays in the voids is termed the *specific retention*. The sum of both must equal the porosity.

The velocity at which water will move through a saturated soil media is a function of the hydraulic resistance that the media provides and the pressure head across the media. Point velocities are generally low and depend on the size of the spatial openings between particles and the particular point on the path that the water must negotiate in and around the inter-connected solid particles. Rather than concern oneself with the point velocity at which the water is moving, one can determine the average velocity of the water passing through the soil media. Figure 5-23 illustrates the classic way in which flow resistance through soil is determined. A conduit of known cross-section is filled with soil, as shown. A pressure head is created across the soil media, which will cause water to flow from the left to right, or from high pressure to low pressure. If the energy equation is applied, it can be shown that the head loss will be proportional to the average velocity V and the length of soil column, inversely proportional to a term known as the *hydraulic conductivity*, K, which is dependent on the size, shape, and surfaces of the porous media (analogous to pipe and channel roughness). In porous media flow, the velocities are sufficiently low and the resistance sufficiently high, so that the flow structure is *laminar*.

Figure 5-23
Darcy experiment to determine flow dependence through a porous media

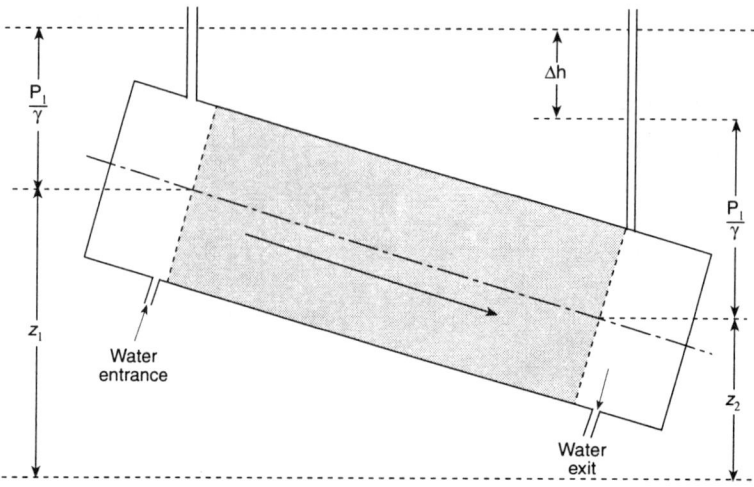

Applying the energy equation between points 1 and 2 shown on Figure 5-23, it is noted that the velocity head terms will be small by comparison to the other terms (and, therefore, can be dropped) and that p_1 is equal to p_2 since both will be at atmospheric pressure, the result is:

$$h_L = z_1 - z_2$$

Using Equation 4-16 to define the head loss, the result is:

$$h_L = Vl / K \qquad [5\text{-}35]$$

Equation 5-35 can be further modified by using the continuity equation resulting in:

$$Q = -(KA)dh / dl \qquad [5\text{-}36]$$

Equation 5-36 is *Darcy's Law*; named for Henri Darcy who first proposed it. It can also be expressed in terms of the Darcy flux, $q = Q/A$, as:

$$q = -Kdh / dl \qquad [5\text{-}37]$$

Darcy's Law relates flow rate in a porous media to *head loss per unit length, dh /dl*, in the flow direction. For porous media the head $h = p / \gamma + z$. For Equation 5-36 to be valid, the flow must be lami-

TIP FOR THE P.E. EXAMINATION CANDIDATE

Darcy's Law is applicable in analyzing well-based water supplies and in the effects of groundwater pollution for municipal solid waste landfills and hazardous waste site remediation (see Chapter 9).

nar. Whether the flow is laminar or not will depend on the ratio of viscous to inertial forces, which can be defined by the Reynolds number. For porous media, a convenient means for expressing the Reynolds number is:

$$N_R = q d_{50} / v \qquad [5\text{-}38]$$

The characteristic length, d_{50}, is the diameter of the average particle found by means of a sieve analysis measured in m. For Darcy's Law to be valid, Equation 5-38 must be ≤ 1. Table 5-6 provides representative values of the hydraulic properties of several porous media.

Table 5-6
Typical properties of porous media

Material	Porosity	Specific yield	Specific retention	Hydraulic conductivity (m/day)
Soil	0.55	0.40	0.15	10^{-3} - 5
Clay	0.50	0.02	0.48	10^{-9} - 10
Sand	0.25	0.22	0.03	0.06 - 120
Gravel	0.20	0.19	0.01	100 - 7000
Limestone	0.20	0.18	0.02	10^{-4} - 5000
Sandstone	0.11	0.06	0.05	10^{-5} - 0.5
Basalt	0.11	0.08	0.03	10^{-8} - 1000
Granite	0.001	0.0009	0.0001	10^{-8} - 5

Source: Roberson, 1988.

5.7.2 Flow Nets To analyze the motion of water in saturated soils, the construction of a *flow net* is a valuable analytical tool. A flow net is developed by constructing the streamlines that define the velocity pattern in the saturated soil (streamlines) and a set of orthogonal lines, which are termed *equipotential lines*. Equipotential lines connect points that have *equal energy or head*. When drawn correctly, they will form lines that intersect streamlines at right angles. An equipotential line may also be thought of as a contour line defining points of equal energy.

Since the flow in the porous media is almost always laminar (exceptions are found in the flow through course gravel), the behavior of the fluid will correspond to that of an *ideal fluid*. This paradox, that an extremely viscous flow will behave like a nonviscous flow, is possible because of the relative smallness of the inertial forces in comparison to the viscous and pressure forces that are present. With this point of reference, both streamline and equipotential line patterns obey the *Laplace* equation:

$$\nabla^2 h = 0 \qquad [5\text{-}39]$$

The Laplace equation may be solved by a number of methods ranging from direct or numerical integration to graphical construction by trial-and-error. Figure 5-24 illustrates the development of a flow net.

Figure 5-24
Steps in constructing a flow net

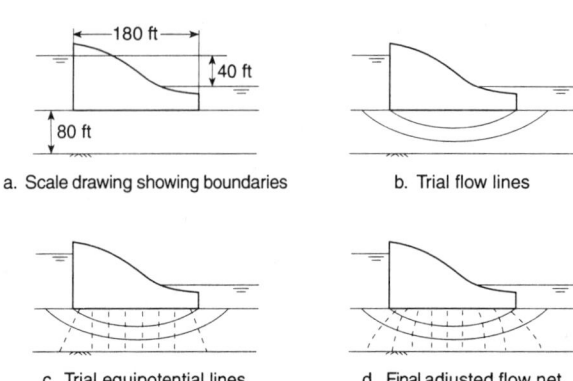

5.7.3 One-Dimensional Porous Media Flow
Where the flow can be considered one-dimensional, Darcy's Equation can be applied without modification. Figure 5-25 illustrates this situation. Here the flow is *confined* and the streamlines are essentially parallel. When the flow is *unconfined*, as is the case in Figure 5-26, analysis is more complicated, in that the location of the phreatic surface must be determined. The procedures for doing this are outside the scope of this coverage. One-dimensional flow analysis is illustrated by the following example.

Figure 5-25
Confined flow

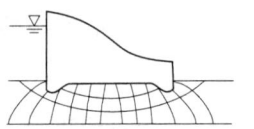

Figure 5-26
Unconfined flow

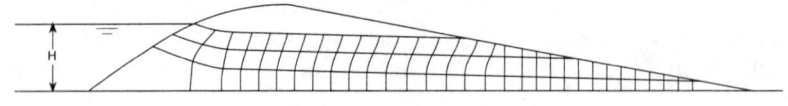

Example 5-12

A 130 foot long concrete settling tank is supported on a layer of silty sand, which has a $K = 1 \cdot 10^{-6}$ m/sec. Below this layer lies a layer of clay which can be considered impervious. A construction trench is cut to facilitate repairs to the tank. Determine the minimum pumping capacity in gpm required to dewater the trench to the level of the clay layer.

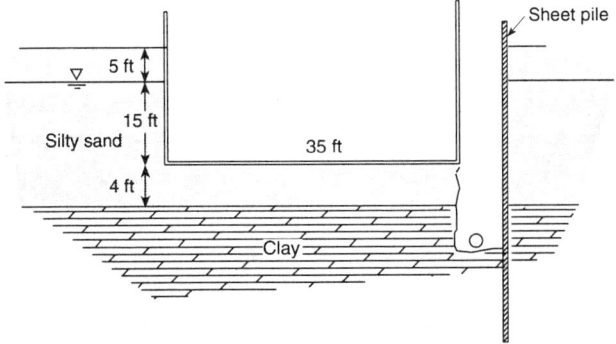

Solution:

Step 1. Use Darcy's Equation in which we first determine the following:

$dh/dl = -19/35 \quad A = (4)(130) = 520 \text{ ft}^2 \quad K = 1 \cdot 10^{-6} \text{ m/sec} = 3.28 \cdot 10^{-6} \text{ ft/sec}$

Step 2.

$Q = -(KA) dh/dl = -(3.28 \cdot 10^{-6})(520)(-19/35) = 9.25 \cdot 10^{-4} \text{ ft}^3/\text{sec}$

$Q_{gal/min} = 9.25 \cdot 10^{-4} \text{ ft}^3/\text{sec} / 2.23 \cdot 10^{-3} = 0.415 \text{ gal/min}$

5.7.4 Radial Porus Media Flow

A common groundwater situation is the flow in an aquifer to a *well point*. This is illustrated in Figure 5-27.

In this situation, the flow will be radial inward along streamlines toward the well. The cross-sectional flow area will be the surface of a cylinder of radius *r*. As the flow gets closer to the well, the velocity must increase since the cross-sectional area is being reduced. This results in a drawdown of the water table, defined by the upper phreatic surface. The limit of this effect is known as the *cone of depression*. The greater the rate of pumping, the larger the drawdown will be. Once pumping

Figure 5-27
Flow in an aquifer to a well point

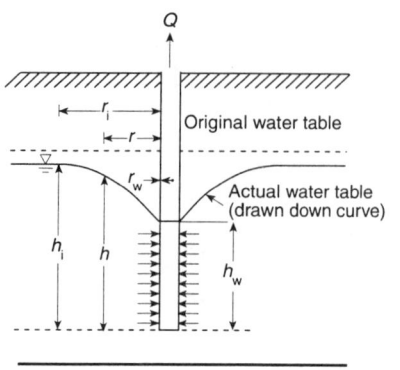

has ceased and sufficient time has passed for recharge, the phreatic surface will return to its original level provided there is a source for replenishment.

The relationship for calculating the discharge from a well can be determined by applying Darcy's Law with modifications to define the cross-sectional area. Since $Q = AV$, the relationship may be written as:

$$Q = -(KA)dh/dl = -(K2\pi rh)dh/dr$$

[5-40]

If the differential equation is integrated between the limits of the radius of the well r_w and the radius that defines the zone of influence r_i, the result is:

$$Q = -(K\pi)[h_i^2 - h_w^2 / \ln(r_i/r_w)]$$

[5-41]

Equation [5-41] assumes that a continuous source of water resides outside of the radius of influence, unaffected by the flow toward the well.

5.8 Standard Hydraulics/ Hydrology Notations

Symbol Units	Definition	
a	Acceleration	L/T²
A	Area	L²
d	Depth of Flow	L
d	Pipe Diameter	L
d_I	Impeller Diameter	L
d_{50}	Mean Soil Particle Diameter	L
C	Runoff coefficient	
C	Hazen-Williams Friction Coefficient	none
C_c	Chezy Coefficient	none
D	Hydraulic Depth	L
D	Duration of Rainfall	T
d	Diameter	L
e	Efficiency	none
e_m	Motor Efficiency	none
e_p	Pump Efficiency	none
E	Energy	none

Chapter 5: Hydraulics & Hydrology

Symbol	Description	Units
E_k	Kinetic Energy	FL
E_p	Pressure Energy	FL
E_z	Potential Energy	FL
E_M	Mechanical Energy	FL
E_T	Thermal Energy	FL
f	Friction Factor	none
F	Force = ma^2	F or ML/T^2
g	Gravitational Constant	L/T^2
H_L	Head Loss	L
h_f	Head Loss	L
i	intensity	
I	Inflow	
l	Length	L
n	Manning Friction Factor	none
N_F	Froude Number	none
N_R	Reynolds Number	none
N_s	Specific Speed	
O	Outflow	
q	Q/A	
Q	Volumetric Flow Rate	L^3/T
Q_{gpm}	Volumetric Flow Rate in gpm	L^3/T
Q_m	Mass Flow Rate	M/T
P	Power	L
p	Pressure	F/L^2
P_R	Hydraulic radius	
R	Rainfall depth	
S	Storage	
S_f	Slope of Energy Grade Line	none
S_o	Slope of Channel Bottom	none
S_w	Slope of Free Surface	none
t	Time	T
T	Top Width of Channel	L
T	Torque	
v	Velocity Vector	L/T
u	Velocity Component in x - direction	L/T
v	Velocity Component in y - direction	L/T
V	Average Velocity for One-dimensional Flow	L/T
W_p	Wetted perimeter	L
W_x	Weight	M
y	Depth	
Z_s	Section factor	$L^{5/3}$
Z	Side slope run	L

<u>Greek Symbols</u>

α	Energy Correction Coefficient	none

ß	Momentum Correction Coefficient	none
γ	Specific Weight	F/L^3
∇	Del Operator $\partial^2/\partial x^2 + \partial^2/\partial y^2$	none
ν	Kinematic Viscosity	L^2/T
μ	Dynamic Viscosity	FT/L^2
π	3.14159	none
ρ	Mass Density	M/L^2
τ_o	Shear Stress	F/L^2
ω	Angular velocity	T^{-1}
η	Porosity	

5.9 References

Albertson, M.L., J.R. Barton and D.B. Simons. 1960. *Fluid Mechanics for Engineers*. Upper Saddle River, New Jersey: Prentice Hall.

Bird, R.B., W.E. Stewart and E.N. Lightfoot. 1960 *Transport Phenomena*. New York: Wiley & Sons.

Chow, Ven Te. 1959. *Open-channel Hydraulics*. New York: McGraw-Hill.

Chow, Ven Te, David R. Maidment and Larry W. Mays. 1988. *Applied Hydrology*. New York: McGraw Hill.

Davis, C.V. 1952. *Handbook of Applied Hydraulics*. New York: McGraw-Hill.

Domenico, Patrick A., and Franklin W.Schwartz. 1990. *Physical and Chemical Hydrogeology*. New York: John Wiley & Sons.

Freeze, R.A. and J.A. Cherry. 1979. *Groundwater*. Upper Saddle River, New Jersey: Prentice Hall.

French, R.H. 1985. *Open-channel Hydraulics*. New York: McGraw-Hill.

Henderson, F.M. 1966. *Open Channel Flow*. New York: Macmillan.

King, H.W., and E.F. Brater. 1963. *Handbook of Hydraulics*. 5th ed. New York: McGraw-Hill.

Roberson, J.A., and C.T. Crowe. 1985. *Engineering Fluid Mechanics*. 3rd ed. Boston: Houghton-Mifflin.

Roberson, J.A., et. al. 1988. *Hydraulic Engineering*. Boston: Houghton-Mifflin.

Shames, I.V. 1982. *Mechanics of Fluids*. 2nd ed. New York: McGraw-Hill.

Simon, A.L. 1981. *Practical Hydraulics*. 2nd ed. New York: John Wiley & Sons.

Strack, Otto, D.L. 1989. *Groundwater Mechanics*. Prentice Hall.

Viessman, Warren, Jr., Gary L.Lewis and John W. Knapp. 1989. *Introduction to Hydrology*. New York: Harper & Row Publishers.

Chapter 6

WATER TREATMENT

by W. Christopher King, P.E., DEE

6.1 Introduction

The need for water of adequate quantity and quality for human use gave rise to the discipline of environmental engineering. Providing safe water continues to be a fundamental service of the profession. This is an essential area of knowledge for environmental engineers because of its public health implications. Until recently, water treatment focused on pathogenic organisms, solids, and chemicals affecting taste, odor, hardness, etc. To address these concerns, processes developed in the early part of the 20th century have been continuously refined ever since. These processes and their proper applications are the foundation of water treatment.

Today, there are new emerging areas of concern regarding the quality of water supplied to consumers. Many of these relate to the degradation of source-water quality caused by the volume and toxicity of materials discharged to the environment, some of which are relatively immune to standard water treatment processes. A second group of problems involve removing microorganisms, such as chyrptospiridia, giardi, etc. from waters previously presumed to be free of pathogenic organisms. A third area of concern is the dissolution of lead and copper from the distribution system into the water due to corrosivity and their affect on public health. Corrosivity also affects the structural integrity of the distribution system from the plant to tap. Finally, an ever-increasing population, often in areas of the country lacking adequate source capacity, e.g., the Southwestern U.S. and the world, dictates that drinking water quality and water resource management will continue to be a key area of environmental engineering practice. Water is essential for life.

As with all of the major subject areas of environmental engineering, water treatment has numerous subtopics — more than will be covered here. Additional information can be acquired from such references as Viessman and Hammer (1993), Corbitt (1990), Salvato (1992), and Montgomery (1995), with the last being the most complete, primarily because water treatment is the book's sole topic.

6.2 Water Quality

The first principle of water supply is that drinking water is not pure, even when it comes from a fancy bottle labeled "artesian spring water." Drinking water that is safe to consume will still contain a variety of chemicals which can impart an unpleasant taste or be corrosive to the distribution system. Therefore, it is important to determine what chemicals are found in water and to identify safe levels of consumption for them. Defining the materials found in water is normally done by characterizing water according to the following categories of quality parameters:

> Physical — turbidity, color, taste, suspended solids, total dissolved solids, and odor
>
> Chemical —
> Organics — benzene, toluene, etc.
> Synthetic organics — PCBs, pesticides, etc.
> Inorganics — lead, copper, sulfate, chloride, etc.
>
> Biological Parameters — total coliform, *Giardia lamblia,* fecal streptococci, viruses, etc.
>
> Radiological Parameters — gross alpha, beta & photon, radon-222 (see Chapter 10 for an explanation of radiological hazards)

Because water will always contain some level of foreign material (contaminants), it is necessary to determine the "safe" levels for each. This is a yet uncompleted task because of the number of chemicals already produced and the number being developed annually. While much is still being learned about the health effects of most all chemicals, synthetic organic chemicals represent the greatest challenge.

6.2.1 Water Quality Regulations To ensure public health and safety, the "safe" level of an extensive list of potential water constituents are prescribed by federal and state regulations. Generally, these are distinctly separate from regulations governing the quality of water *discharged* to the environment which consider a broader range of uses.

Federal drinking water regulations are promulgated under the authority of the Public Health Service Act as amended by the Safe Drinking Water Act (SDWA) of 1965. Additional amendments have been enacted since, the most recent being the Safe Drinking Water Act Amendments of 1986. These rules (codified as 40 CFR 141-149) prescribe safe levels of chemicals by name or by class, establish monitoring and reporting rules, specify treatment methods for pollutants when standards are exceeded, and in some cases specify protection to water sources. The health-based criteria included in the National Primary Drinking Water Regulations are referred to as *maximum contaminant levels* (MCLs). The MCLs are the enforceable standards of the SDWA. These are generally based on risk assessment for a lifetime exposure or a shorter period, depending on the human response to a chemical. (Risk assessment is discussed in Chapter 10). MCLs generally apply to all public water systems serving more than 25 people or 15 service connections. MCLs established for inorganic, organic, synthetic organic, and radiological chemicals are presented in Tables 6-1 through 6-4, respectively. MCLs are normally applied to water at the point of entrance into the distribution system. However, lead and copper standards apply at the tap. Total trihalomethanes (TTHMs) pose a unique problem and are controlled using a different regulatory approach because they are created primarily as a byproduct of the disinfection process.

TIP FOR THE P.E. EXAMINATION CANDIDATE

The PE examination typically avoids testing on regulatory details because they vary across the country and change frequently, making it difficult to incorporate them into a national exam. Code of Federal Regulations (CFR) references are provided for the candidate to locate additional information on the federal standards.

40CFR 141-149 translates as Title 40 of the Code of Federal Regulations, Parts 141-149.

Below MCLs in the hierarchy of the drinking water regulations are se*condary maximum contaminant levels* (SMCLs) and *maximum contaminant level goals* (MCLGs). SMCLs are primarily aesthetic criteria for public water supplies (as described in 40 CFR 143). MCLGs (as described in 40 CFR 141.50 -141.52) could be considered good treatment practice criteria, i.e., things the public would like to have (or, more appropriately, not like to have) in drinking water. MCLGs can be viewed as the best-case conditions for public water supply systems. Although generally not enforced, they reflect the principle that, where possible, obtaining the highest quality drinking water should be an objective.

Table 6-1
Maximum contaminant levels (MCLs) for inorganic chemicals

Chemical	MCL (mg/L)
Flouride	4
Asbestos	7 million fibers/L (longer than 10μm)
Barium	2
Cadmium	0.005
Chromium	0.1
Mercury	0.002
Nitrate	10 (as nitrogen)
Nitrite	1 (as nitrogen)
Total Nitrate and nitrite	10 (as nitrogen)
Selenium	0.05
Antimony	0.006
Beryllium	0.004
Cyanide	0.2 (as free Cyanide)
Nickel	0.1
Thallium	0.002

Source: 40 CFR 141.62. 1994.

Lead and copper are regulated at the consumer's tap, in addition to the treatment plant, because corrosion in the distribution system may cause these minerals to enter the water before consumption. Lead and copper levels measured at the tap are compared to *action* levels of 0.015 and 1.3 mg/L, respectively. Action levels are not MCLs, but are standards that, when exceeded, require some response to control the health risk. Responses typically include corrosion control and treatment upgrades.

Another exception to the norm in the drinking water regulations is the regulation of a group of chemicals called *trihalomethanes* (TTHMs). Chemically, these are derivatives of methane that can be formed by combining a halogen (usually chlorine added to the water as a disinfectant) with naturally-occurring organics in the water. Standards for TTHMs require a balance between the disinfection benefits and residual protection provided by chlorine and the health concern posed by substantial concentrations of TTHMs in drinking water. The MCL for TTHMs is 0.1 mg/L. This standard has drastically affected the way microorganisms are controlled in drinking water. Many new treatment approaches are now being developed to reduce the production of TTHMs while retaining disinfection and residual protection of the distribution system.

A unique feature of the water regulations introduced in recent amendments is the *Surface Water Treatment Rule* (SWTR). This rule was added because of growing problems and concerns with microorganisms from public water supplies produc-

Table 6-2
Maximum contaminant levels (MCLs) for organic chemcials

Chemical	MCL (mg/L)
Vinyl chloride	0.002
Benzene	0.005
Carbon tetrachloride	0.005
1,2-Dichloroethante	0.005
Trichloroethylene	0.005
para-Dichlorobenzene	0.075
1,1-Dichloroethylene	0.007
1,1,1-Trichloroethane	0.2
cis-1,2-Dichloroethylene	0.07
1,2-Dichloropropane	0.005
Ethylbenzene	0.7
Monochlorobenzene	0.1
o-Dichlorobenzene	0.6
Styrene	0.1
Tetrachloroethylene	0.005
Toluene	1
trans-1,2-Dichloroethylene	0.1
Xylenes (total)	10
Dichloromethane	0.005
1,2,4-Trichloro-benzene	0.07
1,1,2-Trichloro-ethane	0.005

Source: 40CFR 141.61. 1994.

ing disease outbreaks. Organisms of particular concern are *Giardia lamblia, Legionella, Chryptospiridia,* viruses, and other bacteria. The SWTR establishes threshold criteria for the organism content of water sources. Where source quality is below the standards, i.e., there are too many organisms in the raw water source, then the water must be filtered and disinfected. Compliance with the treatment standards requires adequate removal of microorganisms. The SWTR also specifies the turbidity of water must be less than 1 turbidity unit. The rule has required adding filtration in many plants, even smaller systems using groundwater which is often influenced by surface water. Specifics of the SWTR as they relate to design of disinfection processes are presented in a subsequent section.

6.2.2 Water Quality Parameters The parameters discussed here in the context of water quality are equally applicable to wastewater systems. The text by Sawyer, McCarty, and Parkin (1994) is a good supplemental reference for an explanation of all of these terms.

Table 6-3
Maximum contaminant levels (MCLs) for synthetic organic chemicals

Chemical	MCL (mg/L)
Alachlor	0.002
Aldicarb	0.003
Aldicarb sulfoxide	0.004
Aldicarb solfone	0.002
Atrazine	0.003
Carbofuran	0.04
Chlordane	0.002
Dibromochloropropane	0.0002
2,4-D	0.07
Ethylene dibromide	0.00005
Heptachlor	0.0004
Heptachlor epoxide	0.0002
Lindane	0.0002
Methoxychlor	0.04
Polychlorinated biphenyls	0.0005
Pentachlorophenol	0.001
Toxaphene	0.003
2,4,5-TP	0.05
Benzp[a]pyrene	0.0002
Dalapon	0.2
Di(2-ethylhexyl) adipate	0.4
Di(2-ethylhexyl) phthalate	0.006
Dinoseb	0.007
Diquat	0.02
Endothall	0.1
Endrin	0.002
Glyphosate	0.7
Hexacholorbenzene	0.001
Hexachlorocyclopentadiene	0.05
Oxamyl (Vydate)	0.2
Picloram	0.5
Simazine	0.004
2,3,7,8-TCDD (Dioxin) (μg/L)	0.00003

Source: 40 CFR 141.61. 1994.

Chapter 6: Water Treatment

Table 6-4
Maximum contamiant levels (MCLs) for radioactive materials

Chemical	MCL
Radium - 226	5 pCi/L
Radium - 228	5 pCi/L
Beta and photon	4 millirem/yr
Gross alpha	15 pCi/L

Source: 40 CFR 141.15 and 16. 1994.

Solids are the dissolved and suspended material present in water. Although the term solids encompasses many different components, only those most important for water systems are discussed here. Suspended material is measured as total suspended solids (TSS). Because most people consider it unacceptable to see particles in their drinking water, the primary concern with TSS is aesthetic. Dissolved material is referred to as total dissolved solids (TDS). In some cases, high concentrations of dissolved materials will give water a taste and/or color that consumers may find disagreeable. In the worst case, dissolved material can be toxic or carcinogenic. Most contaminants subject to MCLs occur in dissolved form. The three solids parameters most important to the water engineer are defined below. *The Standard Handbook for the Examination and Wastewater* published by the American Public Health Association (1990) provides exact definitions and analytical procedures for the measurement for these and other solids parameters.

Total Suspended Solids (TSS) — the concentration of material that exists in a filterable state in a sample. It is the weight of material collected on a filter after drying (103°C), divided by the volume of liquid filtered to produce the sample. Units are mg/L.

Total Dissolved Solids (TDS) — the concentration of solids in a water sample passing through a fine filter. The filtered sample is placed in a pre-weighed dish and dried. The difference in weights divided by the initial volume of the sample yields the concentration of the dissolved material. Units are mg/L.

Settleable Solids — the volume of material that will settle from a suspension under quiescent conditions. A volume of sample is placed in a 1-liter graduated cone (Imhoff Cone) and allowed to stand for one hour. The volume of settled material is observed. The results are reported as ml of settleable solids/ml of sample.

TIP FOR THE P.E. EXAMINATION CANDIDATE

PE examination questions concerning water treatment will assume the candidate has a working knowledge of the parameters that describe the quality of drinking water. The PE examination does not routinely require the candidate to be able to solve questions concerning analytical procedures, but understanding the parameters used in water is necessary to demonstrate competence and apply practical knowledge in solving the problems presented.

Other physical/chemical parameters of primary concern in water treatment include:

Hardness — a measure of the bivalent cations present in a water sample, typically dominated by Ca^{+2} and Mg^{+2}. This parameter is of concern because of its effect on water uses. Hard water interferes with the activity of soaps while water which is too soft makes rinsing away of soaps difficult. Hard water also causes scaling of pipes and heat exchangers, effectively reducing their efficiency and life expectancy. Hardness is typically reported as mg/L as calcium carbonate.

Alkalinity — a measure of the acid-neutralizing capability of water. Alkalinity can be segregated into hydroxide, carbonate, and bicarbonate forms. Alkalinity has a great impact on the corrosivity of a water which can leach elements such as lead and copper into the water as well as reduce the life expectancy of treatment and distribution system components. Alkalinity is reported as mg/L as calcium carbonate.

Turbidity — a surrogate parameter that measures the colloidal solids in water. This is generally an aesthetic parameter for drinking water and water supplies. Turbidity is a measure of the light refracted by the particles in a sample, reported in comparison to the refraction in a standard suspension. In one form of the turbidity test, 1 mg/L of SiO_2 = 1 turbidity unit. Turbidity has increased in importance because of recent changes in water treatment regulations which have set the MCL for turbidity at 1 turbidity unit. For the environmental engineer, turbidity is the solids component that must be removed after sedimentation, typically by some form of filtration.

6.3 Basic Water Treatment

Figure 6-1 presents a typical water treatment plant. Although it is difficult to define a "typical" water treatment plant, each plant is a collection of any of the basic and additional unit processes listed in the figure, e.g., coagulation/flocculation, that are needed to treat the water to acceptable standards. This section discusses the more common unit processes. For each treatment, the unit operations are briefly described followed by discussion of the standard design approach(es) and design examples.

TIP FOR THE P.E. EXAMINATION CANDIDATE

Candidates are expected to understand the design and operation of the units shown in Figure 6-1 and may be required to have some knowledge of the additional unit operations listed.

The key aspects of the water source that are relevant to plant design include quantity (total and rate), water quality, and temporal variations in quality and flow rate. Water sources are primarily groundwater or surface water. Surface water may be from a stream or

Figure 6-1
Typical water treatment plant

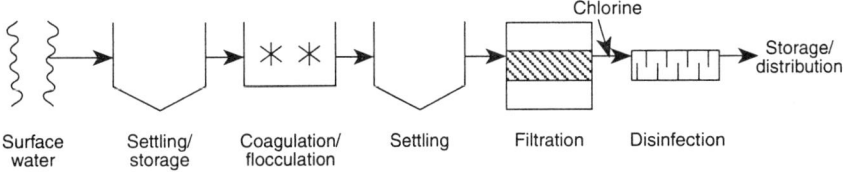

Surface water — Settling/storage — Coagulation/flocculation — Settling — Filtration — Disinfection (Chlorine) — Storage/distribution

Additional unit operations include, water softening, corrosion control, pH adjustment, fluoridation, trihalomethane control

an impoundment, each having different characteristics important to the engineer. Table 6-5 presents an overview of general source characteristics, assuming the absence of heavy man-made pollution.

Typically, source characteristics are obtained from site-specific field studies or relevant published data from prior studies. However, in some cases, such as feasibility studies, it is necessary to estimate one or more characteristics. The use of estimating factors is demonstrated in the solutions to example problems. Reference materials such as Corbitt (1990) and Salvato (1992) are good sources for this type of information.

TIP FOR THE P.E. EXAMINATION CANDIDATE

Normally, the examination question will state whether the source is groundwater or surface water and require an estimate of the quantity of water needed for a specific scenario. The scenario may include a population with commercial, industrial and other quantity requirements. It will be necessary to sum the total daily requirements for the different uses, multiply by the selected daily and seasonal peaking factors, and consider population and demand growth factors to reach a final design flow. For questions in the multiple-choice format, all of the required information will be given, along with some additional information as distracters, thus requiring the examination candidate to identify the correct factors needed to complete the design. In essay questions, it may be necessary to select estimating factors.

6.4 Settling

Gravitational settling is used for many different purposes in water treatment. The standard design factors are *detention time* and *overflow rate*. Detention time is defined as

$$\text{Detention time} = t_d = \frac{\text{Tank volume}}{\text{Flowrate}} \quad [6\text{-}1]$$

$$t_d = \frac{\text{length}(x) \bullet \text{depth}(y) \bullet \text{width}(w)}{Q} \quad [6\text{-}2]$$

The *overflow rate* of a settling chamber is the flow rate into the tank divided by the *surface area* of the tank. Overflow rate is difficult to understand primarily because it may be represented in units of velocity, length/time, or as units such as gal/day-ft^2.

Very little manipulation is required to show that the two units are equivalent, but it is useful for the engineer to understand the underlying concepts for each set of units.

Table 6-5
Characteristic of water sources

Source	Flow	Quality
Surface: Streams	Vary with time. Expect limited supplies.	Quality varies greatly with time. Normal seasonal changes impact treatment. Characteristics: high SS, low hardness, low TDS, high microorganisms Concerns: SS, turbidity microorganisms, organics.
Lakes	Varies over long time periods.	Quality more steady. Characteristics: low hardness, low TDS, low SS, microorganisms. Concerns: turbidity, organics
Groundwater:	Must recharge at rate of use or will be depleted (mined).	Qualtiy constant Characteristics: low SS, low turbidity, high TDS, low microorganisms. Concerns: hardness, pH, TDS, and specific inorganics.

Figure 6-2 shows the active portion of a rectangular sedimentation basin (clarifier). The goal of the clarifier is to have the particle entering the tank (upper left in the figure) reach the bottom of the tank before it reaches the exit point. The side view assumes that water and particles enter from the left and are evenly distributed from the top to the bottom. For a particle to be removed from the flow (i.e., settle out), it must hit the bottom before it reaches the right hand wall of the tank, or it must settle y feet in time t, where t equals the detention time in the basin. By combining Equation 6-2 and *settling velocity* (Y/t)

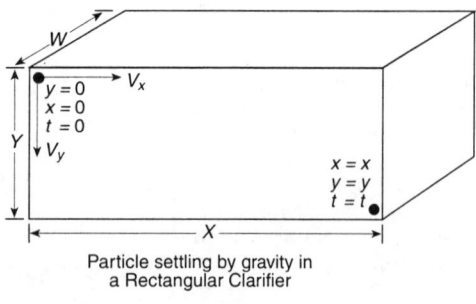

Figure 6-2
Removal of a particle in a clarifier

Particle settling by gravity in a Rectangular Clarifier

Chapter 6: Water Treatment 157

$$t_d = \frac{X \bullet W \bullet Y}{Q} = \frac{A_s \bullet Y}{Q} \qquad [6\text{-}3]$$

$$\text{Overflow rate} = V_o = \text{settling velocity} = \frac{Y}{t_d} \text{ or } \frac{Q}{A_s} \qquad [6\text{-}4]$$

where: Q = Flow rate, typically in gal/day
A_s = length • width

In summary, for settling velocity of a particle to equal the overflow rate, the area required is the tank *surface area* defined by x and w in the diagram.

Within a single sedimentation unit settling occurs in phases and zones. Typically, four classifications for settling are used to describe these phases/zones:

Type I. Discrete particle settling; a single particle falling at a constant velocity.

Type II. Flocculant settling; particles agglomerate and settle at increasing velocity.

Type III. Hindered or zone settling; particles collide and limit settling velocities.

Type IV. Compression; increasing density in the solids blanket at the bottom of a clarifier; settling velocity is very low.

6.4.1 Sedimentation Types *Type I* settling is the simplest form and involves a single particle falling at a constant velocity in a quiescent fluid. This settling is classically modeled by Stoke's Law (See Weber (1972) for a detailed development of Stoke's Law). A sample of a Stoke's Law settling problem is provided in Example 6-1.

$$\text{Stoke's Law} = V_t = \frac{g(\rho_p - \rho)d_{px}^2}{18\mu} \qquad [6\text{-}5]$$

where: g = gravitational constant
= 32.2 ft/sec^2
= 9.8 m/sec^2
r_p = density of particle
r = density of liquid
d_p = diameter of particle
μ = absolute viscosity of liquid
= 1.002 • 10^{-3} kg/m • s @20°C
= 2.050 • 10^{-5} lb-sec/ft^2 @70°F
V_t = terminal settling velocity

Example 6-1 — Type I settling

Determine the overflow rate (gal/ft²-day) and depth for a clarifier designed to achieve 100% removal of a 25μm sand particle with a specific gravity 2.5. Detention time for the tank should be 2 hr minimum at a flow of 1 Mgal/day.

Solution:

Step 1. Determine the settling velocity for a 25 μm particle that must settle the full depth of the lock

$$V_t = \frac{g(P_p - P)d_p^2}{18\mu} \quad \text{(Equation 6-5)}$$

$$V_t = \frac{9.8\frac{m}{sec^2}[(2.5(1000) - 1000)]\frac{kg}{m^3} \bullet (25 \bullet 10^{-6})^2 m^2}{18 \bullet 1.002 \bullet 10^{-3} \frac{kg}{m \bullet sec}}$$

$$= .00051 \frac{m}{sec}$$

$$V_t = 1.83 \frac{m}{hr} \text{ or } 6.02 \frac{ft}{hr}$$

Step 2. Convert V_t into an overflow rate

$$V_t = 6.02 \frac{ft}{hr} \bullet 7.48 \frac{gal}{ft^3} \bullet 24 \frac{hr}{day}$$

$$= 1,080 \frac{gal}{ft^2 - day}$$

The foregoing indicates that 1,080 gal/day-ft² is the surface loading for the clarifer Therefore, for 1 Mgal/day:

$$ft^2 \text{ of Surface Area} = \frac{1 \bullet 10^6 \frac{gal}{day}}{1,080 \frac{gal}{ft^2 - day}}$$

$$\text{Area} = 926 \text{ ft}^2$$

Chapter 6: Water Treatment

Step 3. Determine the depth of the tank based on a 2 hr detention time; use Equation 6-1

$$t_d = \frac{\text{volume}}{\text{flow rate}} \text{(watch units)}$$

$$2\,\text{hr} = \frac{926\,\text{ft}^2 \bullet \text{depth}}{1 \bullet 10^6 \frac{\text{gal}}{\text{day}} \propto \frac{1\,\text{day}}{24\,\text{hr}} \propto \frac{1\,\text{ft}^3}{7.48\,\text{gal}}}$$

$$\text{depth} = 12\,\text{ft}$$

This answer should have been intuitively obvious in that $V_t = 6$ ft/hr, and therefore this particle will settle 12 ft in 2 hr. Use Equation 6-2 and 6-4

$$t_d = \frac{V}{Q} = \frac{\text{surface area \& depth}}{\text{volume / time}}$$

$$t_d = \frac{\text{depth}}{\text{overflow rate}}$$

Preliminary settling in water treatment plants is the unit operation closest to Type I settling. Design for Type I settling may be accomplished by assuming particle properties and then calculating the dimensions required to remove that specific particle as shown in Example 6-1. However, this is very rarely done. Rather, standards of practice, detention times, and overflow rates are used to establish the size of primary clarifiers. Table 6-6 provides typical values for these unit operations.

TIP FOR THE P.E. EXAMINATION CANDIDATE

Stoke's Law has many applications in environmental engineering in addition to water systems. For example, its applications include the design of grit chambers and primary clarifiers for wastewater systems and in air pollution control it is applicable to discrete particle settling.

In Type II settling, the particles initially fall according to Stoke's Law, but collide and adhere as they fall, causing the particles to grow in size and, as a result, their settling accelerates. Stoke's Law establishes why larger particles will overtake smaller particles; as the particles agglomerate, the particle velocity increases because of the increased effective diameter. This process continues with depth throughout the settling basin, and no terminal settling velocity is achieved. Flocculation basins for water treatment and lower sections of primary settling in wastewater plants are the settling unit processes that most closely mimic Type II settling. The scientific approach to design of Type II

Table 6-6
Standard design criteria for gravity settling

Treatment process	Overflow rate (gal/ft^2-day)	Detention time (hr)	Weir loading rate (gal/ft-day)
Water Treatment Plant			
- Preliminary settling[1]	-	3	-
- Settling after coagulation/flow[1]	500-1,000[1]	2 – 4	20,000
Wastewater Treatment[2]			
- Grit removal		1/60	
- Primary settling	400-800		20,000
- Secondary, w/activated sludge	600-800	2-3	
- Secondary, fixed film	400-800		20,000

1 Source: Viessman and Hammer, 1993.
2 Source: Metcalf and Eddy, Inc., 1991.

settling is based on a graphical analysis of settling data. An example of this method is included as Example 6-2. There are two forms to the overall removal efficiency equation (Viessman and Hammer (1993) and Metcalf and Eddy, Inc. (1991)); both forms are valid for all data sets. The method from Metcalf and Eddy, Inc. is included in Example 6-2 because the authors consider it easier to understand. In practice, many designs for primary settlers and flocculation basins are based on typical values for overflow rates and detention times such as those shown in Table 6-6.

Example 6-2 — Type II settling

Settling column data is collected to determine the rates for Type II settling. These data are collected in a column with sample ports at heights over the depth of the tank. At several times during the test, samples are taken from each port and the suspended solids concentrations are reported as shown in the table below. These data are plotted as shown with the lines of equal settling efficiency constructed by interpolating between the data points. These curves can then be used to calculate the treatment efficiency for any time within the range of the data. There are at least two procedures to employ these data in the design of a clarifier. The method presented in Metcalf and Eddy, Inc. (1991) will be used here.

Chapter 6: Water Treatment

Time (min)	Depths (ft)			
	2	4	6	8
0	0	0	0	0
15	41	32	28	25
30	60	46	41	37
45	67	59	52	47
60	71	65	60	55
75	76	69	64	61
90	80	72	68	64

$$\text{Percent Removal} = \frac{\Delta h_1}{h_d}\left(\frac{R_1+R_2}{2}\right) + \frac{\Delta h_2}{h_d}\left(\frac{R_2+R_3}{2}\right) + \ldots \qquad [6\text{-}6]$$

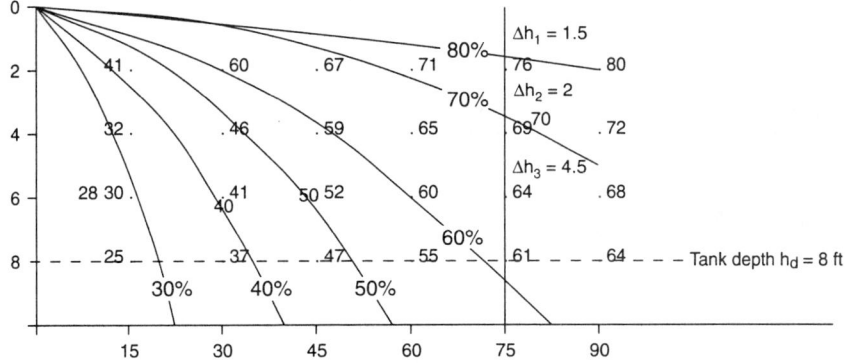

Design an 8-foot deep clairfier that can treat an average flow of 2.5 Mgal/day and achieve a 70% suspended solids removal efficiency. (Note: this problem can be a water or wastewater problem).

Solution:

Step 1. An iterative solution is required by choosing a time and using Equation 6-6 to calculate the efficiency until the 70% removal efficiency is reached. Presented below is the solution for $t = 75$ minutes which does meet this criteria.

$$\text{SS\% Removal} = \frac{\Delta h_1}{h_d}\left(\frac{R_1+R_2}{2}\right) + \frac{\Delta h_2}{h_d}\left(\frac{R_2+R_3}{2}\right) + \frac{\Delta h_3}{h_d}\left(\frac{R_3+R_4}{2}\right)$$

$$= \frac{1.5}{8}\left(\frac{100+80}{2}\right) + \frac{2}{8}\left(\frac{80+70}{2}\right) + \frac{4.5}{8}\left(\frac{70+61}{2}\right)$$

$$= 16.88 + 18.75 + 36.84$$

$$= 72.47\%$$

Step 2. Determine overflow rate

$$V_o = \frac{h_d}{time} = \frac{8\,ft}{75\,min}$$

$$= 0.1067\frac{ft}{min} \cdot \frac{7.48\,gal}{ft^3} \cdot \frac{60\,min}{hr} \cdot \frac{24\,hr}{day}$$

$$= 1,149\frac{gal}{ft^2 - day}$$

Adjust to account for imperfections of real world system

$$(1,149)0.75 = 862\frac{gal}{ft^2 - day}$$

Step 3. Size the clarifier

$$Surface\,area = \frac{Q}{overflow\,rate}$$

$$= \frac{2.5 \bullet 10^6\,gal/day}{862\,gal/day - ft^2}$$

$$= 2,900\,ft^2$$

Volume = area • depth

$$= 2,900 \bullet 8$$

$$= 23,202\,ft^3$$

$$t_d = \frac{Volume}{Q}$$

$$= \frac{23,202\,ft^3}{2.5 \bullet 10^6\frac{gal}{day} \cdot \frac{1\,ft^3}{7.48\,gal} \cdot \frac{1\,day}{24\,hr}}$$

$$= 1.67\,hr$$

Type III settling occurs when the particle concentration is sufficiently high that collisions and particle interactions begin to limit the settling velocities of the individual particles. For settling to occur, the particles tend to move downward as a mass or in a zone; hence Type III settling is sometimes referred to as *zone settling*. Also, this type of settling will sweep or capture small discrete particles, improving the overall treatment efficiencies. This process typifies the settling that occurs in clarifiers where high suspended solids concentrations occur such as secondary clarifiers in activated sludge treatment. There are two forms of graphical analyses of data employed to design clarifiers based on zone settling. The simplest uses interface height data to calculate the required detention time for the basin (See Example 6-3.) The second method, solids flux analyses, is considerably more complex. It is described in detail in Viessman and Hammer (1993) and numerous other texts, but is not be included here.

Compression settling, *Type IV*, occurs at the bottom of the settling basin and in sludge thickeners. This process is slow relative to all other types of settling. It is a complex combination of compression from the weight of the sludge, water being forced out of the matrix, and the support caused by the sludge particle lattice. Design is based on graphical data analysis techniques as demonstrated in Example 6-3.

Example 6-3 — Zone and compression settling

Alum sludge from a water treatment plant needs to reach a concentration of 8,000 mg/L before it can be further processed for disposal. The initial concentration of suspended solids entering a settler is 1,500 mg/L. Results of a settling test on this sludge are presented as Curve A in the diagram below. Based on the data in Curve A, determine the required overflow rate. Be sure to check your results to see if they make sense.

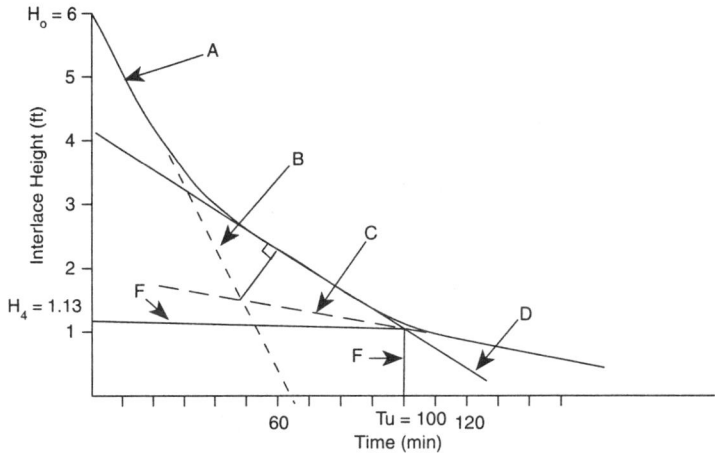

Solution: (This procedure is described in detail in Metcalf and Eddy, Inc. (1991).)

The straight line on the upper part of Curve A represents the zone settling velocity. As the curve becomes nonlinear it represents a transition to compression settling, and the far right of the curve, which is again linear, is the Type IV settling velocity. The first calculation is to determine the settling velocity for the zone settling portion of the curve found by determining the slope of the linear portion of curve A. This is the slope of Line B.

Step 1. Determine size of the clarifier based on zone settling

V_o = settling velocity = slope of line A

$$= \frac{6 \text{ ft}}{65 \text{ min}} = 0.092 \frac{\text{ft}}{\text{min}} = 5.54 \frac{\text{ft}}{\text{hr}}$$

Convert to overflow rate:

$$5.54 \frac{\text{ft}}{\text{hr}} \cdot \frac{24 \text{ hr}}{\text{day}} \cdot \frac{7.48 \text{ gal}}{\text{ft}^3} = 994 \frac{\text{gal}}{\text{ft}^2 - \text{day}}$$

Step 2. Next, a series of steps employing a graphical solution are used to determine the overflow rate based on transition and compression settling. The steps are as follows:

Step 2.1. Construct Line C by extending the linear portion of the right portion of Curve A. Now the tough part, construct a Line D tangent to Curve A at a point perpendicular to the intersection of Lines B and C. The line constructed is labeled as Line D. Now calculate H_u.

$$Hu = \frac{Co\, Ho}{Cu}$$

where: Hu = Height of interface to reach underflow concentration
Co = Initial SS concentration = 1,500 mg/L
Ho = Initial height of interface
Cu = Desired effluent concentration = 8,000 mg/L
Hu = 1,500(6)/8,000
= 1.13 ft

Step 2.2. Construct Line E from H_u=1.13 parallel to the x-axis to the intersection with Line D. Construct Line F vertically from the intersection of Lines E and D. This is the detention time required to reach the desired underflow concentration. It may be converted into a settling velocity and overflow rate as follows:

Time Tu = time required to reach desired underflow concentration

$$\text{Velocity} = \frac{Ho - Hu}{Tu} = \frac{6 - 1.13}{100 \text{ min}} = 0.0487 \text{ ft/min}$$

$$= 525 \frac{\text{gal}}{\text{ft}^2 - \text{day}}$$

The limiting settling rate is based on compression settling. Accordingly, the clarifier design should be based on an overflow rate of 525 gal/ft²-day. Design for Q.

To check the design, compare the calculated results with the typical values reported in Table 6-6 (or any references preferred). The calculated overflow rate is on the low side (500 to 1,000 gal/ft²-day), but within the ranges reported in this table. The detention time is low, but this is most probable because safety factors have been applied to the data in Table 6-6. In practice, it is likely that the results would also be scaled-up to provide a safety factor for peaking and flow increases with age of the plant.

6.4.2 Clarifier Design In practice, all types of settling occur within a single clarifier. Types I and II occur near the surface, while zone, and then compression, settling develop near the bottom of the tank. The size of the clarifier is obviously controlled by the slowest process, which is commonly either zone or compression (thickening) which depends partly on the suspended solids concentrations required from the clarifier. Design can be based on standard loading factors, ideally derived from processes similar to the design application, pilot tests, or from laboratory settling data. Table 6-6 presents a compilation of loading factors and detention times. Analysis of settling data for Types III and IV settling is generally based on graphical techniques shown in Examples 6-2 and 6-3 as applied to the specific type of data collected.

TIP FOR THE P.E. EXAMINATION CANDIDATE

For examination purposes, standing loading factors and/or detention times data must be given for multiple-choice problems and are generally provided for essay format problems. Table 6-6 could be used if these factors were not provided.

6.5 Coagulation/Flocculation

Coagulation/flocculation is a process that produces a chemical floc in the water being treated to enhance the removal of small suspended solids that could not otherwise be removed by gravity. The coagulation portion of the process is where one or more chemicals are added to the water and their reaction, together or with chemicals already in the water, forms a precipitate in the water. The correct chemical dosages, coupled with slow, mechanical stirring, cause the precipitate particles to grow large enough that they tend to settle. Particle growth is an esssential goal of the flocculation process. These floc particles, either while forming or as they begin to settle, capture any small colloidal particles they collide with in the raw water. The settling chamber that follows the coagulation and flocculation basin removes these particles thereby reducing the turbidity of the treated water. Example 6-5 provided later in this section presents the basic calculations for chemical dosages and sludge volumes in the coagulation/flocculation process. The two most common water treatment coagulants are aluminum sulfate (alum) and ferrous sulfate. Alum reacts with either natural alkalinity shown as calcium bicarbonate below in Equation 6-7 or added lime, Equation 6-8.

$$Al_2(SO_4)_3 + 3Ca(HCO_3)_2 \rightarrow 2Al(OH)_3 \downarrow + 3CaSO_4 + 6CO_2 \quad [6\text{-}7]$$

$$Al_2(SO_4)_3 + 3Ca(OH)_2 \rightarrow 2Al(OH)_3 \downarrow + 3CaSO_4 \quad [6\text{-}8]$$

[Note: Often these reactions are shown including 14 waters of hydration with the alum. This may need to be accounted for in the molecular weight used in any calculations.]

The ferrous sulfate reactions are:

$$2FeSO_4 \bullet 7H_2O + 2Ca(HCO_3)_2 + 0.5O_2$$
$$\rightarrow 2Fe(OH)_3 \downarrow + 2CaSO_4 + 4CO_2 + 13H_2O \quad [6\text{-}9]$$

$$2FeSO_4 \bullet 7H_2O + 2Ca(OH)_2 + 0.5O_2$$
$$\rightarrow 2Fe(OH)_3 \downarrow + 2CaSO_4 + 13H_2O \quad [6\text{-}10]$$

Design of a coagulation/flocculation unit involves selecting the correct dosages of chemicals, providing adequate mixing of the chemicals with the water (usually through rapid or flash mixing), and finally providing the correct energy to mix the solution in the flocculator to obtain optimum floc formation. Too much mixing shears particles apart, while too little mixing does not produce the number of particle collisions necessary to grow large floc particles. Rapid mixing can be achieved with high-energy mechanical mixers and detention times in a range from 10 to 60 seconds, depending on the tank configuration. Often, mixing is accomplished by injecting the chemicals into a turbulently flowing pipe. With sufficient residence time, no additional mixing is required. For flocculation, Viessman and Hammer (1993) report optimum floc formation occurs in a paddle flocculator where the velocity gradient ranges from 10-75 (sec^{-1}) using Equation 6-11.

$$G = (P/\mu V)^{0.5} \qquad [6\text{-}11]$$

where: G = velocity gradient, ft/sec-ft or (sec^{-1})
P = power input, ft-lb/sec
m = dynamic viscosity of water, lb-sec/ft^2
V = tank volume, ft^3

For paddle mixers, power can be calculated by:

$$P = 0.5\,(C_d\,A\,\rho\,v^3) \qquad [6\text{-}12]$$

where: C_d = Drag coefficient, 1.8 for flat paddles
A = Paddle area, ft^2
r = Density of water, lb-sec/ft^4
v = paddle velocity, = 2p • (radius) • (revolutions per second)

6.6 Filtration

Filtration is the second most common unit operation in water treatment (after disinfection) and is required for most public water supplies. In filtration, small particles in the water are removed as the water moves through a bed of sand, or sand with other media, such as garnet or anthracite coal. There are gravity and pressure filtration units, but gravity filtration is most common for large-scale water treatment plants. In gravity systems, the filters are operated as slow or rapid filters depending on the velocity through the filter. Obviously, filters can remove only a certain amount of solid materials before the pore spaces in the filtration media become clogged with solids. The process of cleaning the filters is called *backwashing* because it entails reversing the flow of water through the filter to partially suspend or lift the bed thereby freeing the trapped particles which are carried off by the backwash water suspension. Properly-designed filters operate from hours to days between backwashing cycles.

Most references explain the filtration process as a combination of three to four processes occurring simultaneously inside the filter. *Straining* occurs when a suspended solid larger than the opening in the filtration media is trapped in the opening between two media granules. *Interception* occurs when the flow through the media brings a particle in contact with a media grain and the particles adhere. *Sedimentation* occurs when a particle settles onto a piece of the media. Finally, where coagulation/flocculation has preceded filtration, the flocculation process can continue inside the filter to a point where a particle grows sufficiently larger to enable one of the other three mechanisms to remove it.

There are numerous operational configurations for filters and many ways to configure filtration within the treatment scheme. Filtration design normally involves sizing the filters based on a given flow, calculating headloss through a clean filter, and estimating the backwash flow rates. Sizing filters is normally based on standardized loading criteria. Media selection and filter grain size are parameters that affect these calculations.

> **TIP OF THE P.E. EXAMINATION CANDIDATE**
>
> Operational configuration details are beyond the scope of this text and are not typically addressed in the P.E. examination.

Filtration media typically are sand, coal, garnet or some combinations of these. The simpliest, most common filter consists of a single type of sand media. This type has a uniform pore size throughout the filter bed and, as a result, the pore space near the top of the filter fills and clogs before the bottom of the filter is fully utilized. To overcome this problem, multi-media filters are used which are comprised of a less dense media with a larger pore space overlying a more dense material with a smaller grain size, normally sand. The density difference is necessary to keep the two layers separated during the backwashing process. How the density produces this separation can be ascertained by applying Stoke's Law. Stoke's Law reveals that larger particles will tend to settle more rapidly than smaller particles. Likewise, more dense particles will tend to settle more rapidly than lighter particles of the same size. Therefore, for a given backwash flow rate, the less dense upper layer will remain above the more dense lower layer during the backwashng process. Table 6-7 provides details of design factors for different filters.

Table 6-7
Water filtration rates

Plant Type	Range of Rates (gal/ft^2-min)	Typical Rate (gal/ft^2-min)
Conventional Plant Confguration		
Slow sand filter[1]	0.05-0.1	-
Slow sand filter[2]	0.5-0.15	-
Slow sand filter[3]	0.015-0.06	0.04
Rapid sand filter[2]	1-2.5	2
Rapid sand filter[3]	3-4	
Rapid sand filter[1,4]	2-10	5
Direct filtration[1]	1-6	-
Direct filtration[3]	4-5	-

1 Viessman and Hammer, 1993.
2 Corbitt, 1990.
3 Salvato, 1992.
4 Montgomery, 1985.

6.7 Disinfection

Disinfection inactivates pathogenic bacteria rendering the water suitable for consumption. This process is now changing after years of standardized design and operation following its original development in the early part of the 1900s. Typically, the last step in a treatment system, such as shown in Figure 6-1, is a sufficient dose of chlorine to kill the remaining organisms and provide a chlorine residual for protection of the water as its moves through the distribution system to the consumer. If the water contained chlorine-demanding substances or significant numbers of organisms, the operator simply increased the amount of chlorine because it was effective and inexpensive.

Developments occuring since the late 1980s have produced significant changes in the disinfection process. First, studies leading to the implementation of the new Surface Water Treatment Rule (SWTR) regulations established that too many organisms have been allowed to pass through treatment systems, causing an incidence of disease attributable to public water supplies. Second, it has been established that use of chlorination coupled with low levels of organics causes the formation of trihalomethanes, which pose long-term health risks at low levels. The maximum contaminant level (MCL) for total trihalomethanes (TTHMs) currently is 0.1 mg/L. This MCL poses a problem for standard chlorination disinfection, requiring other methods be used. Ozone and other non-halogenated oxidants are strong disinfectants, but these generally don't provide residual protection in the distribution system. Removal of the organics that form TTHMs requires adding other expensive unit processes. As previously discussed, if raw water is found to have significant concentrations of microorganisms, filtration is required to remove bacteria to a level where reasonable chlorine dosages can kill the remaining organisms. Microorganism standards at the plant and in the distribution system require monitoring of organic activity in the product water, with the goal being zero. Many researchers note that there are disease-producing organisms, such as *Giardia lamblia, Legionella, Chryptospiridia* and viruses, that can survive these processes and are not measured with coliform tests, thus exposing the public to disease through public water supplies. The approach that has evolved is a compromise which trys to adhere to both the bacterial standards and the trihalomethane MCL. It involves a combination of filtration and disinfection with the proper dosage and residence time to achieve the quality required by the SWTR standards.

The requirements for disinfection under the SWTR are 99.9% removal of Giardia cysts (3-log reduction) and 99.99% removal for viruses (4-log reduction). Removal is achieved through a combination of filtration and disinfection. For filtration, Table 6-8 shows the expected removal efficiencies for various treatment processes. Also shown in Table 6-8 are the additional removals required through disinfection. This removal is accomplished by meeting the required CT value, which is the concentration of free residual of disinfectant in mg/L times the deten-

Table 6-8
Log-removal/inactivation requirements based on treatment technique

	Expected Log Removal		Additional Log Inactivation Required for Compliance	
Filtration	Giardia	Viruses	Giardia	Viruses
Conventional	2.5	2.0	0.5	2.0
Direct	2.0	1.0	1.0	3.0
Slow Sand	2.0	2.0	1.0	2.0
Diatomaceous Earth	2.0	1.0	1.0	3.0

Source: U.S. Army Enviromental Hygiene Agency, 1994.

tion time in minutes. The CT values required for levels of removal, reported as log inactivations, are shown in Table 6-9 (chlorine as a disinfectant). This table shows that removal is also a function of pH and chlorine dosage. Not evident from the table is the impact of factors relating to the effectiveness of the contact chamber used. Example 6-4 provides a sample calculation for selection of disinfection level and contact chamber size.

Example 6-4 — Design of disinfection process

A water plant provides disinfection with chlorine following standard water treatment with the processes shown in Figure 6-1. Studies of the water source indicate the chlorine dosage needs to be below 1.2 mg/L to avoid problems with TTHMs. Based on the data below, determine the required CT time and design the chlorine contact chamber.

Data:

Non corrosive water
Design flow = 2.5 Mgal/day
Temperature = 10-25°C
Chlorine demand = 50% of dosage
Design criteria from SWTR = 4 log reduction in viruses
= 3 log removal of *Giardia*

Removal obtained by the treatment plant (Table 6-8)

Giardia = 2.5 log-removal
Viruses = 2.0 log

Required to be removed by disinfection (Table 6-8)

Table 6-9
CT Values for inactivation of *Giardia* cysts by free chlorine at 20°C[1]

Chlorine Concentrat-ion (mg/L)	pH ≤6 Log Inactivations						pH =6.5 Log Inactivations						pH=7.0 Log Inactivations						pH=7.5 Log Inactivations					
	0.5	1.0	1.5	2.0	2.5	3.0	0.5	1.0	1.5	2.0	2.5	3.0	0.5	1.0	1.5	2.0	2.5	3.0	0.5	1.0	1.5	2.0	2.5	3.0
≤0.4	6	12	18	24	30	36	7	15	22	29	37	44	9	17	26	35	43	52	10	21	31	41	52	62
0.6	6	13	19	25	32	38	8	15	23	30	38	45	9	18	27	36	45	54	11	21	32	43	53	64
0.8	7	13	20	26	33	39	8	15	23	31	38	46	9	18	28	37	46	55	11	22	33	44	55	66
1	7	13	20	26	33	39	8	16	24	31	39	47	9	19	28	37	47	56	11	22	33	45	56	67
1.2	7	13	20	27	33	40	8	16	24	32	40	48	10	19	29	38	48	57	12	23	34	46	57	69
1.4	7	14	21	27	34	41	8	16	25	33	41	49	10	19	29	39	48	58	12	23	35	47	58	70
1.6	7	14	21	28	35	42	8	17	25	33	42	50	10	20	30	39	49	59	12	24	36	48	60	72
1.8	7	14	22	29	36	43	9	17	26	34	43	51	10	20	31	41	51	61	12	25	37	49	62	74
2	7	15	22	29	37	44	9	17	26	35	43	52	10	21	31	41	52	62	13	25	38	50	63	75
2.2	7	15	22	30	37	44	9	18	27	35	44	53	11	21	32	42	53	63	13	26	39	51	64	77
2.4	8	15	23	30	38	45	9	18	27	36	45	54	11	22	33	43	54	65	13	26	39	52	65	78
2.6	8	15	23	31	38	46	9	18	28	37	46	55	11	22	33	44	55	66	13	27	40	53	67	80
2.8	8	16	24	31	39	47	9	19	28	37	47	56	11	22	34	45	56	67	14	27	41	54	68	81
3	8	16	24	31	39	47	10	19	29	38	48	57	11	23	34	45	57	68	14	28	42	55	69	83

Chlorine Concentrat-ion (mg/L)	pH ≤8.0 Log Inactivations						pH =8.5 Log Inactivations						pH=9.0 Log Inactivations					
	0.5	1.0	1.5	2.0	2.5	3.0	0.5	1.0	1.5	2.0	2.5	3.0	0.5	1.0	1.5	2.0	2.5	3.0
≤0.4	12	25	37	49	62	74	15	30	45	59	74	89	18	35	53	70	88	105
0.6	13	26	39	51	64	77	15	31	46	61	77	92	18	36	55	73	91	109
0.8	13	26	40	53	66	79	16	32	48	63	79	95	19	38	57	75	94	113
1	14	27	41	54	68	81	16	33	49	65	82	98	19	39	59	78	98	117
1.2	14	28	42	55	69	83	17	33	50	67	83	100	20	40	60	80	100	120
1.4	14	28	43	57	71	85	17	34	52	69	86	103	20	41	62	82	103	123
1.6	15	29	44	58	73	87	18	35	53	70	88	105	21	42	63	84	105	126
1.8	15	30	45	59	74	89	18	36	54	72	90	108	21	43	65	86	108	129
2	15	30	46	61	76	91	18	37	55	73	92	110	22	44	66	88	110	132
2.2	16	31	47	62	78	93	19	38	57	75	94	113	23	45	68	90	113	135
2.4	16	32	48	63	79	95	19	38	58	77	96	115	23	46	69	92	115	138
2.6	16	32	49	65	81	97	20	39	59	78	98	117	24	47	71	94	118	141
2.8	17	33	50	66	83	99	20	40	60	79	99	119	24	48	72	95	119	143
3	17	34	51	67	84	101	20	41	61	81	102	122	24	49	73	97	122	146

[1] CT = CT for 3 log inactivation.
Source: U.S. Army Enviromental Hygiene Agency, 1994.

Giardia = 3 - 2.5 = 0.5 log removal
Viruses = 4 - 2.0 = 2.0 log removal

From Table 6-9 for 10°C (worst case)

Chlorine = 0.5 (1.2) = 0.6 free
pH = 8.0 (for non corrosive water)
$CT = 26$
C = concentration in mg/L
T = retention time in minutes

for free available chlorine of 0.6 mg/L

$$\frac{26}{(0.6)} = T$$

43.3 minutes = Time

$$2.5 \bullet 10^6 \frac{\text{gal}}{\text{day}} \bullet \frac{1 \text{ day}}{24 \text{ hr}} \bullet \frac{1 \text{ hr}}{60 \text{ min}} \bullet \frac{1 \text{ ft}^3}{7.48 \text{ gal}} = 232 \frac{\text{ft}^3}{\text{min}}$$

$232.1 \bullet 43.3 = 10,050 \text{ft}^3$

The last consideration is the actual effectiveness of the chlorine contact chamber. Baffle efficiency can be used to adjust the size of the contact chamber. Assuming superior baffle construction, use 0.7,

$$\text{Actual size} = \frac{10,050 \text{ ft}^3}{0.7} = 14,357 \text{ft}^3$$

size into standard length, width, and depth dimensions.

6.8 Solids Management

The sources of solids from a water treatment plant are the sedimentation basins and the filter backwash. There is no standard for how water treatment plant solids are managed. The primary options available for treatment include clarifiers/thickeners, drying beds or lagoons, pressure filtration, centrifugation, and chemical recovery. Disposal options include direct discharge of the suspension to the sanitary sewer, landfilling of dried

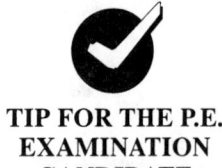

TIP FOR THE P.E. EXAMINATION CANDIDATE

For P.E. examination purposes, the candidate should understand where water treatment solids are generated, how much is generated, how to process these solids (which are still relatively dilute suspensions), and how to dispose or recycle them. The primary calculations employed in the examinations are quantification of the sludge mass and volume, as demonstrated in Example 6-5.

Chapter 6: Water Treatment

sludge, or land application. For each plant, economic and regulatory analyses are required to select the best option(s).

Example 6-5 — Water treatment waste solids

Suspended solids are to be removed through coagulation/flocculation followed by gravity settling. Alum and lime are added at stoichometric rates sufficient to form 25 mg/L of aluminum hydroxide sludge. For the following data, calculate the alum and lime dosages, the dry weight mass of solids generated per day, and the wet volume per day of sludge.

Data:

Flow- 1.0 Mgal/day
TSS - 5.0 mg/L
Thickened Sludge - 20,000 mg/L

Solution:

Alum reacts lime as follows

$$Al_2(SO_4)_3 + 3Ca(OH)_2 \rightarrow 2Al(OH)_3 \downarrow + 3CaSO_4 \qquad [6\text{-}8]$$

Using molecular weights to achieve 25 mg/L of the $Al(OH)_3$.

$$342 + 222 = 156 + 408$$

To base all reaction components on $Al(OH)_3$, divide by 156

$$2.19 + 1.42 = 1 + 2.62$$

Therefore:

Alum dosage = 25(2.19) = 54.81 mg/L
Lime as $Ca(OH)_2$ = 25(1.42) = 35.5 mg/L

For 1 Mgal/day:

Alum = 54.81 • 8.34 • 1 = 467 lb/day
Lime = 35.5 • 8.34 • 1 = 296 lb/day

(Remember: mg/L • Mgal/day•8.34 = lb/day)

The dry mass of sludge produced is:

Floc + suspended solids = (25 + 5) • 8.34 • 1 = 250 lb/day dry solids
The corresponding Volume of sludge is:
250 lb is in a concentration of 20,000 mg/L (2% by weight)
volume in gal = (250/0.02) [lb/day]/ 8.34 [lb/gal] = 1,500 gal/day

---

6.9 Additional Water Treatment Processes

6.9.1 Fine Particulate Removal
Reverse Osmosis (RO) is a membrane process that can remove colloidal particulate and dissolved ions from water by forcing clean water through a water-permeable membrane (the reverse of the natural osmosis process). Raw water enters a chamber separated by a cellulose acetate or similar membrane. When pressure applied to the dirty side of the membrane exceeds the osmotic pressure produced by the ion imbalance across the membrane, clean water flows from the dirty side of the membrane to the clean side. The water on the reject side of the membrane is continually discarded. RO removes most ions with a high efficiency. One drawback is that it not only removes target ions, but most other ions as well. RO water is low in TDS, but is also low in pH and therefore is highly corrosive. RO units routinely require pretreatment to remove suspended solids, and care must be taken that bacterial contamination does not coat the membrane and reduce the flow rates. RO can be used for processing brackish and salt water into drinking water and for specialized treatment of high TDS waters. It is also used to soften water in small-scale applications.

The *Ion Exchange* process employs a natural or synthetic medium to selectively remove unwanted ions from water by exchanging the unwanted ions with a lower activity ion from a solid medium. The resin can be anionic or cationic and, when properly designed, can be very specific for the ion to be removed. The medium is usually contained in a fixed-bed or column, and the water moves through the medium much like a filter. Typical uses of ion exchange include selective ion removal such as softening and ion recovery in waste treatment processes.

6.9.2 Water softening
In locations where excessively hard water exists (typically considered to start in the 150 to 200 mg/L range), water softening can be accomplished at the central plant or through the use of in-home treatment units. Unit processes capable of removing calcium and magnesium include ion exchange, reverse osmosis and coagulation/flocculation, with the last being impractical for in-home use.

Coagulation/flocculation involves a series of reactions which are dependent upon the specific chemicals present in the water. The reactions for removal of calcium and magnesium from water using lime/soda ash treatment are shown below. Example 6-6 demonstrates how to employ these equations.

Carbon dioxide must be removed before any of the hardness reactions can proceed.

$$CO_2 + Ca(OH)_2 \rightarrow CaCO_3 \downarrow + H_2O \qquad [6\text{-}13]$$

Calcium in the presence of natural alkalinity reacts with lime to form calcium carbonate

$$Ca(HCO_3)_2 + Ca(OH)_2 \rightarrow 2CaCO_3 \downarrow + 2H_2O \qquad [6\text{-}14]$$

Magnesium in the presence of natural alkalinity reacts with lime to form magnesium carbonate (not insoluble) and calcium carbonate, but with no loss in hardness

$$Mg(HCO_3)_2 + Ca(OH)_2 \rightarrow MgCO_3 + 2CaCO_3 \downarrow + 2H_2O \qquad [6\text{-}15]$$

To remove magnesium hardness, additional lime must be added

$$MgCO_3 + Ca(OH)_2 \rightarrow CaCO_3 \downarrow + Mg(OH)_2 \downarrow \qquad [6\text{-}16]$$

Magnesium hardness in the form of a sulfate requires lime and soda ash

$$MgSO_4 + Ca(OH)_2 \rightarrow CaSO_4 + Mg(OH)_2 \downarrow \qquad [6\text{-}17]$$

$$CaSO_4 + Na_2CO_3 \rightarrow CaCO_3 \downarrow + Na_2SO_4 \qquad [6\text{-}18]$$

The goal of softening is not zero hardness, but reasonable hardness, generally a range from 50 to 80 mg/L. Based on the reactions above and considering that soda ash is more expensive than lime, the most economical approach generally is to maximize the removal of calcium. However, economics and approach will, of course, depend on the actual quality of the water to be treated.

Example 6-6 — Water softening example

Raw Water Data

CO_2 = 6.0 mg/L   Na = 5 mg/L
Ca = 50 mg/L   Alkalinity = 120 mg/L as $CaCO_3$
Mg = 20 mg/L SO_4 = 94 mg/L

Determine the quantity of lime and soda required for water softening per million gallons to achieve a hardness of 40 mg/L as $CaCO_3$. (Equivalent weights are available in Table 3-3.)

Step 1. Develop the millequivalents diagram to define the water quality.

Chemical	Conc (mg/L)	Equivalent wt.	meq/L
CO_2	6.0	22	.027
Ca	50	20	2.5
Mg	20	12.2	1.64
Na	5	23	0.22
Alka.	120	50 ($CaCO_3$)	2.4
SO_4	94	48	1.96

```
    .27        2.5              4.14              4.36
   ┌────┬───────────────┬──────────────────┬──────────┐
   │    │      Ca       │        Mg        │    Na    │
   │CO₂ ├───────────────┴──────────┬───────┴──────────┤
   │    │          HCO₃            │        SO₄       │
   └────┴──────────────────────────┴──────────────────┘
                   2.4                              4.36
```

Results are:
$Ca(HCO_3)_2 = 2.4$ meq/L
$MgSO_4 = 1.64$ meq/L
$CaSO_4 = 0.1$ meq/L
$Na_2SO_4 = 0.22$ meq/L

Step 2. By examining the stoichiometric relations it is noted that the following chemical reactions occur:

Compound	meq/L	Lime	Soda Ash	Stiochiometric Equation
$Ca(HCO_3)_2$	2.4	2.4		[6-13]
$CaSO_4$	0.1		0.1	[6-17]
$MgSO_4$	1.64	1.64	1.64	[6-16]
CO_2	0.27	0.27		[6-12]
Total		4.31	1.74	

Reaction rates for $Mg(HCO_3)_2$ require 2 meq of lime plus 1 meq of soda ash for each meq of magnesium bicarbonate.

$$\text{Lime required} = \left(4.31 \frac{\text{meq}}{\text{L}} \cdot \frac{28 \text{ mg}}{\text{meq}}\right) + 35 \frac{\text{mg}}{\text{L}}$$

$$= 155.7 \frac{\text{mg}}{\text{L}}$$

Note: $28 \frac{\text{mg Lime}}{\text{meq}}$, $53 \frac{\text{mg Soda ash}}{\text{meq}}$

Note: An excess 35 mg/L of lime is added to provide a driving force to move the reaction toward completion (an express assumption necessary to the problem solution).

$$\text{Soda Ash} = \left(1.74 \frac{\text{meq}}{\text{L}} \bullet 53 \frac{\text{mg}}{\text{meq}}\right) = 92.2 \frac{\text{mg}}{\text{L}}$$

The quantities of lime and soda ash are:

$$\text{Lime: } 155.7 \frac{\text{mg}}{\text{L}} \bullet 8.34 \bullet 1 \text{ Mgal/day} = 1299 \frac{\text{lb}}{\text{d} - \text{Mgal/day}}$$

$$\text{Soda Ash: } 92.2 \frac{\text{mg}}{\text{L}} \bullet 8.34 \bullet 1 \text{ Mgal/day} = 769 \frac{\text{lb}}{\text{d} - \text{Mgal/day}}$$

Experience based on solubilities and reaction rates suggests the maximum hardness removal achievable is approximately 35 to 50 mg/L. pH adjustment and recarbonation are normally required to stabilize lime/soda-softened water.

6.9.3 Corrosion/Scaling Control Corrosion and scaling have historically been operational parameters that related to the aesthetic value of water and equipment life expectantcy, but with no known health implications. In the past, many operators attempted to keep water slightly corrosive to prevent scaling in the distribution system and in heat exchangers, such as home water heaters, because it decreases the useful life of piping and reduces the efficiency of heat exchangers. The establishment of MCLs for lead and copper has drastically changed this approach to scaling/corrosion control. Current operation strategies attempt to maintain waters in the slight scaling range.

Corrosion control is specified in the regulations for lead and copper (40 CFR 141.80). The scaling or corrosion potential of water can be measured with the Langelier Index. A positive Index value indi-

TIP FOR THE P.E. EXAMINATION CANDIDATE

The Langelier Index is presented to demonstrate the variables which affect corrosion/scaling. From this equation, it can be seen that raising the pH is a primary way of reducing corrosion. This fact is reflected in the regulatory strategy for control of lead and copper which specifies pH treatment standards, where copper and/or or lead are determined to be above the action levels at the point of consumption.

cates scaling and a negative Index value suggests corrosive water (Viessman and Hammer (1993)):

$$\text{Langelier Index} = pH - pH_s = pH - [(pK_2 - pK_s) + pCa + pAlk] \qquad [6\text{-}19]$$

where:
- pH = Actual measured pH
- pH_s = pH at calcium carbonate saturation
- $pK_2 - pK_s$ = constants, 1.9-2.34 for most waters
- pCa = neg log of calcium ion concentration in moles/L
- $pAlk$ = neg log of alkalinity in equivalents/L

6.9.4 Fluoridation Fluoride in water is a two-edged sword; fluoride is added to water to control tooth decay, but at levels to avoid dental fluorosis. Reeves (1986) recommends fluoride levels of 0.7 to 1.2 mg/L, depending on temperature. The MCL for fluoride is 4.0 mg/L, with an MCLG of 2.0 mg/L. Fluoride is easily added to water through injection as sodium fluoride, sodium silicofluoride or hydrofluosilicic acid. Removal of high levels of naturally-occurring fluoride can be accomplished with bone char or activated alumina (Viessman and Hammer (1993)).

6.9.5 Total Trihalomethanes Control The regulations that apply to disinfectants and disinfectant byproducts are presently in an uncertain state. Total trihalomethanes (TTHMs) are formed by the reaction of naturally-occurring organic material, often generated by the decomposition cycle, with chlorine or other disinfecting halogens. Chloroform is a common TTHM. Haloacetic acids are also byproducts of natural organic material and halogens that pose health concerns in drinking water systems.

Controlling the formation of these chemicals can be accomplished in a variety of ways, but must be accomplished while also complying with the disinfection standards. It has been shown that eliminating the practice of prechlorination of water as it enters the plant can reduce the levels of TTHMs in the water. Prechlorination is an operational practice used to keep plant equipment free of biological growth. Obviously, this practice of dosing the raw water with chlorine provides the opportunity for TTHM formation before any of the natural organics can be removed. Standard water treatment systems will not remove high levels of dissolved organics. Where dissolved organics appear only seasonally in the water source, additional treatment processes, such as carbon adsorption, can be applied to remove organics and thus prevent the formation of TTHMs. Other operational schemes used to control TTHMs include step-wise chlorination and use of non-halogenated disinfectants to supplement halogens. Longer detention times in the disinfection process make it possible to meet the CT standard with much lower chlorine dosages and, thus, lower TTHMs.

6.10 Operations

Key operating parameters in water treatment are headloss through the filter and the rate of backwash water flow. Headloss and backwash rates can be determined from empirically-derived formulas or standard operating factors. Montgomery (1985) presents standard factors data for backwash rates, Table 5-8, but it does not address headloss.

Determining the rate required for filter backwashing can be accomplished from either mathematical relationships based on media parameters or through standardized loading factors (Viessman and Hammer (1993)). Corbitt (1990) reports typical backwash rates as 15 to 25 gal/ft^2-min for 8 to 10 min. The mathematical methods are not presented because there is no generally-accepted method. Viessman and Hammer (1993) describe in detail one mathematical method for determination of backwash rates.

6.11 Standard Water Treatment Notations

Symbol	Definition
A	Area
A_s	Surface area
C_d	Drag coefficient
CFT	Code of federal regulations
C_i	Curie (see chapter 10)
G	Velocity gradient
MCL	Maximum contaminant level
MCLG	MCL goal
meq	milli equivalent
mg	milligram
P	Power
PCB	Polychlorinated biphenols
Q	Flowrate
ρ	Density
SDWA	Safe drinking water act
SMCL	Secondary MCL
SWTR	Surface water treatment rules
TDS	Total dissolved solids
TSS	Total suspended solids
TTHM	Total trihalomethane
V	Volume
v	Velocity
V_x	Velocity in horizontal direction
V_y	Velocity in vertical direction
W	Width

x	Length
Y	Depth
μ	Viscosity
μm	Micrometers

6.12 References

40 Code of Federal Regulations 141-149. 1994. Drinking Water.

American Public Health Association. 1990. *Standard Handbook for the Examination of Water and Wastewater*. 17th ed. Washington, DC: APHA.

Corbitt, Robert A. 1990. *Standard Handbook of Environmental Engineering*. New York: McGraw-Hill.

Metcalf and Eddy, Inc. 1991. *Wastewater Engineering Treatment, Disposal, and Reuse*. 3rd ed. New York: McGraw-Hill.

Montgomery, James M. Inc. 1985. *Water Treatment Principles and Design*. New York: Wiley-Interscience.

Reeves, Thomas G. 1986. *Water Fluoridation A Manual for Engineers and Technicians*. Washington, DC: US Department of Health and Human Services, Sept.

Salvato, Joseph A. 1992. *Environmental Engineering and Sanitation*. 4th ed. New York: Wiley-Interscience.

Sawyer, Clair N., Perry L. McCarty, and Gene F. Parkin. 1994. *Chemistry fot Environmental Engineers*. 4th ed. New York: McGraw-Hill.

U.S. Army Environmental Hygiene Agency. 1994. *USAWHA TG No. 199 Surface Water Treatment Rule*.

Viessman, Warren, and Mark Hammer. 1993. *Water Supply and Pollution Control*. 5th ed. New York: Harper-Collins.

Weber, Walter W. 1972. *Physicochemical Processes*. New York: Wiley-Interscience.

Chapter 7

WASTEWATER TREATMENT

by John D. Dietz, Ph.D., P.E.

7.1 Introduction

A specialty of environmental engineering corresponding to water treatment for consumption discussed in Chapter 6 is the treatment of wastewater. Many of the same unit processes such as sedimentation, filtration, disinfection, etc. are employed. Because both water and wastewater treatment involve water and have similar chemical characteristics, the principles discussed in the preceeding chapters on Chemistry. Fluids and Hydraulics are equally-applicable.

The development of wastewater treatment paralleled water treatment. However, its sophistication did not evolve at a corresponding rate. While the first major activated sludge wastewater treatment plants were constructed in the early part of this century (Houston, 1918; Chicago, 1923), most communties employed only primary treatment, if any, until after World War II. Even then, secondary treatment was not universal. With the passage of the Water Pollution Control Act Amendments of 1972, the Nation set a goal of universal secondary treatment; a goal which has been largely achieved. Today, the focus is on increasingly higher degrees of treatment and water quality-based standards. Lacking the water supply industry's consumer-driven impetus for quality water, wastewater treatment has been primarily driven by government-imposed treatment standards.

7.2 Treatment Standards

Modern wastewater treatment is based on the Water Pollution Control Act Amendments of 1972 (PL 92-500). The goals prescribed by this legislation include:

1. Elimination of the discharge of pollutants;

2. Provision of surface water quality suitable for sustaining fish, shellfish and wildlife and for recreational purposes whenever possible; and

3. Prohibition of the discharge of toxic pollutants.

Water-quality criteria were established pursuant to PL 92-500 to protect beneficial uses of surface waters and are classified as follows:

1. Maintenance of adequate dissolved oxygen to support desirable aquatic life forms. A minimum dissolved oxygen of 5 mg/L is commonly accepted for support of sport-fish species. Certain cold-water species (trout) may require higher dissolved oxygen levels (6 mg/L).

2. Reduction of plant and algal nutrient levels to avoid cultural eutrophication problems. Eutrophication is more likely to be a concern for discharge to lakes and reservoirs than for discharges to free-flowing rivers. Effluent standards are imposed for total nitrogen and phosphorus in specific regions of the country, e.g., the Great Lakes Basin, the Chesapeake Bay, and Florida.

3. Maintenance of concentrations of toxic substances at values that do not pose a threat to aquatic species and/or other potential uses, e.g., potable water supply. Toxic substances are more commonly associated with industrial sources and are typically regulated for each industrial category. Examples of regulations governing a variety of toxic materials, including heavy metals, synthetic organic compounds and acids/bases are discussed in this chapter. Whole effluent, toxicity-bioassay testing may be required to verify the absence of toxic agents in effluents. This test exposes different aquatic species to the particular industrial effluent to determine if they can survive in the effluent at different dilutions with the receiving waters.

4. Elimination of pathogens to control transmission of waterborne diseases. Disinfection of wastewaters that pose a risk of disease transmission is prescribed prior to discharge.

5. Maintenance of suitable aesthetic qualities to foster recreational use of surface water resources. Accordingly, requirements exist to limit settleable solids, floatable materials and oily materials.

Section 402 of PL 92-500 established procedures for issuance of discharge permits for municipal and industrial wastewaters, the National Pollutant Discharge Elimination System (NPDES) (Nemerow and Dasgupta (1991)). These permits identify maximum allowable concentrations of pollutants that may be present in a facility discharge. Monitoring and reporting requirements are also prescribed in the permits. Certain states, delegated authority by EPA to enforce NPDES permits, have developed their own permit systems which may contain more stringent (than NPDES) requirements.

The level of treatment that must be provided for municipal wastewaters will vary as necessary to maintain receiving water quality standards. In those cases where dissolved oxygen (DO) concentrations can be maintained above applicable criteria (typically 5 mg/L), secondary treatment is adequate, as defined in Table 7-1.

Table 7-1
Secondary treatment standards

Parameter	Maximum concentration (mg/L)
BOD_5	30
SS	30

Secondary treatment requires a maximum effluent of five-day biological oxygen demand (BOD_5) and suspended solids (SS) of 30 mg/L. In addition, the pH must be maintained within a range of 6 to 9. There is often a minimum DO level specified for the effluent to assure that the in-stream DO does not fall below established limits in the mixing zone. These standards may be achieved by many biological treatment processes. Disinfection would typically also be required. Dechlorination may be practiced prior to discharge if necessary to mitigate any toxic effects associated with the disinfectant residual.

Table 7-2
Effluent standards for nitrification

Parameter	Maximum concentration (mg/L)
BOD_5	15
SS	15
TKN	5

In those cases where discharge of a secondary effluent will not satisfy receiving water quality criteria, greater removal of oxygen-demanding materials is required. Specific numerical limits are site-specific, depending on receiving water characteristics. Removal of additional BOD_5 and nitrogen compounds (total Kjeldahl nitrogen, or TKN) may be specified to maintain adequate dissolved oxygen levels. One possible set of effluent standards is reported in Table 7-2. These limits could be achieved by nitrification within a biological treatment process.

Table 7-3
Effluent Standards for Nutrient Removal

Parameter	Maximum concentration (mg/L)
BOD_5	5
SS	5
Total N	3
Total P	1

For wastewater discharge into lakes or impoundments, removal of nutrients (nitrogen and phosphorus) may be prescribed to minimize algal growth potential. Numerical limits will be site-specific. One possible set of effluent standards for a nutrient removal system is shown in Table 7-3. Various combinations of biological processes (nitrification/denitrification, biological phosphorus uptake) and/or chemical processes (phosphorus precipitation) can be employed to achieve these limits. Terminal filtration would also typically be provided.

Industrial facilities with a direct surface water discharge are subject to specific discharge requirements established in an NPDES permit issued to the industry (USEPA (1985)). In addition to regulation of conventional pollutants (pH, BOD_5, SS, oil and grease), categorical industry standards have been established for each industry to define the best available technology (BAT) for reduction of toxic compounds (129 specific compounds corresponding with 65 recognized toxic substance classes). For industries that discharge to a publicly-owned treatment works (POTW), Pretreatment Standards are established for the toxic substances. The federal standards represent minimum pretreatment requirements. Individual municipalities may establish pretreatment standards that are more restrictive than the federal mandates. The specific effluent standards imposed on industrial dischargers are highly variable because of different wastewater characteristics and local sewer use ordinances.

7.3 Receiving Stream Water Quality

An essential step in wastewater treatment design is the determination of the waste assimilative capacity of the waters to which a treated effluent will be discharged. This analysis detemines if the minimum treatment, secondary, is sufficient or if a higher degree of treatment is required. In modern engineering practice sophisticated, dynamic mathematical models are employed for this analysis. These are based on determining the impact of the effluent on the DO in the receiving stream which varies according to temperature in Table 7-4 presents DO concentration at saturation for variations in temperature.

A simplified model (Metcalf & Eddy (1991)) for determining the dissolved oxygen in a river is presented below to illustrate the foundation of these models. This model is developed from a mass balance on a reach of a river that receives a single input of wastewater. The model considers only two factors that influence the oxygen level: decomposition of organics and atmospheric reaeration.

Table 7-4
Saturation values for dissolved oxygen (mg/L) in water at 1 atmosphere pressure as a function of temperature (°C)

Temp	DO_{sat}	Temp	DO_{sat}	Temp	DO_{sat}	Temp	DO_{sat}
1	14.20	11	11.02	21	8.90	31	7.41
2	13.81	12	10.77	22	8.73	32	7.29
3	13.45	13	10.53	23	8.56	33	7.17
4	13.09	14	10.29	24	8.40	34	7.05
5	12.76	15	10.07	25	8.24	35	6.93
6	12.44	16	9.86	26	8.09	36	6.82
7	12.13	17	9.65	27	7.95	37	6.72
8	11.83	18	9.45	28	7.81	38	6.61
9	11.55	19	9.26	29	7.67	39	6.51
10	11.28	20	9.08	30	7.54	40	6.41

Adapted from: Metcalf & Eddy, 1991.

$$D = \frac{k_1 L_o}{k_2 - k_1}[\exp(-k_1 t) - \exp(-k_2 t)] + D_o \exp(-k_2 t) \qquad [7\text{-}1]$$

where:
- D = Dissolved oxygen deficit (mg/L)
- k_1 = Deoxygenation coefficient (base e) (days^{-1})
- L_o = Ultimate BOD in mixing zone at the point of discharge (mg/L)
- k_2 = Reaeration coefficient (base e) (days^{-1})
- t = Time measured downstream from the point of discharge (days)
- D_o = Dissolved oxygen deficit in mixing zone at the point of discharge (mg/L)

The time to achieve the maximum dissolved oxygen deficit, t_{max}, in days, is determined by:

$$t_{max} = (\frac{1}{k_2 - k_1})\ln[\frac{k_2}{k_1}(1 - \frac{D_o(k_2 - k_1)}{k_1 L_o})] \qquad [7\text{-}2]$$

The maximum dissolved oxygen deficit, D_{max}, in mg/L, is determined by:

$$D_{max} = \frac{k_1}{k_2} L_o \exp(-k_1 t_{max}) \qquad [7\text{-}3]$$

To correct the deoxygenation coefficient for 20°C to the actual stream temperature, the following formula is used:

$$k_{1_T} = k_{1_{20}} (1.135)^{(T-20)} \qquad [7\text{-}4]$$

where: k_{1_T} = Deoxygenation coefficient at a general temperature = T (days^{-1})
$k_{1_{20}}$ = Deoxygenation coefficient at a reference temperature = 20°C (days^{-1})
T = Temperature (°C);

Equation 7-4 is valid for temperatures from 4° to 20°C. If the stream temperature exceeds 20°C, then a modification of Equation 7-4 is required. The modified formula is:

$$k_{1_T} = k_{1_{20}} (1.056)^{(T-20)} \qquad [7\text{-}5]$$

The reaeration coefficient must also be adjusted for temperature:

$$k_{2_T} = k_{2_{20}} (1.024)^{(T-20)} \qquad [7\text{-}6]$$

where: k_{2_T} = Reaeration coefficient at a general temperature = T (days^{-1})
$k_{2_{20}}$ = Reaeration coefficient at a reference temperature = 20°C (days^{-1})

The oxygen deficit in the stream is determined by:

$$D = DO_{sat} - DO \qquad [7\text{-}7]$$

where: DO_{sat} = Dissolved oxygen saturation concentration (mg/L)
DO = Dissolved oxygen concentration (mg/L)

Conversion of ultimate BOD (mg/L) to 5-day BOD (mg/L) is accomplished by:

$$BOD_5 = BOD_{ult}[1 - \exp(-5k_{1_{20}})] \qquad [7\text{-}8]$$

Determination of the dissolved oxygen in a river downstream from the point of discharge of a wastewater requires completion of the following steps:

Step 1. Characterization of the conditions in the mixing zone at the point of discharge. A mass balance must be completed to determine the initial ultimate BOD and the initial dissolved oxygen. A thermal balance must be completed to determine the initial temperature. All BOD_5 values must be converted to ultimate BOD_{ult} values using Equation 7-8. Flow, temperature, dissolved oxygen and BOD_{ult} must be known (or assumed) for the river upstream from the point of discharge and for the effluent.

Step 2. Based on the temperature in the river after discharge, values can be determined for the dissolved oxygen saturation concentration (Table 7-4), deoxygenation coefficient (Equation 7-4 or 7-5), and the reaeration coefficient (Equation 7-6). The initial deficit is then calculated (Equation 7-7).

Step 3. For a known value of time, the deficit is calculated (Equation 7-1).

Step 4. To determine the minimum dissolved oxygen concentration, the maximum deficit is determined using Equations 7-2 and 7-3. The minimum dissolved oxygen is determined with Equation 7-7.

The value for the deoxygenation coefficient may be specific to the wastewater because of differences in the inherent rate of degradation of the organic compounds present. For domestic wastewaters at 20°C, a range of values from 0.10 to 0.17 days^{-1} has been reported (Sawyer, McCarty, and Parkin (1994)). Values for industrial wastewaters would be expected to exhibit even greater variability.

The reaeration coefficient is highly dependent on local river conditions (depth and velocity), with a range of values reported from 0.10 days^{-1} for small ponds to 1.15 days^{-1} for swift streams (Metcalf & Eddy (1979)). Site-specific values for these coefficients should be obtained for accurate calculation of receiving water dissolved oxygen.

In many practical situations, dissolved oxygen sag calculations are completed to determine an allowable concentration of BOD in an effluent discharge that will preserve water quality criteria (typically 5 mg/L of dissolved oxygen). In the foregoing procedure, the concentration of BOD in the effluent is unknown. Thus, it requires an iterative calculation in which successive values of the effluent BOD are assumed until the calculated minimum dissolved oxygen concentration satisfies the receiving stream DO standard.

Example 7-1 — Stream water quality

For a wastewater with a flow rate of 4 Mgal/day, determine the minimum dissolved (DO) oxygen concentration downstream from a treatment facility that achieves secondary treatment. The characteristics of the receiving stream upstream from the point of discharge are:

Minimum flow	=	20 cfs
Temperature	=	32°C
Ultimate BOD	=	3 mg/L
Dissolved oxygen	=	7 mg/L

The wastewater temperature is 25°C and the wastewater dissolved oxygen is 5 mg/L. The deoxygenation coefficient for the wastewater (base e) is 0.17 day^{-1} at 20°C. The reaeration coefficient for the stream (base e) is 0.40 day^{-1} at 20°C.

Solution:

The initial conditions (BOD$_{ult}$, DO, and temperature) in the stream at the point of discharge must be determined using mass or energy balances based on the defined inputs. The concentration and temperature of the mixture would be a flow-weighted average of the inputs:

$$BOD_{ult\ mixture} = \frac{(BODult_{eff})(Flow_{eff}) + (BODult_{stream})(Flow_{stream})}{Flow_{eff} + Flow_{stream}}$$

Secondary treatment would be expected to produce an effluent with a BOD$_5$ < 30 mg/L (see Table 7-1). All BOD$_5$ values must be converted to BOD$_{ult}$ using Equation 7-8:

$$BOD_{ult} = \frac{BOD_5}{[1-\exp(-5k_1)]}$$

$$BOD_{ult} = \frac{30\ mg/L}{[1-\exp((-5\ day)(0.17/))]} = 52.4\ mg/L$$

The resulting inputs and output mixture are summarized below:

Effluent	Stream	Mixture
Flow = 4 Mgal/day	Flow = 20 cfs = 12.93 Mgal/day	Flow = 16.93 Mgal/day
BOD$_{ult}$ = 52.4 mg/L	BOD$_{ult}$ = 3 mg/L	BOD$_{ult}$ = 14.7 mg/L
DO = 5 mg/L	DO = 7 mg/L	DO = 6.5 mg/L
Temp = 25°C	Temp = 32°C	Temp = 30.3°C

The rate coefficients (assumed values) must be corrected for temperature using Equations 7-5 and 7-6:

$$k_1 = (0.17\ day^{-1})(1.056)^{(30.3-20)} = 0.30\ day^{-1}$$

$$k_2 = (0.40\ day^{-1})(1.024)^{(30.3-20)} = 0.51\ day^{-1}$$

The DO saturation concentration at 30.3°C is 7.50 mg/L (see Table 7-4). The initial deficit is calculated with Equation 7-7:

$$D_o = 7.5 - 6.5 = 1\ mg/L$$

Chapter 7: Wastewater Treatment

The time of occurrence of the maximum deficit is calculated with Equation 7-2:

$$t_{max} = (\frac{1}{k_2 - k_1})\ln\left(\frac{k_2}{k_1}(1 - \frac{D_o(k_2 - k_1)}{k_1 L_o})\right)$$

$$t_{max} = (\frac{1}{0.51 - 0.30})\ln\left(\frac{0.51}{0.30}(1 - \frac{1.0(0.51 - 0.30)}{(0.30)(14.7)})\right) = 2.30 \text{ days}$$

The maximum deficit is calculated with Equation 7-3:

$$D_{max} = \frac{k_1}{k_2} L_o \exp(-k_1 t_{max}) = \frac{0.30}{0.51}(14.7)\exp(-(0.30)(2.30)) = 4.3 \text{ mg/L}$$

The minimum DO is calculated with Equation 7-7:

$$DO_{min} = DO_{sat} - D_{max} = 7.5 - 4.3 = 3.2 \text{ mg/L}$$

Example 7-2 — Stream Water Quality

For the situation described in Example 7-1, determine the maximum allowable effluent BOD_5 concentration that can be discharged while maintaining compliance with a water quality standard of 5 mg/L dissolved oxygen in the receiving stream.

Solution:

The solution requires a trial-and-error approach in which successive values of effluent BOD_5 are assumed and a corresponding minimum DO is determined. The methodology is identical to the procedure presented in the solution to Example 7-1. Results are tabulated below. The assumed value for the third trial was determined by linear interpolation based on the results of the first two trials:

$$BOD_5 = \left(\frac{5.00 \text{ mg/L} - 3.19 \text{ mg/L}}{6.46 \text{ mg/L} - 3.19 \text{ mg/L}}\right)(0 \text{ mg/L} - 30 \text{ mg/L}) + 30 \text{ mg/L}$$
$$= 13 \text{ mg } BOD_5/L$$

Trial	BOD_5 mg/L	BOD_{ult} mg/L	L_o mg/L	t_{max} days	D_{max} mg/L	DO_{min} mg/L
1	30	52.4	14.7	2.30	4.31	3.19
2	0	0.0	2.3	0.84	1.04	6.46
3	13	22.7	7.7	2.08	2.40	5.10
4	14	24.5	8.1	2.11	2.51	4.99

The maximum allowable BOD_5 in the effluent is 13 mg/L.

7.4 Wastewater Treatment

7.4.1 Characterization of Wastewaters

Wastewater treatment objectives may be quite varied for either municipal or industrial wastewater applications because of differing effluent standards and/or variant raw wastewater characteristics. Untreated municipal wastewaters exhibit consistency in composition, so generalizations can be offered regarding quantity and quality. Industrial wastewaters are extremely variable in composition, so site-specific sampling is necessary for characterization. A review of the literature (e.g., Nemerow and Dasgupta (1991)) may be useful for approximation of industrial wastewater characteristics.

Description of municipal wastewater characteristics may be achieved by quantifying per-capita pollutant generation rates and per-capita wastewater volume generation rates. Alternatively, typical wastewater concentrations may be defined. Caution must be exercised, however, because of the variability in flow rates that results from differing water use and infiltration/inflow. Details may be obtained from Metcalf & Eddy (1991) or ASCE/WPCF (1969). Average per-capita flows are commonly estimated at 100 gal/day (Ten States Standards (1978)), including an allowance for normal infiltration/inflow. Modification of this value (100 gal/day) is often appropriate to account for local infiltration/inflow conditions and water consumption.

Example 7-3 — Wastewater Flow Characteristics

Estimate wastewater flow characteristics for a community with a population of 12,000. Report values for average flow, and peak hourly flow.

Solution:

The average flow rate is based on a typical per-capita wastewater generation rate of 100 gal/day (Ten States Standards (1978)). The resulting average flow is:

Chapter 7: Wastewater Treatment   191

$$\text{Average flow rate} = \frac{(12{,}000 \text{ pop})(100 \text{ gpcd})}{10^6 \text{gal/MGAL}} = 1.2 \text{ MGD}$$

The peak hourly flow rate is determined using historical peak flow measurements. For a population of 12,000, the ratio of peak to average wastewater flow is reported to be 2.9 to 3.3, or approximately 3 (ASCE/WPCF (1969)). Thus:

$$\text{Peak flow rate} = (3)(1.2 \text{ Mgal/day})$$
$$= 3.6 \text{ Mgal/day}$$

Typical municipal wastewater quality is reviewed in Tables 7-5 and 7-6. Site-specific flow and quality data should be used whenever available because of the significant local variability that may be encountered.

TIP FOR THE P.E. EXAMINATION CANDIDATE

Typical wastewater characteristics may be given (typically in multiple-choice problems) or the candidate may be required to select and document the validity of the values selected. If data is supplied, the candidate is expected to use the supplied data.

Table 7-5
Per-capita waste loading factors (lb/capita-day) — normal domestic wastewater with complete grinding of food wastes

Component	Range	Typical
BOD_5	0.18 to 0.26	0.22
Suspended solids	0.20 to 0.33	0.26
Total Kjeldahl nitrogen	0.022 to 0.040	0.033
Total phosphorus	0.007 to 0.013	0.009

Adapted from: Metcalf & Eddy, 1979.

Example 7-4 — Domestic Wastewater Characteristics

Estimate wastewater characteristics for a community with a population of 12,000. The sewer system is known to suffer from excessive infiltration, with a resulting average wastewater flow rate of 2.5 Mgal/day. Report values for concentration of BOD_5, suspended solids, total nitrogen and total phosphorus. Cite references for all data sources.

Solution:

The information provided is sufficient to determine per-capita wastewater generation rates:

$$\text{Per - capita flow rate} = \frac{(2.5 \text{ Mgal / day})(10^6 \text{ gal / Mgal})}{12{,}000 \text{ pop}}$$

$$= 208 \text{ gal / capita - day}$$

The resulting per-capita flow rate exceeds typical values (approximately 100 gal/capita-day) because of the excessive infiltration. Consequently, the wastewater concentrations would be expected to be less than the typical values reported in Table 7-6. The wastewater characteristics may be determined using the per-capita waste loading factors reported in Table 7-5. The daily mass generation rates are equal to the product of the population and the per-capita loading factor:

Mass generation rate = (Population) (Per-capita loading factor)

The concentration is obtained by dividing the mass generation rate by the average flow rate:

$$\text{Concentration} = \frac{(\text{Population})(\text{Per - capita loading factor lb / day})}{(\text{Flow rate})(8.34 \frac{\text{lb / Mgal}}{\text{mg / L}})}$$

$$\text{Concentration} = \frac{(12{,}000)(\text{Per - capita loading factor lb / day})}{(2.5 \text{Mgal / day})(8.34 \frac{\text{lbs / Mgal}}{\text{mg / L}})}$$

The results are tabulated below:

Parameter	Population	Per-capita loading factor (lb/day)	Concentration (mg/L)
BOD$_5$	12,000	0.22	127
SS	12,000	0.26	150
N	12,000	0.033	19
TP	12,000	0.009	5

These concentrations are significantly less than the typical values reported in Table 7-6, reflecting the impact of excessive infiltration.

Recently, regulatory emphasis has been placed on modifying manufacturing processes to alter the characteristics of industrial wastewaters e.g., the Pollution Prevention Act of 1990, thereby reducing the quantity (either volume or mass) and the toxicity of materials that require treatment. It is acknowledged that this approach

Table 7-6
Composition of untreated domestic wastewater

Parameter	Units	Weak	Typical	Strong
Total dissolved solids	mg/L	250	500	850
Total suspended solids	mg/L	100	220	350
Fixed suspended solids	mg/L	20	55	75
Volatile suspended solids	mg/L	80	165	275
Settleable solids	mL/L	5	10	20
BOD_5	mg/L	110	220	400
TOC	mg/L	80	160	290
COD	mg/L	250	500	1000
Total nitrogen (as N)	mg/L	20	40	85
Organic nitrogen (as N)	mg/L	8	15	35
Ammonia (as N)	mg/L	12	25	50
Nitrate (as N)	mg/L	0	0	0
Nitrite (as N)	mg/L	0	0	0
Total phosphorus (as P)	mg/L	4	8	15
Organic phosphorus (as P)	mg/L	1	3	5
Inorganic phosphorus (as P)	mg/L	3	5	10
Alkalinity (as $CaCO_3$)	mg/L	50	100	200
Grease	mg/L	50	100	150

Adapted from: Metcalf & Eddy, 1991.

has been widely practiced for many years to minimize costs associated with end-of-pipe wastewater treatment by minimizing water use and recovering valuable by-products from waste streams. These pollution prevention or waste minimization goals are outlined in the following hierarchy (USEPA (1988)):

1. *Source Reduction.* Reduce the amount of waste at the source, through changes in industrial processes.

2. *Recycling.* Reuse and recycle wastes for the original or some other purpose, such as materials recovery or energy production.

3. *Incineration/Treatment.* Destroy, detoxify and neutralize wastes into less harmful substances.

4. *Secure Land Disposal.* Deposit wastes on land using volume reduction, encapsulation, leachate containment, monitoring, and controlled waste releases.

Development of optimal strategies for industrial waste treatment must begin with an evaluation of the manufacturing process to define opportunities for pollution prevention prior to consideration of requirements for end-of-pipe treatment. Although very important, this analysis is beyond the scope of this text. The reader is referred to the literature that is specific to the industry under consideration.

7.4.2 Process Selection Selection of a sequence of unit processes for wastewater treatment depends upon the raw wastewater characteristics and the effluent requirements. Generalizations regarding industrial waste practice are difficult to make because of the variable nature of conventional pollutants (e.g., BOD_5 and suspended solids) and toxic materials (e.g., hexavalent chromium, trichloroethylene or cyanide) that may require removal. However, typical facilities for treatment of municipal wastewaters may be identified based on requirements for secondary treatment, nitrification or nutrient removal. One such flow sheet to achieve secondary treatment is presented in Figure 7-1.

Figure 7-1
Typical activated sludge facility for secondary treatment of municipal wastewater

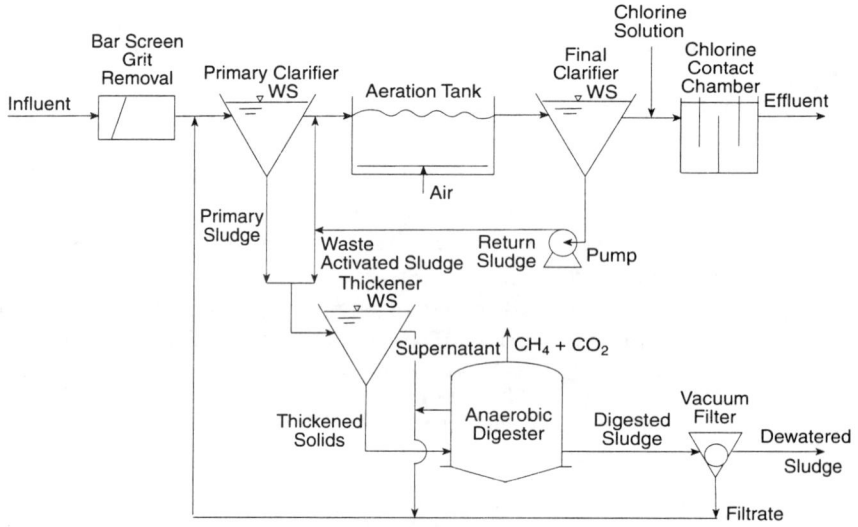

Adapted from: Reynolds, 1982.

There are many alternative configurations that can achieve the same result. A partial summary of unit operations and unit processes that could be substituted for those identified in Figure 7-1 is presented in Table 7-7. Actual process selection will depend on system interactions, client/engineer preferences, operations capability, residuals disposal constraints, and economics.

Similar information is contained in Table 7-8; however, the unit operations and processes are classified according to the pollutant that would be removed by the particular operation/process. Many of the contaminants listed in Table 7-8 are more commonly associated with industrial wastewaters rather than municipal applications (e.g., heavy metals).

Table 7-7
Alternative unit operations/processes

Function	Alternatives
Grit removal	Horizontal flow grit chamber
	Aerated grit chamber
	Vortex grit chamber
Biological treatment	Activated sludge
	Trickling filter
	Rotating biological contactor
Disinfection	Chlorination
	Ozonation
	Ultraviolet radiation
Sludge thickening	Gravity thickening
	Dissolved air flotation thickening
	Centrifugal thickening
	Belt filter press thickening
Sludge stabilization	Anaerobic digestion
	Aerobic digestion
	Composting
	Lime treatment
	Drying
Sludge dewatering	Vacuum filtration
	Belt filter press
	Centrifugation
	Sand drying bed
Sludge disposal	Land application
	Landfill

Other conventional contaminants (e.g., suspended solids) are routinely associated with both municipal and industrial wastewaters. Anticipated removals of conventional pollutants for a municipal wastewater are presented in Table 7-9 for a facility that provides primary and secondary treatment (as illustrated in Figure 7-1). Provision of additional unit operations/processes (e.g., filtration) would be expected to improve removals. Performance data obtained from studies with municipal wastewaters should not be generalized to industrial wastewaters unless supported by pilot studies with the particular wastewater.

7.4.3 Key Design Factors The design of individual unit operations and processes is normally based on a combination of factors, including hydraulic loading, approach velocity, hydraulic residence time, weir loading rate, organic loading, solids loading, and/or solids residence time. Many of the hydraulic and solids loading constraints must be evaluated for both average and peak loading conditions. Relevant information is summarized in Table 7-10 for selected unit operations and processes.

Table 7-8
Unit operation/process objectives

Contaminant	Unit operation or unit process
Suspended solids	Screening and comminution
	Grit removal
	Sedimentation [1]
	Filtration [1]
	Coagulation/sedimentation [1]
Biodegradable organics	Activated sludge
	Trickling filters
	Rotating biological contactors
	Lagoons
Volatile organics	Air stripping
	Carbon adsorption
Pathogens	Chlorination
	Ozonation
	Ultraviolet radiation
	Thermal treatment
	Lime treatment
	Filtration [1]
Kjeldahl nitrogen	Biological nitrification
Nitrate nitrogen	Biological denitrification
Total nitrogen	Nitrification/denitrification
	Ammonia stripping
	Ion exchange [1]
	Breakpoint chlorination
	Natural systems
Phosphorus	Metal salt addition (Fe or Al)
	Lime coagulation/Sedimentation
	Biological phosphorus removal
	Natural systems
Refractory organics	Carbon adsorption
	Ozonation
	Ozonation/ultraviolet radiation
Heavy metals	Chemical precipitation
	Ion exchange [1]
	Membrane processes [1]
Grease	Gravity separation
	Flotation
	Biological processes
	Screens
Dissolved solids	Ion exchange [1]
	Membrane processes [1]

[1] See Chapter 6 for process descriptions
Source: Metcalf & Eddy (1991).

Table 7-9
Secondary treatment of a municipal wastewater

Treatment Unit	Process efficiency expressed as % removal					
	BOD	COD	SS	Total P	Organic N	NH_3-N
Bar Racks	0	0	0	0	0	0
Grit Chambers	0-5	0-5	0-10	0	0	0
Primary Sedimentation	30-40	30-40	50-65	10-20	10-20	0
Activated Sludge	80-95	80-85	80-90	10-25	15-50	8-15
Trickling Filter						
High Rate, Rock	65-80	60-80	60-85	8-12	15-50	8-15
High Rate, Plastic	65-85	65-85	65-85	8-12	15-50	8-15
Rotating Biological Contactor (RBCs)	80-85	80-85	80-85	10-25	15-50	8-15

Organic values provided assume no specific design for nitrogen removal.
Adapted from: Metcalf & Eddy, 1991.

Definition of these key design parameters is provided in Equations 7-9 to 7-17.

$$\text{Overflow Rate or Hydraulic Loading} = \frac{\text{Flow Rate}}{\text{Surface Area}} \quad [7\text{-}9]$$

$$\text{Approach Velocity} = \frac{\text{Flow Rate}}{\text{Cross} - \text{sectional Area}} \quad [7\text{-}10]$$

$$\text{Hydraulic Residence Time} = \frac{\text{Tank Volume}}{\text{Flow Rate}} \quad [7\text{-}11]$$

$$\text{Weir Loading Rate} = \frac{\text{Flow Rate}}{\text{Length of Weir}} \quad [7\text{-}12]$$

$$\text{Organic Loading} = \frac{\text{Mass of Organics per Unit Time}}{\text{Reactor Volume}} \quad [7\text{-}13]$$

$$\text{Organic Loading} = \frac{\text{Mass of Organics per Unit Time}}{\text{Reactor Surface Area}} \quad [7\text{-}14]$$

$$\text{Organic Loading} = \frac{\text{Mass of Organics per Unit Time}}{\text{Mass of Biomass in Reactor}} = \frac{F}{M} \quad [7\text{-}15]$$

$$\text{Solids Loading} = \frac{\text{Mass of Suspended Solids per Unit Time}}{\text{Surface Area}} \quad [7\text{-}16]$$

Table 7-10
Process-design variables for selected unit

Unit operation/process	Design factor	Sizing criteria	Equation
Pumping/piping	Maximum hourly flow	Velocity, head loss	7-10
Screening	Maximum hourly flow	Velocity, head loss	7-10
	Minimum hourly flow	Velocity, solids deposition	7-10
Grit removal	Maximum hourly flow	Overflow rate	7-9
	Maximum hourly flow	Velocity, scour of solids	7-10
	Minimum hourly flow	Velocity, scour of organics	7-10
Primary	Maximum/average flow	Overflow rate	7-9
Sedimentation	Average flow rate	Hydraulic residence time	7-11
	Maximum/average flow	Weir loading	7-12
	Average flow rate	Hydraulic residence time	7-11
Activated sludge aeration basin	Average organic loading	Food to microorganism ratio	7-13 or 7-15
	Average solids waste rate	Solids residence time	7-17
Trickling filters	Average flow rate	Hydraulic loading rate	7-9
	Average organic loading	Organic loading	7-13 or 7-14
Activated sludge	Maximum/average flow	Overflow rate	7-9
Clarification	Maximum/average flow	Weir loading	7-12
	Maximum/average solids loading	Solids loading	7-16
	Average flow rate	Hydraulic residence time	7-11
Trickling filter	Maximum/average flow	Overflow rate	7-9
Clarification	Maximum/average flow	Weir loading	7-12
	Average flow rate	Hydraulic residence time	7-11
Chlorination	Maximum hourly flow	Hydraulic residence time	7-11
Filtration	Average flow rate	Hydraulic loading rate	7-9

Adapted from: Metcalf & Eddy, 1991.

Chapter 7: Wastewater Treatment

$$\text{Solids Residence Time} = \frac{\text{Mass of Biomass in Reactor}}{\text{Mass of Biomass Removed per Unit Time}} \quad [7\text{-}17]$$

TIP FOR THE P.E. EXAMINATION CANDIDATE

Dimensional consistency must be maintained during application of the relationships in Equations 7-9 to 7-17. Familiarization with customary units and probable unit conversions in advance of the examinations is encouraged. Presentations of units with calculations is also encouraged to diagnose errors.

7.5 Preliminary Treatment

Various screening devices may be provided to remove rags and other debris that could damage equipment or clog piping within the treatment facility. Alternately or in combination, grinding devices (comminutors) may be installed to reduce the size of such objects. Typical design criteria for coarse screening equipment are provided in Table 7-11.

Table 7-11
Design criteria for screening equipment

Item	Range	Comment
Trash rack openings	1.5 to 6 in.	Commonly used on combined systems
Manual screens openings	1 to 2 in.	Used in small plants or in bypass channels
Approach velocity	1 to 2 ft/sec	
Mechanically cleaned bar screen openings	0.25 to 1.5 in.	0.75 in. opening satisfactory
Approach velocity		
Maximum	2 to 4 ft/sec	
Minimum	1 to 2 ft/sec	To prevent grit deposition
Continuous screen openings	0.25 to 1.5 in.	Effective in 0.25 to 0.75 in. range
Approach velocity		
Maximum	2 to 4 ft/sec	
Minimum	1 to 2 ft/sec	
Head loss	0.5 to 2 ft	

Table 7-11 (cont.)
Design criteria for screening equipment

Item	Range	Comment
Comminutor (size reduction only) openings	0.25 to 0.5 in.	
Grinder (size reduction only) openings	0.25 to 0.5 in.	
Head loss	1 to 1.5 ft	

Adapted from: WEF/ASCE, 1991.

In addition to screening or grinding, grit removal is usually provided to eliminate inorganic solids (typically sand) that would cause abrasion in mechanical equipment and accumulate in reactors (aeration basins and digesters), thereby reducing the effective volume for treatment. Design criteria are provided for aerated grit chambers and horizontal flow grit chambers in Tables 7-12 and 7-13, respectively.

Table 7-12
Typical design criteria for aerated grit chambers

Item	Range	Comment
Dimensions		
Depth, ft	6 to 16	Varies widely
Length to width ratio	2.5:1 to 5:1	
Width to depth ratio	1:1 to 5:1	2:1 typical
Minimum detention time at peak flow, min	2 to 5	3 typical
Air supply, standard, ft^3/min-ft	3 to 8	5 typical
Diffuser type	Medium to coarse bubble	
Distance from bottom, ft	2 to 3	
Transverse roll velocity, ft/sec	2 to 2.5	

Adapted from: WEF/ASCE, 1991.

Table 7-13
Typical design criteria for horizontal flow grit chambers

Item	Range	Comment
Dimensions		
Depth, ft	2 to 5	Depends on channel area and flow rate
Length, ft	10 to 80	Function of channel depth and grit settling velocity
Width, ft		Compatible with equipment for grit removal
Overflow rate, gal/day-ft^2	20,000 to 45,000	Function of particle size to be removed
Detention time at peak flow, min	0.25 to 1.5	Function of velocity and channel length
Horizontal velocity, ft/sec	0.5 to 1.25	Optimum velocity = 1 ft/sec

Adapted from: WEF/ASCE, 1991 and Metcalf & Eddy, 1991.

Example 7-5 — Grit Chamber Design

Determine the total volume of an aerated grit chamber for a wastewater treatment plant with an average flow of 10 Mgal/day and peak to average flow ratio of 2.2.

Solution:

Aerated grit chamber design is based on the hydraulic residence time (Equation 7-11) at peak flow. The recommended HRT is 3 min (see Table 7-12).

TIP FOR THE P.E. EXAMINATION CANDIDATE

It is generally accepted practice to provide multiple units in parallel to facilitate operation during maintenance. Many essay problems will require specification of the number of units as well as dimensions.

$$\text{Volume} = \frac{(22 \text{ MGD})(10^6 \text{ gal / MGAL})(3 \text{ min})}{(1440 \text{ min / day})(7.48 \text{ gal / ft}^3)} = 6127 \text{ ft}^3$$

Two basins in parallel are recommended, with dimensions of 6 ft (depth), 12 ft (width), and 43 ft (length). Additional basin depth is necessary for freeboard.

7.6 Sedimentation

7.6.1 Objectives and Design Criteria The primary purpose of sedimentation operations in wastewater treatment is to remove particulate material, thereby reducing the suspended solids and organic content of the effluent. Sedimentation is accomplished by units called clarifiers which may be applied early in the treatment train, *primary*, or after a biologic process, *secondary* or *tertiary*. Clarifiers may be either circular in configuration or rectangular and arranged with multiple units, typically in parallel, in each phase. A secondary function is to produce a residual solids stream (sludge) that has sufficient solids content to reduce sludge volumes and minimize sludge processing costs. Additionally, clarifiers provide capacity for limited storage of solids during peak loading events or during periods when sludge processing equipment is not operational.

Particulates that are present in raw wastewaters or produced during biological treatment processes are flocculent in nature. Consequently, the performance of sedimentation operations depends upon the hydraulic residence time and the hydraulic loading (or overflow rate). The design of activated sludge final clarifiers must also consider the solids flux (or solids loading rate). These variables were defined previously (Equations 7-9, 7-11, 7-12, and 7-16); application to specific clarification operations is defined in Equations 7-18 to 7-21.

The hydraulic residence time (HRT) for clarification processes is defined based on the plant influent wastewater flow rate (Q_o):

$$\text{HRT} = \frac{V}{Q_o} \qquad [7\text{-}18]$$

where: HRT = hydraulic residence time
V = primary or secondary clarifier volume
Q_o = plant influent flow rate.

The overflow rate (ORA) and weir loading rate (WLR) for clarification processes are also defined based on the plant influent wastewater flow rate:

$$\text{ORA} = \frac{Q_o}{A} \qquad [7\text{-}19]$$

$$\text{WLR} = \frac{Q_o}{L} \qquad [7\text{-}20]$$

where: ORA = overflow rate
A = clarifier surface area
WLR = weir loading rate
L = total length of weir.

The solids loading rate (SLR) for activated sludge secondary clarifiers is defined based on the total solids loading on the clarifier from the aeration basin:

$$\text{SLR} = \frac{Q_o(1+R)X}{A} \quad [7\text{-}21]$$

where: SLR = solids loading rate
R = recycle ratio (decimal)
   = Q_{return}/Q_o
X = mixed liquor suspended solids concentration

Clarifier design for municipal applications is commonly based on experience. Heuristic guidelines for facility design based on past experience are presented in Tables 7-14 through 7-16. Design procedures have also been established based on experimental pilot-testing programs for non-flocculent slurries (Rich (1961)), flocculent slurries (Rich (1961)), and activated sludge (Keinath et al. (1977)). When available, site-specific settling data should be used for design rather than these general guidelines developed from experience.

7.6.2 Primary Clarification Substantial removals of BOD and suspended solids are an expected result of primary clarification of municipal wastewaters. Marginal removals of organic nitrogen and phosphorus also result. Typical performance data were reported previously in Table 7-9. Typical design guidelines for primary clarifiers are shown in Table 7-14.

Table 7-14
Design guidelines for primary clarification of municipal wastewater

Design criteria	Range	Typical
Hydraulic detention time, hr	1.5 to 2.5	2.0
Overflow rate, gal/day-ft^2		
Average flow	800 to 1200	1000
Peak hourly flow	2000 to 3000	2500
Weir loading, gal/day-ft		
Average flow	10,000 to 40,000	20,000
Peak flow	100,000	

Adapted from: WEF/ASCE, 1991 and Metcalf & Eddy, 1991.

Example 7-6 — Primary clarifier design

Determine the size of clarifier(s) for an activated wastewater treatment plant with an average flow of 10 Mgal/day and a peak to average flow ratio of 2.2.

Solution:

The design of primary clarifiers is based on the overflow rate (peak and average hydraulic loading, (Equation 7-9) and the detention time (Equation 7-11). Empirical design criteria were reported in Table 7-14.

$$\text{Area (peak flow)} = \frac{(22 \text{ Mgal/day})(10^6 \text{gal/Mgal})}{2,500 \text{ gal/day-ft}^2} = 8,800 \text{ ft}^2$$

$$\text{Area (average flow)} = \frac{(10 \text{ Mgal/day})(10^6 \text{gal/Mgal})}{1,000 \text{ gal/day-ft}^2} = 10,000 \text{ ft}^2$$

The larger area (10,000 ft²) must be selected to satisfy both loading criteria. Two circular units are selected with a diameter of 80 ft.

The clarifier depth is determined based on the desired hydraulic residence time of 2 hr at average flow:

$$\text{Depth} = \frac{(10 \text{ Mgal/day})(10^6 \text{gal/Mgal})(2 \text{ hr})}{(24 \text{ hr/day})(7.48 \text{ gal/ft}^3)(2 \text{ units})\frac{(3.14159)(80 \text{ ft})^2}{4}} = 11.1 \text{ ft}$$

The recommended sidewater depth is 11 ft, with an additional 2 ft allowance for freeboard

7.6.3 Secondary Clarification Typical design criteria for clarifiers used with attached-growth processes, Trickling Filters and Rotating Biological Contactors are summarized in Table 7-15. Multiple units are normally specified to sustain plant operation during clarifier maintenance.

Typical design criteria for clarifiers used with several suspended-growth, activated sludge processes are summarized in Table 7-16. The guidelines indicate that sludges from a high-purity oxygen process may exhibit superior settling properties, while sludges derived from an extended aeration process may have degraded settling properties. Multiple units are normally specified to sustain operation during clarifier maintenance.

Table 7-15
Design guidelines for secondary clarification of municipal wastewater (Attached growth processes)

Design factor	Trickling filter	Rotating biological contactor
Overflow rate, gal/day-ft^2		
Average flow	400 to 600	400 to 800
Peak flow	1000 to 1200	1000 to 1200
Hydraulic detention time, hr	3 to 6	2.5 to 6
Weir loading, gal/day-ft		
Average	10,000	10,000
Peak (small clarifiers)	20,000	20,000
Peak (large clarifiers)	30,000	30,000
Depth, ft	10 to 15	10 to 15

Adapted from: WEF/ASCE, 1991 and Metcalf & Eddy, 1991.

Table 7-16
Design guidelines for secondary clarification of municipal wastewater — Activated sludge (suspended growth)

Design factor	Conventional activated sludge	Extended aeration	Pure oxygen activated sludge
Overflow rate, gal/day-ft^2			
Average flow	400 to 800	200 to 400	400 to 800
Peak flow	1000 to 1200	600 to 800	1000 to 1200
Hydraulic detention time, hr	3 to 6	6 to 12	3 to 6
Solids loading, lb/day-ft^2			
Average	20 to 30	5 to 24	24 to 35
Peak	48	33	48
Weir loading, gal/day-ft			
Average	10,000	10,000	10,000
Peak (small clarifiers)	20,000	20,000	20,000
Peak (large clarifiers)	30,000	30,000	30,000
Depth, ft	12 to 20	12 to 20	12 to 20

Adapted from: Metcalf & Eddy, 1991.

Example 7-7 — Secondary clarifier design

For a community of 100,000, determine the size of the secondary clarifier(s) in a municipal activated sludge treatment facility (see Figure 7-1): The wastewater characteristics are:

Average per-capita flow rate = 100 gal/capita-day

Peak factor = 2.2

Solution:

Average flow rate = (100 gal/capita-day)(100,000 pop)/(10^6 gal/Mgal) = 10 Mgal/day

Peak hourly flow rate = (10 Mgal/day)(2.2) = 22 Mgal/day

The design of activated sludge secondary clarifiers is based on the overflow rate (peak and average hydraulic loading, Equation 7-19), the detention time (Equation 7-18), and the peak and average solids loading (Equation 7-21). Empirical design criteria are reported in Table 7-16. The mixed-liquor suspended solids concentration is assumed to be 2,500 mg/L and the average return activated sludge percent recycle is assumed to be 50%.

$$\text{Area (peak flow)} = \frac{(22 \text{ Mgal/day})(10^6 \text{ gal/Mgal})}{1,100 \text{ gal/day-ft}^2}$$

$$= 20,000 \text{ ft}^2$$

$$\text{Area (average flow)} = \frac{(10 \text{ Mgal/day})(10^6 \text{ gal/Mgal})}{600 \text{ gal/day-ft}^2}$$

$$= 16,700 \text{ ft}^2$$

$$\text{Area (peak solids loading)} = \frac{(22 \text{ Mgal/day})(1+0.50)(2500 \text{ mg/L})\left(8.34 \frac{\text{lb/Mgal}}{\text{mg/L}}\right)}{48 \frac{\text{lb}}{\text{day}-\text{ft}^2}}$$

$$\text{Area} = 14,300 \text{ ft}^2$$

$$\text{Area (average solids loading)} = \frac{(10 \text{ Mgal/day})(1+0.50)(2,500 \text{ mg/L})\left(8.34 \frac{\text{lb/Mgal}}{\text{mg/L}}\right)}{25 \frac{\text{lb}}{\text{day}-\text{ft}^2}}$$

$$\text{Area} = 12,500 \text{ ft}^2$$

The largest area (20,000 ft²) must be selected to satisfy all loading criteria. Four circular units with a diameter of 80 ft are selected. The clarifier depth is determined based on the desired hydraulic residence time of 4.5 hr at average flow:

$$\text{Depth} = \frac{(10 \text{ Mgal/day})(10^6 \text{ gal/Mgal})(4.5 \text{ hr})}{(24 \text{ hr/day})(7.48 \text{ gal/ft}^3)(4 \text{ units})\frac{(3.14159)(80 \text{ ft})^2}{4}} = 12.5 \text{ ft}$$

The recommended sidewater depth is 13 ft, with an additional 2 ft allowance for freeboard.

7.7 Biological Treatment

7.7.1 Overview Biological wastewater treatment processes are used for municipal and many industrial wastewater treatment applications. Biological processes have been demonstrated to achieve BOD removal, suspended solids removal, nitrification, nutrient removal, nitrogen removal (nitrification/denitrification), phosphorus removal, and removal of many synthetic organic compounds from municipal and industrial wastewaters.

The design of biological treatment systems may be accomplished with general guidelines developed from extensive experience. Heuristic guidelines are sufficiently established to support design of municipal treatment facilities because of the amount of accumulated data and similarity in domestic wastewater characteristics. For selected industrial waste applications, similar guidelines have been adopted based on past experience. Because of the substantial variability in industrial wastewater characteristics and the scarcity of historical data, the design of many industrial wastewater treatment facilities must include a more fundamental approach based on evaluation of the reaction kinetics for the specific wastewater. The kinetic results are integrated into a design approach using mass balances and reactor engineering principles. This fundamental approach is valid for all applications and should be used whenever site-specific kinetic data are available to support this type of analysis. Procedures for conducting kinetic studies are reviewed in Metcalf & Eddy (1991), Grady and Lim (1980), and Reynolds (1982).

The empirical relationships and kinetic parameter estimates provided in this section are generally sensitive to various environmental factors, e.g., pH and temperature. The values reported may be reasonable for normal conditions e.g., temperature = 20°C; however, adjustment may be necessary for extreme conditions.

7.7.2 Process Kinetics

The design of biological wastewater treatment facilities must address reaction rates for substrate removal and production of biomass. In this context, substrate may be represented by any measure of organic material oxidized by the biomass to supply energy, e.g., BOD, COD, TOC, or any specific organic compound. For nitrification processes, the substrate is ammonia (measured as TKN). Knowledge of the rate of substrate utilization is necessary to evaluate the BOD or TKN removal efficiency relative to effluent standards (see Tables 7-1 to 7-3).

Biological growth reactions in wastewater treatment achieve substrate removal by a combination of conversion of substrate to biomass and oxidation of substrate. The assimilative removal of substrate results in the production of suspended solids. The quantity of suspended solids affects the design of many unit operations and processes, including the clarifiers which follow the reactor and sludge-handling facilities. Knowledge of the rate of biomass production is therefore necessary to estimate residuals (sludge) quantities and complete design of selected unit operations and processes.

Various empirical models have found widespread use for analysis of kinetics. Selected models are defined in Equations 7-22 through 7-28.

Four common models for specific substrate utilization rates are presented in Equations 7-22 through 7-25 representing zero-order, first-order, variable-order (Monod), and inhibitory kinetics, respectively:

Zero order $\quad q = \dfrac{-r_s}{X} = k_o \quad$ [7-22]

where: q = specific substrate utilization rate $\left(\dfrac{\text{mg } S}{\text{mg } X \text{-day}}\right)$

$-r_s$ = substrate removal reaction rate $\left(\dfrac{\text{mg } S}{L - \text{day}}\right)$

X = biomass concentration (either SS or VSS) $\left(\dfrac{\text{mg}}{L}\right)$

k_0 = zero-order reaction rate constant $\left(\dfrac{\text{mg } S}{\text{mg } X \text{-day}}\right)$

First Order $\quad q = \dfrac{-r_s}{X} = k_1 S$ [7-23]

where: k_1 = first-order reaction rate constant $\left(\dfrac{L}{\text{mg } X - \text{day}}\right)$

S = substrate concentration $\left(\dfrac{\text{mg}}{L}\right)$

Variable Order $\quad q = \dfrac{-r_s}{X} = \dfrac{kS}{K_s + S}$ [7-24]

where: k = maximum specific substrate utilization rate $\left(\dfrac{\text{mg } S}{\text{mg } X \text{-day}}\right)$

K_s = half-saturation constant $\left(\dfrac{\text{mg } S}{L}\right)$

Inhibitory $\quad q = \dfrac{-r_s}{X} = \dfrac{kS}{K_s + S + \dfrac{S^2}{K_I}}$ [7-25]

where: K_I = inhibition coefficient $\left(\dfrac{\text{mg } S}{L}\right)$

The variable-order Monod model (Equation 7-24) is widely used because of the flexibility to describe both zero-order (large concentrations) and first-order kinetics (small concentrations). The inhibitory model is appropriate for specific organic compounds where the rate of biological activity is reduced at excessive substrate concentrations. The inhibitory model degenerates to the Monod model as the inhibition constant approaches infinity. Various synthetic organic compounds are reported (Grady (1990)) to exhibit inhibition at high concentrations, including 4-chlorophenol, 1,3-dichlorobenzene and 1,4-dichlorobenzene.

The rate of biomass production is related to the rate of substrate utilization by the reaction stoichiometry. The stoichiometric ratio for a defined reaction is referred to as the observed yield (Y_{obs}):

$$\mu = \frac{r_x}{X} = Y_{obs}\frac{-r_s}{X} \qquad [7\text{-}26]$$

where: μ = specific growth rate (days^{-1})

Y_{obs} = observed yield $\left(\frac{\text{mg } X}{\text{mg } S}\right)$

The observed yield varies with the system operation and is, therefore, not constant. The observed yield may be expressed as a function of the system-specific growth rate and the maximum yield (Y_{max}) as reported in Equation 7-27. The maximum yield is independent of system operation and is evaluated as a constant that is characteristic of the wastewater.

$$Y_{obs} = \frac{Y_{max}}{1 + \frac{k_e}{\mu}} \qquad [7\text{-}27]$$

where: Y_{max} = maximum yield $\left(\frac{\text{mg } X}{\text{mg } S}\right)$

k_e = endogenous decay coefficient (days^{-1})

Equations 7-26 and 7-27 may be combined to yield Equation 7-28:

$$\mu = Y_{max}q - k_e \qquad [7\text{-}28]$$

The kinetic equations for substrate removal and biomass production can be solved simultaneously to establish a relationship between the specific growth rate and the substrate concentration. For first-order substrate removal kinetics (Equation 7-23):

$$S = \frac{\mu + k_e}{Y_{max}k_1} \qquad [7\text{-}29]$$

A similar derivation for variable-order kinetics yields:

$$S = \frac{K_s(\mu + k_e)}{Y_{max}k - \mu - k_e} \qquad [7\text{-}30]$$

It is evident from Equations 7-29 and 7-30 that the effluent quality (represented by the substrate concentration, S) is determined by the experienced specific growth rate (μ). As the growth rate (μ) is decreased, the removal of substrate is improved (Lawrence and McCarty (1970)).

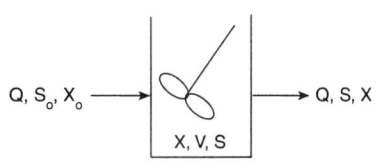

Figure 7-2
Complete-mix suspended growth system without solids recycle

Through development of material balances for biomass, the specific growth rate is easily related to conventional design parameters for suspended growth systems. For complete-mix suspended growth systems without solids recycle (such as aerated lagoons, aerobic digestion or anaerobic digestion) at steady-state, the growth rate is related to the reactor hydraulic residence time (see Figure 7-2).

Input - Output + Generation = Accumulation [7-31]

$$QX_o - QX + Vr_x = 0 \qquad [7\text{-}32]$$

$$0 - QX + Vr_x = 0 \qquad [7\text{-}33]$$

$$\mu = \frac{r_x}{X} = \frac{Q}{V} = \frac{1}{\text{HRT}} \qquad [7\text{-}34]$$

where: Q = influent flow rate
         X_o = influent biomass concentration (assumed negligible)
         V = reactor volume
         HRT = hydraulic residence time

For complete-mix suspended growth systems with solids recycle (such as activated sludge) at steady-state, the growth rate is related to the solids residence time (see Figure 7-3).

Input - Output + Generation = Accumulation [7-35]

$$QX_o - Q_e X_e - Q_w X_w + Vr_x = 0 \qquad [7\text{-}36]$$

$$\mu = \frac{r_x}{X} = \frac{Q_e X_e + Q_w X_w}{VX} = \frac{1}{\text{SRT}} \qquad [7\text{-}37]$$

Figure 7-3
Complete-mix suspended growth system with solids recycle

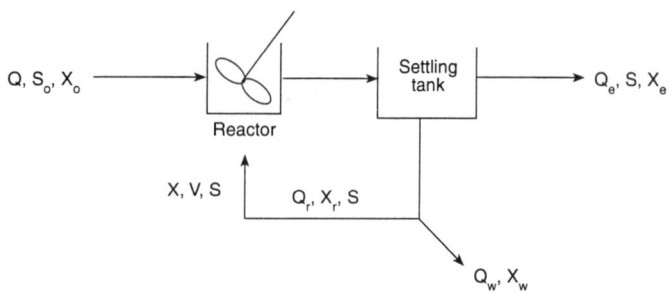

where:
- Q = influent flow rate
- X_o = influent biomass concentration (assumed negligible)
- Q_e = effluent flow rate
- Xe = effluent biomass concentration
- Q_w = waste sludge flow rate
- X_w = waste sludge biomass concentration
- V = reactor volume
- SRT = solids residence time (see Equation 7-17).

The solids residence time is also commonly referred to as the sludge age or the mean cell residence time.

Organic loading can be used as an alternative to the specific growth rate for design of biological treatment systems. The use of organic loading is particularly attractive for attached-growth systems (trickling filters and rotating biological contactors) because of difficulties in measuring biomass concentration in these systems. The organic loading is defined as the rate of supply of substrate/unit of biomass (Equation 7-15 or 7-38):

$$\text{Organic Loading} = \frac{\text{Mass of Organics per Unit Time}}{\text{Mass of Biomass in Reactor}} \qquad [7\text{-}15]$$

$$\text{Organic Loading} = \frac{QS_o}{VX} \qquad [7\text{-}38]$$

where:
- Q = influent flow rate
- S_o = influent substrate concentration
- V = reactor volume
- X = reactor biomass concentration

In systems where direct measurement of biomass is difficult, it is often assumed that the biomass is proportional to the reactor volume or the media surface area.

For these estimates of biomass, the organic loading is defined with Equation 7-13 or 7-14:

$$\text{Organic Loading} = \frac{\text{Mass of Organics per Unit Time}}{\text{Reactor Volume}} \quad [7\text{-}13]$$

$$\text{Organic Loading} = \frac{\text{Mass of Organics per Unit Time}}{\text{Reactor Surface Area}} \quad [7\text{-}14]$$

For all systems that operate at high efficiency, the effluent substrate concentration (S) is much less than the influent (S_o); thus, the organic loading is approximately equal to the specific substrate utilization rate:

$$\text{Organic Loading} = \frac{QS_o}{VX} \cong \frac{Q(S_o - S)}{VX} = \frac{-r_s}{X} = q \quad [7\text{-}39]$$

As previously noted in Equation 7-28, the specific substrate utilization rate can not be specified independently of the specific growth rate. Thus, the organic loading also depends on the growth rate. Therefore, specification of either the specific growth rate or the organic loading rate would determine the substrate removal efficiency for a biological wastewater treatment process. Empirical design guidelines have been developed using indicators of growth rate, e.g., solids residence time and organic loading, e.g., lb BOD_5 /1,000 ft^3-day. Examples to support design of each biological unit process are provided in subsequent sections of this chapter.

7.7.3 Suspended-Growth Processes Numerical values for various kinetic parameters (municipal wastewaters) for oxidation of BOD and for oxidation of ammonia are presented in Table 7-17.

The kinetic values for nitrification in Table 7-17 were developed with a pure culture of nitrifying bacteria. In an operating wastewater treatment facility, the nitrifying bacteria will represent a small fraction of the total biomass (suspended solids). This fraction has been related to the ratio of BOD_5 to TKN in the wastewater as summarized in Table 7-18. Calculation of the specific substrate utilization rate for nitrification (see Equations 7-22 through 7-25) will require adjustment of the total reactor biomass concentration (X) to reflect the actual concentration of nitrifying organisms.

Substrate utilization kinetic data is provided in Table 7-19 for a variety of industrial wastewaters. A typical first-order rate constant for domestic wastewater is included for comparison. It is evident from this data that many industrial wastewaters do not sustain the same magnitude of biological reaction rates as municipal wastewater. Biological treatment would be satisfactory for these industrial wastes, but the necessary growth rate could be very different than for municipal applications. Generalization regarding industrial waste treatment applications is, therefore, very difficult. Site-specific kinetic information should be used for design

Table 7-17
Municipal wastewater kinetic parameter estimates

Parameter	Units	Range	Typical
Maximum specific substrate utilization rate, k			
BOD removal	$\dfrac{\text{mg BOD}_5}{\text{mgVSS} - \text{day}}$	2-10	5
NH_3 removal	$\dfrac{\text{mg NH}_3 - \text{N}}{\text{mgVSS} - \text{day}}$	1.5-15	5
Half-saturation constant, K_s			
BOD removal	$\dfrac{\text{mg BOD}_5}{\text{L}}$	25-100	60
NH_3 removal	$\dfrac{\text{mg NH}_3 - \text{N}}{\text{L}}$	0.2-5.0	1.4
Maximum yield, Y_{max}			
BOD removal	$\dfrac{\text{mgVSS}}{\text{mg BOD}_5}$	0.4-0.8	0.6
NH_3 removal	$\dfrac{\text{mgVSS}}{\text{mg NH}_3 - \text{N}}$	0.1-0.3	0.2
Endogenous decay, k_e			
BOD removal	days^{-1}	0.025-0.075	0.06
NH_3 removal	days^{-1}	0.03-0.06	0.05

Adapted from: Metcalf & Eddy, 1991.

Table 7-18
Nitrifier fraction of the biomass in mixed culture

BOD_5/TKN ratio	Nitrifier fraction
0.5	0.35
1	0.21
2	0.12
3	0.083
4	0.064
5	0.054
6	0.043
7	0.037
8	0.033
9	0.029

Source: Metcalf & Eddy, 1991.

purposes whenever such information is available for municipal or industrial applications.

7.7.4 Attached-Growth Processes

The design of attached-growth processes (trickling filters and rotating biological contactors) is ordinarily based on empirical models developed for the specific wastewater or on guidelines developed from experience. Use of reactor engineering principles is more difficult for these systems because of difficulties in measuring biomass concentration. A diagram of a typical trickling filter configuration is shown in Figure 7-4. Recirculation is provided in all modifications except for the low-rate process. Primary clarification is normally specified for municipal wastewaters prior to trickling filter treatment.

Empirical models have been developed for rock-media trickling filters with municipal wastewaters. Two common models are reported in Equations 7-40 through 7-43 (Reynolds (1982)). These models would not be appropriate for other applications because of differences in degradation rates for alternate organic compounds and differences in mass transfer rates with alternate media.

The Eckenfelder model, developed from reactor theory, assumes plug flow behavior in the filter. The numerical values for the rate constants in Equation 7-40 are specific for municipal wastewaters and stone-media filters. The equation may be used for systems with or without recycle, as long as the influent BOD_5 concentration and the hydraulic loading rate are adjusted to reflect the effects of recycle. The specific units for concentration, depth, and hydraulic loading rate must be utilized in this empirical equation.

Table 7-19
Values for the first-order (Equation 7-23) reaction rate constant for selected industrial wastewaters

Source	$k_1, \dfrac{L}{mg\ VSS-day}$
Domestic wastewater	0.0294
Brewery	0.0053
Coke plant, ammonia liquor	0.0264
Organic chemical	0.0015
Petrochemical	0.0062
Pharmaceutical	0.0094
Phenolic	0.0022
Pulp and paper	0.0100
Refinery	0.0162
Rendering	0.0360
Tetraethyl lead	0.0170
Textile	0.0036
Thiosulfate	0.0038
Vegetable oil	0.0074

Source: Grady and Lim, 1980.

Figure 7-4
High-rate trickling filter recycle configuration

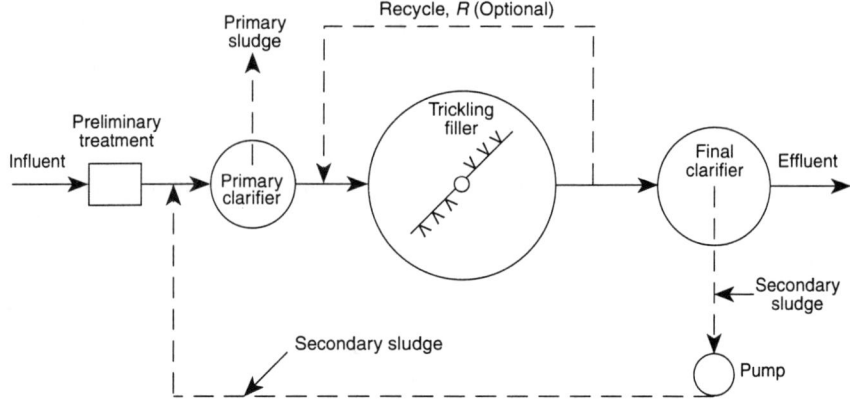

Adapted from: Reynolds, 1982.

$$\frac{S}{S_{in}} = \frac{1}{1 + \frac{2.5 D^{0.67}}{Q_L^{0.50}}} \quad [7\text{-}40]$$

where: S = effluent BOD$_5$ (mg/L)
S_{in} = BOD$_5$ in wastewater input to the filter (mg/L)
D = trickling filter depth (ft)
Q_L = hydraulic loading rate (Mgal/acre-day)

The National Research Council (NRC) collected data from numerous military facilities during the 1940s. The database included low-rate, single-stage high-rate, and two-stage high rate trickling filters with stone media. The empirical equation developed for a single-stage system is reported in Equation 7-41:

$$E_s = \frac{100}{1 + 0.0085 \sqrt{\frac{y_o}{VF}}} \quad [7\text{-}41]$$

where: E_s = BOD$_5$ removal efficiency (%)
y_o = organic loading rate to the filter (lb BOD$_5$/day)
V = stone-media volume (acre-feet)
F = recycle factor

The recycle factor, F, is defined as a function of the recycle ratio $\left(\frac{Q_R}{Q}\right)$ in Equation 7-42:

$$F = \frac{1 + \dfrac{Q_R}{Q}}{\left(1 + 0.1 \dfrac{Q_R}{Q}\right)^2} \qquad [7\text{-}42]$$

where: Q_R = recirculated flow rate

The efficiency of the second-stage of a two-stage system is in Equation 7-43:

$$E_{s2} = \frac{100}{1 + \dfrac{0.0085}{1 - \dfrac{E_{s1}}{100}} \sqrt{\dfrac{y_o'}{V_2 F_2}}} \qquad [7\text{-}43]$$

where:
E_{s2} = BOD$_5$ removal efficiency of the second stage (%)
E_{s1} = BOD$_5$ removal efficiency of the first stage (%)
y_o' = organic loading rate to second stage (lb/BOD$_5$/day)
$= y_o\left(1 - \dfrac{E_{s1}}{100}\right)$
V_2 = stone media volume for second stage (acre-feet)
F_2 = recycle factor for the second stage (see Equation 7-42).

General design criteria for low-rate, intermediate-rate, high-rate and super-high-rate trickling filter systems are presented in Table 7-20.

Table 7-20
Typical design information for trickling filters

Item	Low rate	Intermediate rate	High rate	Super-high rate
Filter medium	Rock	Rock	Rock	Plastic
Hydraulic loading				
gal/ft^2-min	0.02-0.06	0.06-0.16	0.16-0.64	0.2-1.20
Mgal/acre-day	1-4	4-10	10-40	15-90
BOD$_5$ loading				
lb/1000 ft^3-day	5-25	15-30	30-60	30-100
Depth, ft	6-8	6-8	3-6	10-40
Recirculation ratio	0	0-1	1-2	1-2
BOD$_5$ removal efficiency, %	80-90	50-70	65-85	65-80
Nitrification	yes	partial	little	little

Adapted from: Metcalf & Eddy, 1991.

Example 7-8 — Trickling filter design

Determine the size of a low-rate, stone-media trickling filter and clarifier for secondary treatment of a municipal wastewater. The raw wastewater BOD_5 concentration is 250 mg/L, the average flow is 4 Mgal/day, and the peak hourly flow is 10 Mgal/day.

Solution:

It is assumed that primary clarification is provided prior to the trickling filter in order to minimize clogging problems. It is also assumed that 35% of the BOD_5 is removed in primary treatment (see Table 7-9). Because this application involves municipal wastewaters with a stone-media filter, the National Research Council equations are appropriate (Equations 7-41 and 7-42).

The BOD_5 remaining after primary clarification is calculated below:

$$y_o = (4 \text{ Mgal/day})(250 \text{ mg/L})\left(8.34 \frac{\text{lb/Mgal}}{\text{mg/L}}\right)(1-0.35) = 5421 \text{ lb/day}$$

$$BOD_5 = (250 \text{ mg/L})(1-0.35) = 162 \text{ mg/L}$$

The efficiency of the trickling filter is defined by the influent to the filter (162 mg BOD_5/L) and the required effluent concentration of 30 mg BOD_5/L for secondary treatment (see Table 7-1):

$$E_s = \left[\frac{162 \text{ mg BOD}_5/\text{L} - 30 \text{ mg BOD}_5/\text{L}}{162 \text{ mg BOD}_5/\text{L}}\right]100\% = 81.5\%$$

There is no recirculation for low-rate trickling filters; thus the recycle factor (F) =1 (see Equation 7-42). Equation 7-41 is manipulated to solve for the media volume:

$$V = \frac{y_o}{F} \frac{0.0085^2}{\left[\frac{100}{E_s} - 1\right]^2}$$

$$V = \frac{5421}{1} \frac{0.0085^2}{\left[\frac{100}{81.5} - 1\right]^2} = 7.601 \text{ acre-ft} = 331{,}000 \text{ ft}^3$$

The maximum depth of a stone-media filter is limited by structural considerations. Four circular units with a diameter of 120 ft and a depth of 7 ft are recommended for this application (see Table 7-20).

The design of trickling filter secondary clarifiers is based on the overflow rate (peak and average hydraulic loading, Equation 7-9) and the detention time (Equation 7-11). Empirical design criteria are reported in Table 7-15.

$$\text{Area (peak flow)} = \frac{(10 \text{ Mgal / day})(10^6 \text{ gal / Mgal})}{1,100 \text{ gal / day - ft}^2} = 9,100 \text{ ft}^2$$

$$\text{Area (average flow)} = \frac{(4 \text{ Mgal / day})(10^6 \text{ gal / Mgal})}{500 \text{ gal / day - ft}^2} = 8,000 \text{ ft}^2$$

The larger area (9,100 ft²) must be selected to satisfy both loading criteria. Two circular units with a diameter of 80 ft are selected.

The clarifier depth is determined based on the desired hydraulic residence time of 4.5 hr at average flow:

$$\text{Depth} = \frac{(4 \text{ Mgal / day})(10^6 \text{ gal / Mgal})(4.5 \text{ hr})}{(24 \text{ hr / day})(7.48 \text{ gal / ft}^3)(2 \text{ units})\frac{(3.14159)(80 \text{ ft})^2}{4}} = 10.0 \text{ ft}$$

The recommended sidewater depth is 10 ft, with an additional 2 ft allowance for freeboard.

Rotating biological contactors (RBCs) may be used for carbonaceous BOD removal, combined BOD removal and nitrification, and separate-stage nitrification of secondary effluent. A schematic diagram of the process is presented in Figure 7-5. Typical empirical design guidelines are summarized in Table 7-21.

7.7.5 Suspended Growth Process Developing the equations needed to characterize a complete-mix activated sludge facility requires development of equations for conservation of substrate around the biological reactor, for conservation of the mass of biomass around the secondary clarifier, and for conservation of the mass of biomass around the entire system.

Knowledge of the appropriate kinetic relationships for substrate removal and biomass production is also necessary. Development of key design equations for the activated sludge process is represented in Figure 7-6. A steady-state substrate

Figure 7-5
Rotating biological contractor schematic diagram

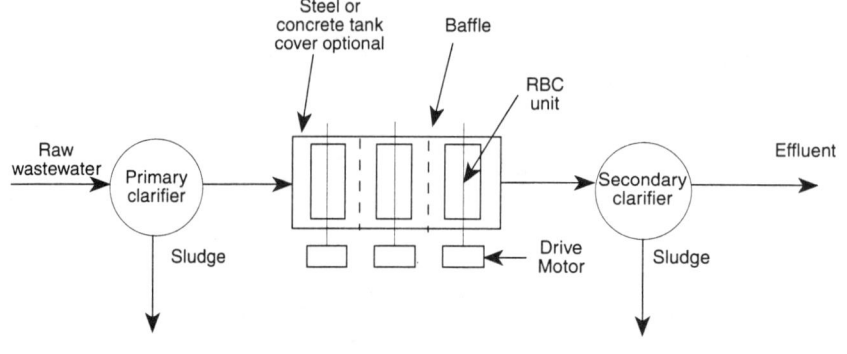

Table 7-21
Typical design information for RBCs

Item	Secondary	Combined nitrification	Separate nitrification
Hydraulic loading gal/ft^2-day	2.0-4.0	0.75-2.00	1.0-2.5
Organic loading *			
lb SBOD$_5$/1000 ft^2-day	0.75-2.0	0.5-1.5	0.1-0.3
lb TBOD$_5$/1000 ft^2-day	2.0-3.5	1.5-3.0	0.2-0.6
Maximum organic loading on the first stage			
lb SBOD$_5$/1000 ft^2-day	4-6	4-6	
lb TBOD$_5$/1000 ft^2-day	8-12	8-12	
NH$_3$ loading			
lb NH$_3$/1000 ft^2-day		0.15-0.30	0.2-0.4
Hydraulic residence time, hours	0.7-1.5	1.5-4.0	1.2-2.9
Effluent BOD$_5$, mg/L	15-30	7-15	7-15
Effluent NH$_3$, mg/L		<2	1-2

*SBOD$_5$ = soluble BOD$_5$, TBOD$_5$ = total BOD$_5$
Adapted from: Metcalf & Eddy, 1991.

Figure 7-6
Schematic representation of complete-mix activated sludge system

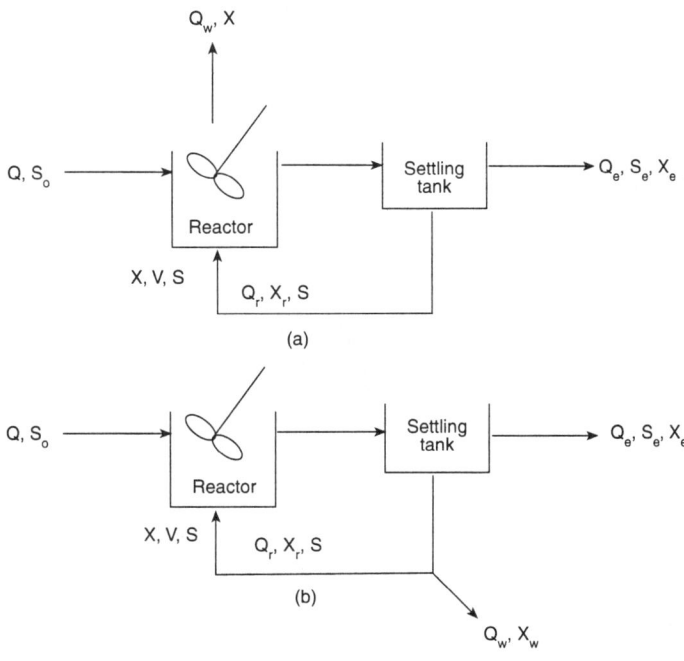

(BOD$_5$ or TKN) mass balance around the treatment facility yields the following equation:

Input - Output + Generation = Accumulation [7-44]

$$QS_o - Q_e S_e - Q_w S_w + Vr_s = 0 \qquad [7\text{-}45]$$

where:
- Q = influent wastewater flow rate
- S_o = substrate concentration in the influent to the activated sludge process
- Q_e = effluent flow rate
- S_e = effluent substrate concentration
- Q_w = waste sludge flow rate
- S_w = waste sludge substrate concentration
- V = aeration basin volume
- $-r_s$ = substrate removal rate

If the reaction in the clarifier is assumed to be negligible, the substrate concentration in the reactor (S) is equal to the concentration in the effluent (S_e) and the waste sludge (S_w). Furthermore, the sum of the effluent flow rate (Q_e) and the waste sludge flow rate (Q_w) is equal to the influent flow rate (Q). Thus:

$$q = \frac{-r_s}{X} = \frac{Q(S_o - S)}{VX} \qquad [7\text{-}46]$$

The aeration basin biomass concentration (X) may be determined by simultaneous solution of Equations 7-28, 7-37 and 7-46 to produce Equation 7-47:

$$\mu = Y_{max} q - k_e \qquad [7\text{-}28]$$

$$\mu = \frac{1}{SRT} \qquad [7\text{-}37]$$

$$X = \frac{Y_{max}}{1 + k_e\, SRT} \frac{Q}{V}(S_o - S)SRT \qquad [7\text{-}47]$$

For first-order kinetics, simultaneous solution of Equations 7-23, 7-28, and 7-37 yields a relationship (Equation 7-49) between the solids residence time and the effluent soluble substrate concentration:

$$q = k_1 S \qquad [7\text{-}23]$$

$$S = \frac{1 + k_e\, SRT}{Y_{max}\, k_1\, SRT} \qquad [7\text{-}48]$$

$$SRT = \frac{1}{S Y_{max}\, k_1 - k_e} \qquad [7\text{-}49]$$

For variable-order kinetics, simultaneous solution of Equations 7-24, 7-28, and 7-37 yields a similar relationship (Equation 7-51) between the solids residence time and the effluent soluble substrate concentration:

$$q = \frac{kS}{K_s + S} \qquad [7\text{-}24]$$

$$S = \frac{K_s(1 + k_e\, SRT)}{(Y_{max} k\, SRT) - (1 + k_e\, SRT)} \qquad [7\text{-}50]$$

$$SRT = \frac{K_s + S}{(Y_{max} k - k_e) S - k_e K_s} \qquad [7\text{-}51]$$

A steady-state biomass balance around the activated sludge treatment facility produces the following equations:

Input − Output + Generation = Accumulation [7-44]

$$QX_o - Q_eX_e - Q_wX_w + Vr_x = 0 \qquad [7\text{-}52]$$

$$0 - Q_eX_e - Q_wX_w + Vr_x = 0 \qquad [7\text{-}53]$$

$$\frac{r_x}{X} = \frac{Q_eX_e + Q_wX_w}{VX} = \frac{1}{\text{SRT}} \qquad [7\text{-}54]$$

where: Q = influent flow rate
X_o = influent biomass concentration (assumed negligible)
Q_e = effluent flow rate
X_e = effluent biomass concentration
Q_w = waste sludge flow rate
X_w = waste sludge biomass concentration
V = reactor volume
SRT = solids residence time (see Equation 7-17)

Equation 7-54 can be solved to define the necessary waste sludge flow rate:

$$Q_w = \frac{VX}{\text{SRT } X_w} - \frac{Q_eX_e}{X_w} \qquad [7\text{-}55]$$

For the special case where sludge is wasted directly from the aeration basin ($X_w = X$) and negligible effluent suspended solids ($X_e = 0$), determination of the necessary waste sludge flow rate is simplified because the SRT may be maintained at the desired value without consideration of the clarifier thickening performance (represented by X_w) or the reactor biomass concentration (X):

$$Q_w = \frac{V}{\text{SRT}} \qquad [7\text{-}56]$$

A steady-state biomass balance around the final clarifier provides a relationship between the recycle flow rate and underflow concentration for systems that practice sludge wasting from the clarifier underflow:

Input - Output + Generation = Accumulation [7-44]

$$(Q + Q_r)X - Q_eX_e - Q_wX_w - Q_rX_r + 0 = 0 \qquad [7\text{-}57]$$

where: Q = influent flow rate
Q_r = recycle flow rate
Q_e = effluent flow rate
X_e = effluent biomass concentration
Q_w = waste sludge flow rate
X_w = waste sludge biomass concentration
X_r = recycle (or underflow) biomass concentration

If the mass of biomass removed in the effluent and waste sludge streams is negligible in comparison to the mass of biomass in the recycled sludge, Equation 7-57 is simplified:

$$X_r = \frac{(Q+Q_r)X}{Q_r} \qquad [7\text{-}58]$$

or

$$Q_r = \frac{XQ}{X_r - X} \qquad [7\text{-}59]$$

General design criteria for selected activated sludge modifications for municipal wastewater applications are shown in Table 7-22. A minimum solids residence time of five days is normally selected to avoid adverse settling characteristics associated with high growth rates (Bisogni and Lawrence (1971)). Higher values of solids residence time (15 to 20 day) may be required to achieve nitrification, particularly at reduced wastewater temperatures.

Example 7-9 — Aeration basin design

For a community of 100,000, determine the size and number of aeration basins in a municipal activated sludge treatment facility (see Figure 7-1): The wastewater characteristics are:

Average per-capita flow rate = 100 gal/capita-day

Peak factor = 2.2

Wastewater concentration

BOD_5 = 220 mg / L
SS = 220 mg / L

Solution:

Average flow rate = (100 gal/capita-day)(100,000 pop)/(10^6 gal/Mgal) = 10 Mgal/day

The size of the aeration basin may be determined based on a typical value for hydraulic residence time of 6 hr for a conventional activated sludge process (see Table 7-22).

Table 7-22
Typical design guidelines for municipal activated sludge systems

Process	SRT (days)	Organic loading $\frac{\text{lb BOD}_5}{\text{lbVSS} - \text{day}}$	Organic loading $\frac{\text{lb BOD}_5}{10^3 \text{ft}^3 - \text{day}}$	MLSS (mg/L)	HRT (hr)	$\frac{Qr}{Q}$
Conventional	5-15	0.2-0.4	20-40	1,500-3,000	4-8	0.25-0.75
Complete-mix	5-15	0.2-0.6	50-120	2,500-4,000	3-5	0.25-1.00
Step-feed	5-15	0.2-0.4	40-60	2,000-3,500	3-5	0.25-0.75
Extended aeration	20-30	0.05-0.15	10-25	3,000-6,000	18-36	0.5-1.5
High-purity oxygen	3-10	0.25-1.00	100-200	2,000-5,000	1-3	0.25-0.50
Oxidation ditch	10-30	0.05-0.30	5-30	3,000-6,000	8-36	0.75-1.50
Single-stage nitrification	8-20	0.10-0.25	5-20	2,000-3,500	6-15	0.50-1.50
Separate-stage nitrification	15-100	0.05-0.20	3-9	2,000-3,500	3-6	0.50-2.00
Contact stabilization	5-15	0.2-0.6	60-75			0.50-1.50
Contact tank				1,000-3,000	0.5-1.0	
Stabilization				4,000-10,000	3-6	
High-rate aeration	5-10	0.4-1.5	100-1,000	4,000-10,000	2-4	1.0-5.0

Adapted from: Metcalf & Eddy, 1991.

$$\text{Volume} = \frac{(10 \text{ Mgal/day})(10^6 \text{ gal/Mgal})(6 \text{ hr})}{(24 \text{ hr/day})(7.48 \text{ gal/ft}^3)} = 334,200 \text{ ft}^3$$

Four parallel basins are recommended with dimensions of 15 ft (water depth), 30 ft (width), and 185 ft (length). An increased basin depth of several feet is recommended for freeboard.

Example 7-10 — Solids residence time-municipal wastewater

Recommend a value of the solids residence time (SRT) for secondary treatment of a municipal wastewater with the following kinetics:

$$m = Y_{max} q - k_e \quad q = k_1 S$$

$$Y_{max} = 0.6 \frac{\text{mg VSS}}{\text{mg BOD}_5} \quad k_e = 0.06 \text{ day}^{-1} \quad k_1 = 0.0294 \frac{L}{\text{mg VSS} - \text{day}}$$

Solution:

To achieve secondary treatment limits, the BOD_5 (S) must not exceed 30 mg/L (see Table 7-1). For first-order kinetics, the SRT is calculated with Equation 7-49:

$$\text{SRT} = \frac{1}{SY_{max} k_1 - k_e}$$

$$= \frac{1}{(30 \text{ mg BOD}_5/L)\left(0.6 \text{ mg}\frac{\text{VSS}}{\text{mg BOD}_5}\right)\left(0.0294 \frac{L}{\text{mg VSS} - \text{day}}\right) - (0.06 \text{ day}^{-1})}$$

$$= 2.13 \text{ day}$$

This SRT is below the minimum value recommended to produce acceptable settling properties (Bisogni and Lawrence (1971)). An increase in SRT to 5 days is recommended. The corresponding effluent BOD_5 concentration would be less than 30 mg/L, achieving compliance with the secondary treatment requirement.

The required solids residence time for industrial wastewater applications is highly variable because of the variability in kinetics (see Table 7-19). Because many industrial wastewaters exhibit lower degradation rates than domestic wastewaters, it may be necessary to operate industrial treatment facilities at much greater solids residence times than indicated in Table 7-22.

TIP FOR THE P.E. EXAMINATION CANDIDATE

Kinetic information will ordinarily be supplied in the examination in some form for industrial waste applications. Use of municipal kinetic values or empirical guidelines developed from municipal applications would be considered a serious flaw in the problem solution due to possible pronounced differences in wastewater characteristics.

Example 7-11 Solids residence time-industrial wastewater

Recommend a value of the solids residence time (SRT) for treatment of a petrochemical wastewater with the following kinetics:

$$m = Y_{max}\, q - k_e \quad q = k_1 S$$

$$Y_{max} = 0.6 \frac{\text{mg VSS}}{\text{mg BOD}_5} \quad k_e = 0.06\ \text{day}^{-1} \quad k_1 = 0.0062 \frac{\text{L}}{\text{mg VSS-day}}$$

The raw wastewater contains 600 mg/L of BOD_5. The process must achieve 95% removal of BOD_5.

Solution:

To achieve the stated effluent condition, the BOD_5 (S) must not exceed 30 mg/L:

$$S = (600\ \text{mg BOD}_5 / \text{L})(1 - 0.95) = 30\ \text{mg BOD}_5 / \text{L}$$

For first-order kinetics, the SRT is calculated with Equation 7-49:

$$SRT = \frac{1}{S\, Y_{max}\, k_1 - k_e}$$

$$= \frac{1}{(30\ \text{mg BOD}_5 / \text{L})(0.6\ \text{mg}\frac{\text{VSS}}{\text{mg BOD}_5})(0.0062\frac{\text{L}}{\text{mg VSS} - \text{day}}) - (0.06\ \text{day}^{-1})}$$

$$= 19.4\ \text{day}$$

Operation of the facility at a solids residence time greater than 20 days is recommended. Note the difference in design between this industrial wastewater application and municipal applications (see Example 7-10). Although both applications require an effluent BOD_5 of 30 mg/L, there is a significant difference in the SRT required because of the different kinetics.

7.8 Gas Transfer - Aeration

7.8.1 Estimation of Oxygen Supply Requirements

Aerobic biological processes require a supply of oxygen to support the desired biological transformations: oxidation of organics (BOD_5) and/or oxidation of ammonia (nitrification). Oxygen supply requirements for biological processes may be determined through analysis with specific kinetic parameters, consideration of specific reaction stoichiometry, or by estimation based on experience for a particular wastewater. Sufficient oxygen must be supplied to maintain aerobic conditions during peak organic loading periods. A minimum peak factor of two (relative to average conditions) has been recommended to account for diurnal loading effects in municipal applications (Metcalf & Eddy (1991)).

For those cases where kinetic parameters are known, the oxygen requirement associated with oxidation of *organics* may be determined with Equations 7-60 and 7-61 (Metcalf & Eddy (1991)):

$$O_2 = \text{Mass of BOD ultimate removed} - 1.42 \text{ Mass of cells wasted} \quad [7\text{-}60]$$

or

$$O_2 = \frac{Q(S_o - S)}{f} - 1.42 \frac{Y_{max}}{1 + \frac{k_e}{\mu}} Q(S_o - S) \quad [7\text{-}61]$$

where:
- O_2 = oxygen supply requirement;
- Q = wastewater flow rate
- S_o = substrate concentration in influent to biological process
- S = effluent substrate concentration
- f = factor to convert S to ultimate oxygen demand
- 1.42 = stoichiometry factor, mg oxygen per mg VSS, nitrification absent
- Y_{max} = maximum cell yield expressed in units of mg VSS per mg S
- k_e = endogenous decay coefficient
- μ = specific growth rate

If substrate is measured as chemical oxygen demand (COD), $f = 1$. If substrate is measured as BOD_5, f can be determined if the BOD reaction rate constant (k_1) is known (see Equation 7-8):

$$f = \frac{BOD_5}{BOD_{ult}} = 1 - \exp(-5k_1) \quad [7\text{-}62]$$

For municipal wastewaters, f is approximately 0.67. Generalization regarding f is difficult for industrial wastewaters. The specific growth rate equals the reciprocal of the solids residence time for activated sludge systems (Equation 7-63) and the reciprocal of the hydraulic residence time for aerated lagoons and other systems without solids recycle (Equation 7-64):

$$\mu = \frac{1}{SRT} \tag{7-63}$$

$$\mu = \frac{1}{HRT} \tag{7-64}$$

Oxygen supply requirements for *nitrification applications* can be estimated based on the reaction stoichiometry (Equation 7-65). Equation 7-61 must be modified to reflect the increased oxygen demand of the waste cell mass (1.98 mg oxygen per mg VSS) for nitrification applications. Total oxygen supply equals the sum of requirements for carbon oxidation (Equation 7-66) and nitrification (Equation 7-67):

$$NH_4^+ + 2O_2 \rightarrow 2H^+ + NO_3^- + H_2O \tag{7-65}$$

$$O_2 = \frac{Q(S_o - S)}{f} - 1.98 \frac{Y_{max}}{1 + \frac{k_e}{\mu}} Q(S_o - S) \tag{7-66}$$

$$O_2 = 4.57 Q(TKN_o - TKN) \tag{7-67}$$

where: O_2 = oxygen supply requirement
4.57 = stoichiometry factor, mg oxygen per mg TKN
Q = wastewater flow rate
TKN_o = TKN concentration in influent to biological process
TKN = effluent TKN concentration

In cases where specific kinetics are not known, Ten States Standards (1978) recommends design oxygen requirements of 1.1 lb of oxygen per lb of BOD_5 for activated sludge modifications other than extended aeration. A value of 1.8 lb of oxygen per lb of BOD_5 is specified for the extended aeration mode of operation. An additional oxygen quantity would be required to satisfy the nitrogenous demand according to Equation 7-67.

7.8.2 Mechanical Aeration Mechanical aeration equipment is rated under standard testing conditions using tap water at 20°C and a dissolved oxygen concentration of 0 mg/L. Actual oxygen transfer rates under field conditions are substantially less than standard rates, so adjustment for field conditions is necessary (Metcalf & Eddy (1991)):

$$N = N_o \left[\frac{\beta \, DO_{sat} - DO}{9.08} \right] 1.024^{(T-20)} \alpha \qquad [7\text{-}68]$$

where:
- N = oxygen transfer rate under field conditions
- N_o = oxygen transfer rate under standard test conditions
- β = ratio of dissolved oxygen saturation concentration in wastewater to concentration in tap water
- DO_{sat} = dissolved oxygen saturation concentration in tap water for the specified temperature and altitude (mg/L) see Table 7-4
- DO = operating dissolved oxygen concentration (mg/L)
- 9.08 = oxygen saturation for standard test conditions (mg/L)
- T = wastewater temperature (°C)
- α = ratio of oxygen mass transfer rate in wastewater to rate in tap water.

For municipal wastewaters, typical values for β and α are 0.95 and 0.82, respectively. A typical dissolved oxygen concentration for activated sludge operation would be 2.0 mg/L. Higher operating DO values may be warranted for nitrification applications. The field and standard transfer rates for various mechanical aeration devices are shown in Table 7-23. Mixing requirements range from 0.75 to 1.50 HP per 1,000 ft^3 of tank volume. Power requirements must be adequate to satisfy both oxygen supply and mixing criteria. Actual mixing requirements and transfer rates will depend on aerator spacing and tank geometry and should be confirmed for the specific equipment used.

Table 7-23
Oxygen transfer rates for mechanical aeration equipment

Aerator Type	Standard, N_0	Field, N
Surface low-speed	2.0 to 5.0	1.2 to 2.4
Surface low-speed with draft tube	2.0 to 4.6	1.2 to 2.1
Surface high-speed	2.0 to 3.6	1.2 to 2.0
Surface downdraft turbine	2.0 to 4.0	1.0 to 2.0
Submerged turbine with sparger	2.0 to 3.3	1.2 to 1.8
Submerged impeller	2.0 to 4.0	1.2 to 1.8
Surface brush and blade	1.5 to 3.6	0.8 to 1.8

Values reported in lb of oxygen per horsepower -hr.
Source: Metcalf & Eddy, 1991.

7.8.3 Diffused Aeration

Diffused aeration equipment may also be used to supply oxygen for biological treatment processes. The standard oxygen transfer efficiency for various devices is shown in Table 7-24. The reported values correspond with a diffuser depth of 15 ft; higher efficiencies would be expected with greater depth. Correction for Alpha, Beta, temperature, and operating dissolved oxygen concentration would be required, as discussed previously for mechanical aeration devices.

Power requirements for blowers, assuming adiabatic compression, may be determined with Equation 7-68 (Metcalf & Eddy (1991)):

$$\text{HP} = \frac{wRT_1}{550\, n\, e}\left[\left(\frac{p_2}{p_1}\right)^n - 1\right] \qquad [7\text{-}69]$$

where: HP = compressor/blower horsepower (HP)
 w = mass air flow rate (lb/sec)
 R = gas constant
  = 53.3
 T_1 = inlet air temperature (°R, note °R = °F + 460)
 p_2 = absolute pressure at blower outlet (psi)
 p_1 = absolute pressure at blower inlet (psi)
 n = 0.283 for air
 e = compressor/blower efficiency (0.70 to 0.90)

Mixing requirements for diffused aeration systems range from 10 to 30 standard ft³/min per 1,000 ft³ of tank volume. Air supply rates must be adequate to satisfy both oxygen supply and mixing requirements. Actual requirements for oxygen supply and mixing will depend on aerator spacing and tank geometry and should be confirmed for the specific equipment used.

The air flow rate (w) in Equation 7-68 must be adjusted to account for the oxygen content of the supply air and the field transfer efficiency of the aeration device:

$$w = \frac{O_2}{(X_{O_2})(\text{FTE})} \qquad [7\text{-}70]$$

where: O_2 = biological oxygen requirement (lb/sec)
 X_{O_2} = oxygen content in make-up air (mass fraction = 0.23)
 FTE = oxygen field transfer efficiency (decimal)

Table 7-24
Oxygen transfer efficiency under standard test conditions

Diffused Aeration Equipment

Diffuser type	Standard transfer efficiency (%)
Ceramic discs - grid	25-40
Ceramic domes - grid	27-39
Ceramic plate - grid	26-33
Rigid porous plastic tubes	
Grid	28-32
Dual spiral roll	17-28
Single spiral roll	13-25
Nonrigid porous plastic tubes	
Grid	26-36
Single spiral roll	19-37
Perforated membrane tubes	
Grid	22-29
Quarter points	19-24
Single spiral roll	15-19
Jet aeration, Side header	15-24
Nonporous diffusers	
Dual spiral roll	12-13
Mid-width	10-13
Single spiral roll	9-12

Source: Metcalf & Eddy, 1991.

Example 7-12 — Aeration equipment design

For a community of 100,000, determine the size of aeration equipment for a ceramic grid diffused aeration system for a municipal activated sludge treatment facility which provides secondary treatment (see Figure 7-1): The wastewater characteristics are:

    Average per-capita flow rate = 100 gal/capita-day
    Peak factor = 2.2
    Wastewater concentration
    BOD_5 = 220 mg / L
        SS = 220 mg / L

The dimensions of the aeration basin were determined in Example 7-9.

Solution:

Average flow rate = (100 gal/capita-day)(100,000 pop)/(10^6 gal/Mgal)

= 10 Mgal/day

Specification of aeration equipment requires determination of peak oxygen supply needs. Average oxygen supply requirements are reported to equal 1.1 lb per lb of BOD_5 removed in the activated sludge process (Ten States Standards (1978)). This value presumes that nitrification will not occur. The organic loading on the activated sludge process must reflect BOD removal in the primary clarifier, estimated to be 35% (see Table 7-9). The average oxygen requirement is therefore:

$$O_2 = (10\,\text{Mgal/day})(220\,\text{mg BOD}_5/\text{L})\left(8.34\frac{\text{lb/Mgal}}{\text{mg/L}}\right)(1-0.35)\left(1.1\frac{\text{lb O}_2}{\text{lb BOD}_5}\right)$$

= 13,120 lb/day

The aeration system must be sized to deliver the peak oxygen requirement, which may be estimated using the peak flow factor of 2.2:

O_2(peak design) = (13,120 lb/day)(2.2) = 28,900 lb/day

The blower horsepower is based on the air supply rate, which can be determined by adjusting the oxygen supply rate by the field transfer efficiency and the oxygen content of the air (Equation 7-70). The standard transfer efficiency of a ceramic grid, diffused air system is estimated to be 30% (see Table 7-24). Under field conditions, the transfer efficiency is estimated to be reduced to 15%. The oxygen content (mass fraction) of air is 0.23. The resulting air flow requirement is calculated below:

$$w = \frac{28,900\,\text{lb O}_2/\text{day}}{(86,400\,\text{sec/day})\left(0.23\frac{\text{lb O}_2}{\text{lb air}}\right)(0.15)} = 9.70\,\text{lb air/sec}$$

Consideration of mixing requirements is also necessary for specification of air supply rates. The density of air is 0.075 lb/standard ft^3 (Reynolds (1982)). The corresponding volumetric air supply rate is calculated below:

$$\text{air} = \frac{(9.70\,\text{lb/sec})(60\,\text{sec/min})}{0.075\,\text{lb/standard ft}^3} = 7,760\,\text{standard ft}^3/\text{min}$$

Mixing requirements are normally satisfied if the air supply rate exceeds 10 to 30 standard ft^3/min per 1,000 ft^3 of aeration basin volume (Metcalf & Eddy (1991)). The aeration basin volume was determined previously (Example 7-9). For this design, the air supply rate is 23 standard ft^3/min per 1,000 ft^3:

$$\text{air supply} = \frac{(7,760 \text{ standard ft}^3/\text{min})(1,000 \text{ ft}^3/1,000 \text{ ft}^3)}{(4 \text{ basins})(15 \text{ ft})(30 \text{ ft})(185 \text{ ft})}$$

$$= 23 \text{ standard ft}^3/\text{min}/1,000 \text{ ft}^3$$

Verification of requirements for mixing based on actual basin geometry and vendor recommendations would be necessary. For purposes of this design, a conservative approach is assumed, with an increase in air supply rate to achieve 30 standard ft³/min per 1,000 ft³:

$$w = \frac{(30 \text{ standard ft}^3/\text{min}/1,000 \text{ ft}^3)(4)(15 \text{ ft})(30 \text{ ft})(185 \text{ ft})(0.075 \text{ lb/standard ft}^3)}{(1,000 \text{ ft}^3/1,000 \text{ ft}^3)(60 \text{ sec/min})}$$

$$= 12.5 \text{ lbs air/sec}$$

The blower horsepower is determined with Equation 7-69. Use of this equation requires assumption of air temperature (assumed to be 100°F = 560°R), compressor efficiency (assumed to be 0.80), absolute pressure of blower output (assumed to exceed static pressure by 3 psi to allow for head loss in the piping system, filter, and diffuser), and absolute pressure at blower inlet (assumed to equal atmospheric pressure = 14.7 psi). The blower output pressure is calculated below:

$$p_2 = 14.7 \text{ psi} + 3 \text{ psi} + \frac{(15 \text{ ft})(62.4 \text{ lb/ft}^3)}{144 \text{ in}^2/\text{ft}^2} = 24.2 \text{ psi}$$

The total blower horsepower is calculated below with Equation 7-70:

$$\text{HP} = \frac{(12.5 \text{ lb/sec})(53.3)(560°R)}{(550)(0.283)(0.80)}\left[\left(\frac{24.2 \text{ psi}}{14.7 \text{ psi}}\right)^{0.283} - 1\right] = 454 \text{ HP}$$

Six units at 100 HP are recommended to provide flexibility to match air supply rates with experienced demand.

7.9 Filtration

Terminal filtration of wastewater is commonly practiced to achieve compliance with effluent suspended solids removal requirements that exceed secondary treatment limitations (30 mg/L). Various types of filter media have been used, including mono-medium (sand), dual media (anthracite and sand) and multi-media (anthracite, sand and garnet). Specifications for these filters are shown in Table 7-25.

Table 7-25
Effluent filtration

Characteristic	Range	Typical
Shallow-bed mono-medium sand filter		
Media depth, in	10-12	11
Effective size, mm	0.35-0.60	0.45
Uniformity coefficient	1.2-1.6	1.5
Hydraulic loading rate, gal/min-ft^2	2-6	3
Minimum backwash velocity, gal/min-ft^2	44-48	
Conventional mono-medium sand filter		
Media depth, in	20-30	24
Effective size, mm	0.4-0.8	0.65
Uniformity coefficient	1.2-1.6	1.5
Hydraulic loading rate, gal/min-ft^2	2-6	3
Minimum backwash velocity, gal/min-ft^2	44-48	
Deep-bed mono-medium sand filter		
Media depth, in	36-72	48
Effective size, mm	2-3	2.5
Uniformity coefficient	1.2-1.6	1.5
Hydraulic loading rate, gal/min-ft^2	2-10	5
Minimum backwash velocity, gal/min-ft^2	44-48	
Dual-media: anthracite and sand		
Anthracite media depth, in	12-30	24
Effective size, mm	0.8-2.0	1.3
Uniformity coefficient	1.3-1.8	1.6
Sand media depth, in	6-12	12
Effective size, mm	0.4-0.8	0.65
Uniformity coefficient	1.2-1.6	1.5
Hydraulic loading rate, gal/min-ft^2	2-10	5
Minimum backwash velocity, gal/min-ft^2	20-30	
Multi-media: anthracite, sand, garnet		
Anthracite media depth, in	8-20	16
Effective size, mm	1-2	1.4
Uniformity coefficient	1.4-1.8	1.6
Sand media depth, in	8-16	10
Effective size, mm	0.4-0.8	0.5
Uniformity coefficient	1.3-1.8	1.6
Garnet media depth, in	2-6	4
Effective size, mm	0.2-0.6	0.3
Uniformity coefficient	1.5-1.8	1.6
Hydraulic loading rate, gal/min-ft^2	2-10	5
Minimum backwash velocity, gal/min-ft^2	20-30	

Source: Metcalf & Eddy, 1991.

The filter area is determined with the hydraulic loading rate (HLR):

$$A = \frac{Q}{\text{HLR}} \qquad [7\text{-}71]$$

where: A = filter surface area
Q = flow rate to filter
HLR = hydraulic loading rate

Filter backwash water requirements are determined with the backwash velocity:

$$Q_{BW} = V_{BW} A \qquad [7\text{-}72]$$

where: Q_{BW} = filter backwash flow rate;
V_{BW} = backwash velocity.

7.10 Disinfection

The most common unit process to achieve disinfection is chlorination. Approximate chlorine doses are summarized for various applications in Table 7-26. Ten State Standards (1978) recommends a minimum contact period of 15 min at peak hourly flow. Contact basins should be designed to minimize short-circuiting and longitudinal dispersion. Reactors that approach plug-flow conditions are preferred. Actual requirements are often dictated by state regulatory agencies or discharge permits that may specify a range of acceptable chlorine residuals and hydraulic residence times.

Table 7-26
Chlorine doses for wastewater disinfection

Application	Dosage Range (mg/L)
Untreated wastewater (prechlorination)	6-25
Primary effluent	5-20
Chemical precipitation effluent	2-6
Trickling filter effluent	3-15
Activated sludge effluent	2-8

Source: Metcalf & Eddy, 1991.

Removal of ammonia may be achieved by breakpoint chlorination:

$$2NH_3 + 3HOCl \rightarrow N_2 + 3H_2O + 3HCl \qquad [7\text{-}73]$$

Dechlorination, if required, is often achieved by addition of sulfur dioxide:

$$SO_2 + HOCl + H_2O \rightarrow Cl^- + SO_4^{2-} + 3H^+ \qquad [7\text{-}74]$$

Alternative disinfection processes include treatment with alternative chlorine compounds (e.g., chlorine dioxide, sodium or calcium hypochlorite, or bromine chloride), ozonation, and UV radiation.

Example 7-13 Chlorination Basin Design

For a community of 100,000, determine the size of the chlorine contact basin in a municipal activated sludge treatment facility (see Figure 7-1): The wastewater characteristics are:

Average per-capita flow rate = 100 gal/capita-day

Peak factor = 2.2

Solution:

Average flow rate = (100 gal/capita-day)(100,000 pop)/(10^6 gal/Mgal)
= 10 Mgal/day

Peak hourly flow rate = (10 Mgal/day)(2.2) = 22 Mgal/day

The disinfection basin is sized to provide a specified hydraulic residence time at peak flow conditions (Equation 7-11). Ten States Standards (1978) recommends a design value of 15 min residence time.

$$\text{Volume} = \frac{(22 \text{ Mgal / day})(10^6 \text{ gal / Mgal})(15 \text{ min})}{(1,440 \text{ min / day})(7.48 \text{ gal / ft}^3)} = 30,600 \text{ ft}^3$$

Baffled basins (serpentine configuration) are desired to approach plug flow conditions. Two parallel basins are recommended with dimensions of 5 ft (depth), 10 ft (width), and 310 ft (channel length).

Example 7-14 Chlorine requirements

Determine the quantity of chlorine gas (lb/day) that must be supplied for breakpoint chlorination of municipal wastewater that contains 3 mg/L of ammonia-nitrogen. The wastewater flow rate is 1 Mgal/day. The relevant reactions are provided below:

$$Cl_2 + H_2O \rightarrow HOCl + HCl$$
$$2NH_3 + 3HOCl \rightarrow N_2 + 3H_2O + 3HCl$$

Solution:

Determine the net reaction to relate the quantity of Cl_2 and NH_3:

$$3[Cl_2 + H_2O = HOCl + HCl]$$
$$+[2NH_3 + 3HOCl = N_2 + 3H_2O + 3HCl]$$

combines to yield:

$$3Cl_2 + 2NH_3 \rightarrow N_2 + 6HCl$$
$$3(71) + 2(14) \rightarrow 2(14) + 6(36.5)$$
$$213 + 28 \rightarrow 28 + 219$$
$$7.61 + 1 \rightarrow 1 + 7.82$$

> Notice that NH_3-N translates to NH_3 reported as nitrogen, therefore, the molecular weight for NH_3 is 14 rather than 17.

Determine the mass of ammonia-nitrogen removed; this quantity becomes the basis for the stoichiometric calculation to determine the required chlorine dose:

$$\text{mass } NH_3 - N = (1 \text{ Mgal/day})(3 \text{ mg/L})\left(8.34 \frac{\text{lb/Mgal}}{\text{mg/L}}\right) = 25 \text{ lb } NH_3 - N/\text{day}$$

Determine the chlorine dose:

$$Cl_2 = 25 \text{ lb/day} \bullet 7.61 \frac{\text{lb } Cl_2}{\text{lb } NH_3 - N} = 190 \text{ lb } Cl_2/\text{day}$$

7.11 Waste Solids Management

7.11.1 Solids Characterization   It is necessary to define the solids content (% solids) and wet sludge specific gravity in order to relate the sludge dry weight and wet sludge volume:

$$VOL_S = \frac{(\text{Dry weight})100\%}{(\%\text{ solids})(\text{specific gravity})(\text{density of water})} \quad [7\text{-}75]$$

where: VOL_S = sludge wet volume
% solids = sludge weight % solids (%)
Dry weight = dry weight of the sludge (suspended solids)

For sludges with a density approximately equal to the density of water, the solids content (% solids) and the suspended solids concentration may be related by Equation 7-76:

$$SS = (\%\text{ solids})\left(10,000\,\frac{mg/L}{\%}\right) \quad [7\text{-}76]$$

where: SS = suspended solids concentration (mg/L)

Typical values for the sludge specific gravity and % solids are summarized in Tables 7-27 and 7-28, respectively. The specific gravity is normally very close to one; consequently, the wet sludge density may ordinarily be assumed equal to the den-

Table 7-27
Typical values for wet sludge specific gravity

Process	Specific gravity
Primary sedimentation	1.02
Waste activated sludge	1.005
Trickling filter sludge	1.025
Waste activated sludge (extended aeration)	1.015
Aerated lagoon	1.01
Filtration	1.005
Algae removal	1.005
Roughing filters	1.02
Suspended-growth denitrification	1.005
Primary sludge with chemical addition	
Low lime addition (350-500 mg/L)	1.04
High lime addition (800-1600 mg/L)	1.05

Source: Metcalf & Eddy, 1991.

Table 7-28
Expected sludge concentrations (% dry solids) for selected treatment processes

Process	Range	Typical
Primary sludge	4.0-10.0	5.0
Primary plus waste activated sludge	3.0-8.0	4.0
Primary plus trickling filter sludge	4.0-10.0	5.0
Primary with iron precipitation of phosphorus	0.5-3.0	2.0
Primary with lime precipitation of phosphorus	2.0-16.0	7.0
Scum	3.0-10.0	5.0
Waste activated sludge	0.5-2.5	0.8
High-purity oxygen activated sludge	1.3-4.0	2.0
Trickling filter sludge	1.0-3.0	1.5
Rotating biological contactor sludge	1.0-3.0	1.5

Source: Metcalf & Eddy, 1991.

sity of water without substantial error. Chemical precipitate sludges and dewatered sludges will exhibit the greatest specific gravity; these cases may warrant use of a specific gravity greater than one. Procedures for determining specific gravity as a function of volatile solids content, non-volatile solids content and moisture content are presented in Metcalf & Eddy (1991). Solids content varies as a function of the treatment process used and the type of sludge processed. Site-specific values should be used instead of these general values, when available.

7.11.2 Sludge Generation Rates Several methods are available for estimating sludge quantities:

- Gross estimates based on historical information related to wastewater flow rate.

- Gross estimates based on historical information related to population served.

- Mass balances based on site-specific flow and concentration data.

- Analysis based on site-specific kinetics and concentration data.

- Stoichiometry for defined reactions and site-specific concentration data.

Use of site-specific information is preferred, when available. Typical sludge quantities for municipal wastewaters, based on historical data, are summarized in Table 7-29.

Chapter 7: Wastewater Treatment

Mass balances may be completed for operations where inputs and removal efficiencies are defined. For example, the primary sludge quantity (dry weight) may be calculated for an application with known input flow and concentration, and suspended solids removal efficiency (see Table 7-9):

$$\text{Sludge} = Q\, X_o\, \text{RemEff} \qquad [7\text{-}77]$$

where: Sludge = generation rate, dry weight
Q = wastewater flow rate
X_o = influent suspended solids concentration
RemEff = suspended solids removal efficiency (decimal).

Secondary sludge quantities may be estimated with knowledge of the observed yield and the quantity of substrate removed. When the yield is expressed in mass of suspended solids per mass of substrate, a single calculation may be completed to define the total (volatile plus non-volatile) amount of sludge:

Table 7-29
Sludge quantities (lb dry solids/1,000 gal) from municipal wastewaters

Process	Range	Typical
Grit	0.03-0.86	<0.4
Scum	<0.01-0.14	0.04
Screenings	0.01-0.32	varies with screen size
Primary sedimentation	0.9-1.4	1.25
Activated sludge, following primary treatment	0.6-0.8	0.7
Extended aeration activated sludge, without primary sedimentation	0.7-1.0	0.8
Trickling filter	0.5-0.8	0.6
Aerated lagoon, without primary sedimentation	0.7-1.0	0.8
Filtration	0.1-0.2	0.15
Algae removal	0.1-0.2	0.15
Primary sedimentation with lime addition, depends on lime dose	2-11	4.5
Suspended-growth nitrification	negligible	
Suspended-growth denitrification	0.10-0.25	0.15

Source: Metcalf & Eddy, 1991 and USEPA, 1979.

$$\text{SSludge} = Q(S_o - S)\, Y_{obs} \qquad [7\text{-}78]$$

where:

- SSludge = secondary sludge generation rate, dry weight
- Q = wastewater flow rate
- S_o = substrate concentration in influent to secondary process
- S = effluent substrate concentration
- Y_{obs} = observed yield (units mass SS/mass S).

If the observed yield is defined in terms of volatile suspended solids, the secondary sludge must be determined (USEPA (1979)) as the sum of the volatile (Equation 7-79) and the non-volatile sludge (Equation 7-80):

$$\text{VolSSludge} = Q(S_o - S)\, Y_{obs} \qquad [7\text{-}79]$$

$$\text{NonVolSSludge} = Q\, XNV_o \qquad [7\text{-}80]$$

or

$$\text{SSludge} = \text{VolSSludge} + \text{NonVolSSludge} \qquad [7\text{-}81]$$

where:

- VolSSludge = volatile secondary sludge generation rate
- Q = wastewater flow rate
- S_o = substrate concentration in influent to secondary process
- S = effluent substrate concentration
- Y_{obs} = observed yield (units mass VSS/mass S)
- NonVolSSludge = non-volatile secondary sludge generation rate
- XNV_o = non-volatile suspended solids in influent to secondary process

Stoichiometric methods are reviewed in Chapter 3. The stoichiometric approach is best-suited for chemical precipitation processes where the precise reaction is defined.

Example 7-15 — Sludge production and oxygen requirements

Estimate sludge production rates (lb dry weight/day) and oxygen supply requirements (lb/day) for an industrial wastewater with the following characteristics:

Flow = 5 Mgal/day BOD_5 = 700 mg/L   TKN = 70 mg/L
SS = 300 mg/L   NVSS = 200 mg/L   VSS = 100 mg/L

Chapter 7: Wastewater Treatment

The kinetics were defined by laboratory testing as follows:

$$m = Y_{max} q - k_e \quad q = \frac{kS}{K_s + S} \quad k_e = 0.04 \text{ day}^{-1}$$

$$Y_{max} = 0.4 \frac{\text{mg VSS}}{\text{mg BOD}_5} \quad k = 3 \frac{\text{mg BOD}_5}{\text{mg VSS} - \text{day}} \quad K_s = 80 \text{ mg BOD}_5/L$$

The BOD rate constant k_e at 20°C was determined to be 0.1 day^{-1}.

A complete-mix activated sludge facility is operated at a solids residence time (SRT) of 15 days. The effluent BOD$_5$ and effluent TKN concentrations are negligible.

Solution:

Sludge production rates are defined using Equations 7-79 and 7-80 for the volatile and non-volatile fraction, respectively. The effluent BOD$_5$, TKN, SS concentrations are assumed to equal zero to obtain a conservative design approach.

$$\text{Volatile} = Q(S_o - S) Y_{obs} = Q(S_o - S) \frac{Y_{max}}{1 + k_e \text{ SRT}}$$

$$= (5 \text{ Mgal/day})(700 \text{ mg BOD}_5/L)\left(8.34 \frac{\text{lb/Mgal}}{\text{mg/L}}\right) \frac{0.4 \text{ mg VSS/mg BOD}_5}{1 + (0.04 \text{ day}^{-1})(15 \text{ day})}$$

Volatile sludge = 7,300 lb/day

$$\text{Non-volatile} = Q \ XNV_o = (5 \text{ Mgal/day})(200 \text{ mg NVSS/L})\left(8.34 \frac{\text{lb/Mgal}}{\text{mg/L}}\right)$$

Non-volatile sludge = 8,340 lb/day

The total sludge mass is 15,640 lb/day.

Oxygen supply requirements are determined with Equations 7-66 and 7-67 for the carbonaceous and nitrogenous demand, respectively. The ratio of 5-day to ultimate BOD is calculated with Equation 7-62 for the stated value of the BOD reaction rate constant.

$$f = 1 - \exp[-(5 \text{ days})(0.1 \text{ day}^{-1})] = 0.39$$

Note that "f" is substantially less for this industrial wastewater than would be typical of municipal wastewaters.

Carbonaceous $O_2 = \dfrac{Q(S_o - S)}{f} - 1.98$ Volatile Sludge

$$= \dfrac{(5 \text{ Mgal/day})(700 \text{ mg BOD}_5/\text{L})\left(8.34 \dfrac{\text{lb/Mgal}}{\text{mg/L}}\right)}{0.39 \dfrac{\text{lb BOD}_5}{O_2}} - \left(1.98 \dfrac{\text{lb } O_2}{\text{lb VSS}}\right)(7{,}300 \text{ lb VSS/day})$$

Carbonaceous $O_2 = 60{,}400$ lb O_2 / day

Nitrogenous $O_2 = 4.57 \, Q(\text{TKN}_o - \text{TKN})$

Nitrogenous $O_2 = \left(4.57 \dfrac{\text{lb } O_2}{\text{lb TKN}}\right)(5 \text{ Mgal/day})(70 \text{ mg TKN/L})\left(8.34 \dfrac{\text{lb/Mgal}}{\text{mg/L}}\right)$

Nitrogenous $O_2 = 13{,}300$ lb O_2 / day

The total oxygen demand is 73,700 lb/day.

7.11.3 Sludge Pumping Design of facilities for pumping sludge must take into account the rheological properties of the sludge. Head loss relationships may be evaluated based on specific measurements of sludge viscosity (USEPA (1979)) (Metcalf & Eddy (1991)); however, discussion of this approach is beyond the scope of this text. A simplified method considers the head loss of sludge in ratio to the head loss for water. For laminar flow, this ratio increases as the solids content of the sludge increases. A distinct relationship has been defined for digested and undigested sludges as illustrated in Figure 7-7.

To minimize clogging problems, velocities of 4 to 6 ft/sec and a minimum pipe diameter of 6 in. are warranted. This combination of velocity and pipe size results in a relatively large volumetric flow rate. It is often necessary to operate pumps intermittently to maintain minimum velocities. Details on pump selection may be found in USEPA (1979) and Metcalf & Eddy (1991).

Example 7-16 Return activated sludge pumping

For the activated sludge application described in Examples 7-7 and 7-9, determine the requirements for return sludge pumping.

Solution:

The return activated sludge pumping capacity must be adequate to provide flexibility for operation during elevated flow conditions and/or adverse settling episodes.

As noted in Table 7-22 normal operating values for recycle flow rate range from 25% to 75% of influent flow rate for the conventional activated sludge process. The design capacity for these pumps should be sufficient to operate at greater recycle flow rates; a capacity of 100% is recommended for design. Five pumps are recommended, each with a rated capacity of 2.5 Mgal/day. This configuration allows dedication of one pump to each secondary clarifier, with one stand-by pump. Pumps should have variable-speed capability to permit operational flexibility.

Figure 7-7
Relative head loss for sludge pumping applications

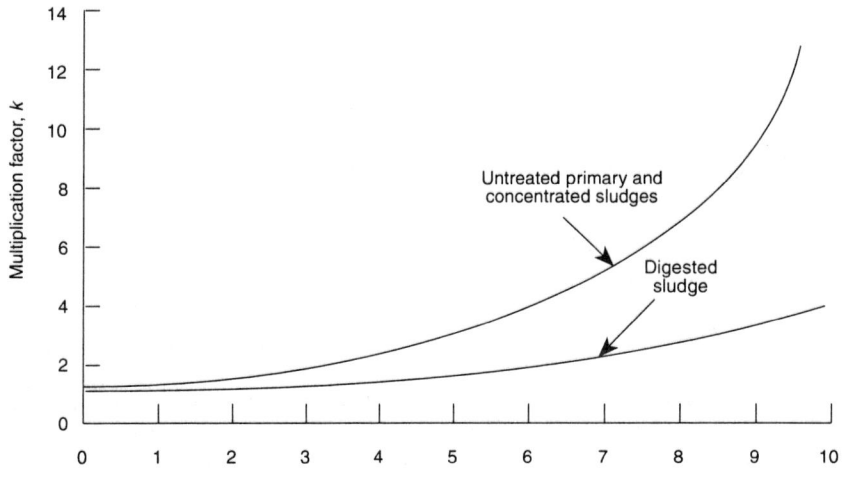

Adapted from: Metcalf & Eddy, 1991.

7.11.4 Stabilization *Aerobic sludge digestion* is commonly used for stabilization of secondary sludges. The economics of this approach are often attractive for small treatment facilities without primary clarification. Most digesters are operated as complete-mix reactors without solids recycle, thus performance would be expected to depend on the hydraulic residence time, as developed in Equation 7-34. The process stoichiometry for digestion of cellular material typical of a secondary sludge is provided in Equation 7-82:

$$C_5H_7O_2N + 7O_2 \rightarrow 5CO_2 + NO_3^- + 3H_2O + H^+ \qquad [7\text{-}82]$$

Theoretically, the reaction requires two pounds of oxygen per pound of volatile suspended solids removed. The oxidation also consumes alkalinity, so a drop in pH may be experienced if the buffering capacity of the sludge is insufficient. Typical design criteria for aerobic sludge digestion are shown in Table 7-30. A hydraulic residence time of 10 to 20 days is recommended based on the type of sludge. Longer times are needed for temperatures below 20°C or for applications where significant reduction of pathogens is required to permit land disposal of sludge.

Table 7-30
Design criteria for aerobic digesters

Parameter	Value
Hydraulic residence time, days, for 20°C	
Waste activated sludge only	10-15
Activated sludge without primary sedimentation	12-18
Primary plus waste activated or trickling filter	15-20
Oxygen requirements, lb O_2/lb organics	
Secondary sludges, ratio to VSS destroyed	2.3
Primary sludge, ratio to BOD_5 in sludge	1.6-1.9
Organic loading, lb VSS per ft^3-day	0.1-0.3
Reduction in volatile suspended solids, %	40-50

Source: Metcalf & Eddy, 1991.

The complex microbiology associated with *anaerobic processes* is reviewed in McCarty (1964). Anaerobic processes are vulnerable to upsets associated with variation in pH, temperature, and a large number of toxic agents. Details regarding reaction stoichiometry, kinetics, and toxicity are provided in McCarty (1964).

Anaerobic digesters may be either high- or low-rate systems. High-rate digesters have continuous mixing and either continuous or intermittent sludge introduction and withdrawal. Low-rate systems have intermittent mixing, feed, and withdrawal. The anaerobic digestion process is particularly well-suited for primary sludges, although all secondary sludges may also be stabilized with anaerobic digestion. Temperature control is usually practiced and may be achieved by combustion of the methane gas generated by the anaerobic process. The digesters are normally operated as complete-mix reactors without recycle; the design is related to the hydraulic residence time. Design parameters and operating characteristics are summarized in Tables 7-31 through 7-33. An increase in the hydraulic residence time would be warranted for lower temperatures or to achieve greater reduction in pathogens.

Table 7-31
Typical design criteria for anaerobic digestion

Parameter	Low-rate	High-rate
Hydraulic residence time, days	30-60	10-20
Organic loading, lb VSS/ft^3-day	0.04-0.10	0.15-0.40
Volumetric requirements, ft^3/capita		
Primary sludge only	2-3	1.5-2.0
Primary plus trickling filter	4-5	2.5-3.5
Primary plus waste activated	4-6	2.5-4.0

Source: Reynolds, 1982.

Table 7-32
Typical operating conditions for mesophilic anaerobic digestion

Parameter	Range
Temperature, °C	
Optimum	35
Usual operating range	29-35
pH	
Optimum	7.0-7.2
Usual operating range	6.7-7.4
Suspended solids reduction, %	
Volatile	50-75
Total	35-50
Gas production, per pound of VSS destroyed	
ft^3 per pound	12-18
lb per pound	1.05-1.75
Gas fuel value, BTU/ft^3	530-730
Gas composition, %	
Methane	55-75
Carbon dioxide	25-45
Hydrogen sulfide	trace
Hydrogen	trace
Nitrogen	trace

Source: Reynolds, 1982.

Table 7-33
Recommended hydraulic residence time for high-rate anaerobic digestion

Temperature (°C)	Hydraulic residence time (day)
18	28
24	20
30	14
35	10
40	10

Source: Metcalf & Eddy, 1991.

7.11.5 Conditioning Chemicals are often added during thickening or dewatering operations to improve the solids-liquid separation. This conditioning can improve the % solids output, increase allowable solids loading on equipment, and/or increased solids capture efficiency. Proper conditioning may also improve the dewatered sludge cake release characteristics. Failure to achieve high solids-capture efficiencies will result in the return of large quantities of suspended solids in the filtrate, supernatant, or centrate side-streams generated during thickening or dewatering operations. This solids loading may impair sedimentation and terminal filtration operations during wastewater processing. The solids-capture efficiency is defined as the fraction of input suspended solids that is recovered in the thickened or dewatered sludge:

$$\text{SCE} = \frac{Q_{out} X_{out}}{Q_{in} X_{in}} \bullet 100\% \quad\quad [7\text{-}83]$$

where: SCE = solids capture efficiency (%)
Q_{out} = output sludge flow rate
X_{out} = output sludge solids concentration
Q_{in} = input sludge flow rate
X_{in} = input sludge solids concentration.

Organic polymers are favored for conditioning biological sludges in most thickening or dewatering operations. Typical doses are indicated in Tables 7-34 and 7-35. Inorganic chemicals (lime and ferric chloride) are often used in conjunction with vacuum filtration and plate and frame filter presses. Sand bed dewatering and gravity thickening are normally conducted without chemical conditioning (USEPA (1979)).

Table 7-34
Polymer conditioning dose (lb of dry polymer per ton of dry sludge) for thickening

Sludge type	Dissolved air flotation	Solid bowl centrifuge	Basket centrifuge	Gravity belt filter
Waste Activated	4-10	0-8	2-6	6-14

Source: Metcalf & Eddy, 1991.

Table 7-35
Polymer conditioning dose (lb of dry polymer per ton of dry sludge) for dewatering

Type of sludge	Vacuum filter	Belt filter press	Solid bowl centrifuge
Primary	2-10	2-8	1-5
Primary + waste activated	10-20	4-16	4-10
Primary + trickling filter	2.5-5.0	4-16	
Waste activated	15-30	8-20	10-16
Anaerobically digested primary	7-14	4-10	6-10
Anaerobically digested primary + waste activated	3-17	3-17	4-10
Aerobically digested primary + waste activated	15-20	4-16	

Source: Metcalf & Eddy, 1991.

7.11.6 Thickening Thickening during sludge processing reduces the volume of sludge for subsequent processing. Thickening is often performed prior to digestion to reduce the volume requirements (and capital cost) of digestion facilities. Primary sludges (and mixtures of primary-plus-secondary sludges) are commonly thickened by gravity sedimentation. The design of such thickeners is based on the solids loading rate:

$$\text{SLR} = \frac{Q_s X_s}{A} \qquad [7\text{-}84]$$

where: SLR = solids loading rate
Q_s = unthickened sludge flow rate
X_s = unthickened sludge suspended solids concentration
A = thickener surface area

Typical design and performance data for gravity thickeners are reported in Table 7-36. Systems that operate at higher solids-loading rates would be expected to produce lower concentrations of thickened sludge.

Table 7-36
Design and operational criteria for gravity thickening

Type of sludge	Unthickened sludge (% solids)	Thickened sludge (% solids)	Solids-loading rate (lb/ft^2-day)
Primary	2-7	5-10	18-28
Trickling filter	1-4	3-6	7-10
Rotating biological contactor (RBC)	1.0-3.5	2-5	7-10
Air activated sludge	0.5-1.5	2-3	2.5-7.0
High-purity oxygen activated sludge	0.5-1.5	2-3	2.5-7.0
Extended aeration activated sludge	0.2-1.0	2-3	5-7
Anaerobically digested primary sludge	8	12	25
Primary + trickling filter	2-6	4-9	12-20
Primary + RBC	2-6	4-8	10-16
Primary + waste activated sludge	2-5	2-8	8-16
Trickling filter + waste activated sludge	0.5-2.5	2-4	2.5-7.0
Anaerobically digested primary + waste activated	4	8	14

Source: Metcalf & Eddy, 1991.

Example 7-17 — Gravity thickener

Determine the size of gravity thickeners for the municipal treatment facility (secondary treatment) described in Figure 7-1. Raw wastewater characteristics were described in Example 7-9.

Solution:

The design of gravity thickeners is based on the solids-loading rate (Equation 7-84). Determination of the mass of solids input to the thickener is necessary to use this equation.

The quantity of primary sludge is determined with knowledge of the raw wastewater characteristics (220 mg/L of suspended solids) and assumption of the suspended solids removal efficiency of the primary clarifiers (assumed to equal 60%).

$$\text{Primary sludge} = (10 \text{ Mgal/day})(220 \text{ mg/L})(0.60)\left(8.34 \frac{\text{lb/Mgal}}{\text{mg/L}}\right)$$

$$= 11,000 \text{ lb/day}$$

The mass of secondary sludge (waste activated sludge) is determined with a manipulated form of Equation 7-55:

$$\text{Waste activated sludge (WAS)} = Q_w X_w = \frac{VX}{\text{SRT}} - Q_e X_e$$

The solids residence time is assumed to equal 5 days for secondary treatment applications (see Table 7-22). The aeration basin dimensions were determined previously (Example 7-9), and the MLSS is assumed to be 2,500 mg/L (see Table 7-22). If the mass of effluent suspended solids is assumed to be negligible in comparison to the mass of secondary sludge, the quantity of waste activated sludge is calculated as follows:

$$\text{WAS} = \frac{4(15 \text{ ft})(30 \text{ ft})(185 \text{ ft})(7.48 \text{ gal/ft}^3)(2,500 \text{ mg/L})\left(8.34 \frac{\text{lb/Mgal}}{\text{mg/L}}\right)}{(5 \text{ day})(10^6 \text{ gal/Mgal})}$$

$$= 10,400 \text{ lb/day}$$

The total input sludge to the thickener is the sum of the primary and secondary sludge:

Input = 11,000 + 10,400 = 21,400 lb/day

For a mixture of primary and waste activated sludge, a solids-loading rate of 12 lb/ft²-day is recommended (see Table 7-36). The necessary area is determined with Equation 7-84:

$$\text{Area} = \frac{\text{Mass Input Sludge}}{\text{Solids Loading Rate}} = \frac{21,400 \text{ lb/day}}{12 \text{ lb/ft}^2 - \text{day}} = 1,783 \text{ ft}^2$$

Two circular units with a diameter of 35 ft are recommended.

Example 7-18 — Anaerobic Digester

Determine the required volume of a high-rate anaerobic digester for the thickened sludge (primary and waste activated) application in Example 7-17.

Solution:

The volume of high-rate anaerobic digesters is determined by specification of the hydraulic residence time. Designs are typically based on a criteria of 20 days (see Table 7-33). Calculating the volume requires determination of the volumetric flow rate of thickened sludge input to the digesters. Equation 7-76 is used to relate the sludge dry weight and wet volume. The thickened sludge solids content is assumed to be 4% for a mixture of primary and waste activated sludge (see Table 7-28). The specific gravity of the wet sludge is assumed to equal 1.01 (see Table 7-27). The volumetric thickened sludge flow rate is calculated with Equation 7-75:

$$\text{Volume / day} = \frac{(\text{Dry weight / day})(100\%)}{(\%\ \text{solids})(\text{specific gravity})(\text{density of water})}$$

$$\text{Flow rate} = \frac{(21,400\ \text{lb / day})(100\%)}{(4\%)(1.01)(62.4\ \text{lb / ft}^3)} = 8,500\ \text{ft}^3\ /\ \text{day}$$

The digester volume is determined with Equation 7-11:

$$\text{Volume} = (8,500\ \text{ft}^3\ /\ \text{day})(20\ \text{days}) = 170,000\ \text{ft}^3$$

Other thickening methods are commonly used for secondary sludges, especially for waste activated sludge. Typical performances for these alternate thickening methods are compared in Table 7-37 for waste activated sludge. Gravity thickening methods do not generally achieve the same degree of volume reduction as alternative methods for waste activated sludge.

7.11.7 Dewatering Sludge dewatering is performed to reduce the volume of sludge prior to transportation for disposal or to reduce the water content prior to incineration or composting. Additionally, land disposal regulations may require a specific solids content before a sludge can be landfilled. Several methods are commonly used, including sand drying beds, centrifugal devices, and filtration systems. Performance data for selected mechanical dewatering processes are shown in Table 7-38. Dewatered sludges exhibit the characteristics of a solid (as opposed to a liquid).

Table 7-37
Typical thickener performance for waste activated sludge

Thickening process	Thickened sludge (% solids)
Gravity	2-3
Dissolved air flotation	3-6
Imperforate basket centrifuge	8-10
Solid bowl centrifuge	4-6
Gravity belt thickener	3-6
Rotary drum thickener	5-9

Source: Metcalf & Eddy, 1991.

Table 7-38
Typical dewatering performance for various sludge types (% solids)

Type of sludge	Vacuum filter	Solid bowl centrifuge	Basket centrifuge	Belt filter press
Primary	27-35	25-35	25-30	28-44
Primary + waste activated	18-30	12-20	12-14	20-35
Primary + trickling filter	20-30	20-25	7-11	20-35
Waste activated	13-25	5-20	8-14	12-20
Anaerobically digested:				25-35
Primary	25-35	25-35	8-14	20-25
Primary + waste activated	18-25	15-20		
Primary + trickling filter	20-27	18-25		
Waste activated				12-20
Aerobically digested waste activated		8-10		12-20

Source: Metcalf & Eddy, 1991.

Example 7-19 — Vacuum Filter

Determine conditioning requirements and vacuum filter area requirements for dewatering the digested sludge in Example 7-18. The filter solids loading rate is 4 lb/ft^2-hr.

Solution:

The solids loading rate for the filter is specified at 4 lb/ft^2-hr for this mixture of digested primary and waste activated sludge. Polymer conditioning is recommended prior to dewatering; the polymer dose is expressed as a fraction of the feed dry sludge mass. A dose of 10 lb of polymer per dry ton of sludge is recommended (see Table 7-35). The mass of sludge remaining after digestion is calculated based on an assumed solids destruction efficiency of 40% during digestion (see Table 7-32):

$$\text{Digested sludge mass} = (21,400 \text{ lb / day})(1 - 0.40) = 12,840 \text{ lb / day}$$

The required polymer dose would be 64 lb/day:

$$\text{Polymer dose} = \frac{(12,840 \text{ lb / day})(10 \text{ lb polymer / ton})}{2,000 \text{ lbs / ton}} = 64 \text{ lb / day}$$

The total mass of solids input to the vacuum filter is 12,900 lb/day.

It is customary to operate mechanical dewatering equipment on an intermittent basis, except for very large facilities where a complete staff is maintained on a 24-hr basis. It is assumed that the filters will be operated only during the lead shift, with an expected weekly operation of 32 hr/unit. Two units are recommended. The necessary size of each unit is based on the allowable solids-loading rate (Equation 7-16):

$$\text{Area} = \frac{(12,900 \text{ lb / day})(7 \text{ day / wk})}{(2 \text{ units})(32 \text{ hr / wk})(4 \text{ lb / ft}^2 \text{ - hr})} = 353 \text{ ft}^2$$

Two units with a width of 12 ft and a diameter of 10 ft are recommended.

Sand drying beds are very popular in small communities where large amounts of land are available. Water removal is accomplished first by gravity drainage and compression followed by evaporation. Sand drying beds are also attractive for aerobically digested waste activated sludge that does not dewater as readily by mechanical means as other sludges. The performance of drying beds is sensitive to climatic factors (temperature, humidity, rainfall) and to the amount of time the sludge is allowed to remain on the bed. The dewatered sludge may range from 15 to 40 % solids (Metcalf & Eddy (1991)).

7.11.8 Disposal Federal regulations for the disposal of sewage sludge were issued in 1992 (40 CFR Part 503). The regulations establish requirements for allowable metal content, pathogen reduction, and vector attractiveness reduction prior to the application of sludges to agricultural land.

7.12 Physical Chemical Treatment

Treatment of industrial wastewaters often requires the use of physical and/or chemical methods as an alternative to biological treatment. A brief discussion of selected physical and chemical treatment processes is provided to indicate their potential applications. Further information may be obtained from the literature (e.g., Montgomery (1985); Weber (1972); and WPCF (1990)).

7.12.1 Oxidation There is increasing interest in oxidative processes for treating synthetic organic compounds. Ozonation, particularly in combination with ultra-violet radiation, has been demonstrated to effectively destroy a variety of organic compounds. Other oxidants used for this purpose include hydrogen peroxide, potassium permanganate, chlorine and chlorine dioxide (WPCF (1990)).

7.12.2 Air Stripping Removal of volatile contaminants from aqueous media may be achieved by air stripping processes, which transfer contaminants from the liquid to a gas medium. The feasibility of using these processes may be evaluated by considering Henry's constant for the particular contaminant. Air stripping has been used to remove many halogenated solvents and hydrocarbons (WPCF (1990)).

7.12.3 Ion Exchange Ion exchange processes have been used to treat potable water (for softening), industrial process water (for demineralization), and industrial wastewater (for material recovery). An extensive history of ion exchange treatment exists in the metal finishing industry, where its uses include recovery of precious metals (gold, silver), recovery of chromic acid, purification of acids for reuse, recovery of common metals, and recovery of a high-quality (low total dissolved solids) rinsewater source from wastewater.

7.12.4 Neutralization Adjustment of the pH of industrial wastewaters is often required to comply with discharge standards. The chemical principles in Chapter 3 may be used to determine the appropriate chemical requirements. The characteristics of chemicals commonly used for pH adjustment are reviewed in Table 7-39.

7.12.5 Precipitation Chemical precipitation may be used to remove a wide range of inorganic compounds. Lime-soda softening to remove calcium and magnesium is routinely employed in potable water treatment. Iron and manganese may also be removed by oxidation/precipitation in groundwater treatment. Waste-

Table 7-39
Compounds used for pH adjustment

Chemical	Purity	Relative Cost [1]
For reduction of pH		
Sulfuric acid, H_2SO_4	93%	1.00
Hydrochloric acid, HCl	32%	1.63
Carbon dioxide, CO_2		
For elevation of pH		
Limestone, $CaCO_3$	93%	0.17
Quicklime, CaO	90%	0.58
Hydrated lime, $Ca(OH)_2$	93%	0.71
Soda ash, Na_2CO_3	50%	4.10
Caustic soda, NaOH	50% liquid	2.46
Ammonia, NH_3	liquid	1.17

1) Costs per equivalent expressed relative to sulfuric acid.
Source: WPCF, 1990

waters from point sources in the metal finishing industry would be expected to contain various metals (copper, lead, silver, zinc, cadmium, nickel, chromium) that can be removed by hydroxide or sulfide precipitation. Chemical precipitation processes generate a sludge that must receive further processing. These sludges may be a listed hazardous material.

7.12.6 Carbon Adsorption Granular activated carbon may be used to remove organic and organo-metallic compounds. Adsorption is often used to treat wastewaters that contain a variety of organic contaminants since the process is relatively non-selective, as evidenced by the range of compounds listed in Table 7-40 that can be adsorbed. Extensive data have been published in Montgomery (1985) regarding the adsorption characteristics of specific organic solutes. Economic feasibility must be evaluated for specific applications through pilot- and/or bench-scale testing.

7.12.7 Membrane Processes Membrane processes may be used to separate dissolved species based (roughly) on molecular size. Membrane products with a range of pore sizes are available, providing a corresponding selectivity for removal of specific solutes (WPCF (1990)). Reverse osmosis has the smallest particle size cutoff (an approximate molecular weight cutoff of 100 to 200), while dialysis and ultrafiltration operate with a larger particle size cutoff (a molecular weight cutoff range up to 100,000). Membrane processes may be used to produce a water of high purity (low dissolved solids) or for recovery of valuable solutes from wastewaters. Operation of membrane systems generates two output streams (permeate and reject); recovered material may be concentrated in the reject stream in certain applications.

Table 7-40
Examples of compounds adsorbed on carbon

Organic chemical class	Examples
Aromatic hydrocarbons	Benzene, toluene, xylene
Polynuclear aromatics	Naphthalene, anthracenes, biphenyls
Chlorinated aromatics	Chlorobenzene, polychlorinated biphenyls, Aldrin, endrin, toxaphene, ddt
Phenolics	Phenol, cresol, resorcinol
Aliphatic hydrocarbons	Gasoline, kerosene
Chlorinated aliphatic hydrocarbons	1,1,1-trichloroethane, trichloroethylene, carbon tetrachloride, perchloroethylene
Aliphatic acids and aromatic acids	Tar acids, benzoic acid
Aliphatic amines and aromatic amines	Aniline, toluene diamine
Ketones, esters, ethers, and alcohols	Hydroquinone, polyethylene glycol
Surfactants	Alkyl benzene sulfonates
Soluble organic dyes	Methylene blue, indigo carmine

Source: WPCF, 1990.

7.13 Standard Wastewater Treatment Notations

Symbol	Definition	Units
A	Area	L^2
Alpha	Ratio of oxygen mass transfer rate in wastewater to rate in tap water	none
Beta	Ratio of dissolved oxygen saturation concentration in wastewater to concentration in tap water	none
BOD_5	Five-day biological oxygen demand	m/L^3
BOD_{ult}	Ultimate biological oxygen demand	m/L^3
D	Depth	L
D	Dissolved oxygen deficit in mixing zone at the point of discharge	m/L^3
Do	Dissolved oxygen deficit in mixing zone at the point of discharge	m/L^3
D_{max}	Maximum dissolved oxygen deficit	m/L^3
DO	Dissolved oxygen concentration	m/L^3
DO_{sat}	Dissolved oxygen concentration at saturation	m/L^3
E_s	BOD removal efficiency (%)	none
e	Compressor/blower efficiency	none
F	Recycle factor for NRC trickling filter equations	none
f	Factor to convert S to ultimate oxygen demand	none
FTE	Oxygen field transfer efficiency (decimal)	none
HLR	Hydraulic loading rate	$L^3/L^2\text{-}T$
HRT	Hydraulic residence time	T
HP	Horsepower	FL/T

Symbol	Description	Units
K_I	Inhibition coefficient	m/L^3
K_s	Half-saturation constant	m/L^3
k	Maximum specific substrate utilization rate	$m/m\text{-}T$
k_e	Endogenous decay cofficient	T^{-1}
K_0	Zero-order reaction rate constant	$m/m\text{-}T$
k_1	First-order reaction rate constant	$L^3/m\text{-}T$
k_1	Deoxygenation cofficient (base e)	T^{-1}
k_1	BOD rate constant (base e)	T^{-1}
k_2	Reaeration coefficient (base e)	T^{-1}
L	Length	L
L_o	Ultimate BOD in mixing zone at the point of discharge	m/L^3
N	Oxygen transfer rate under field conditions	$m/F\text{-}L$
N_o	Oxygen transfer rate under standard test conditions	$m/F\text{-}L$
O_2	Oxygen requirement	m/T
ORA	Overflow rate	$L^3/L^2\text{-}T$
p_1	Absolute pressure at lower inlet	F/L^2
p_2	Absolute pressure at blower outlet	F/L^2
Q	Volumetric flow rate	L^2/T
Q_L	Hydraulic loading rate	$L^3/L^2\text{-}T$
q	Specific substrate utilization rate	$m/m\text{-}T$
R	Universal gas law constant	none
R	Recycle ratio	none
$-r_s$	Substrate removal rate	$m/L^3\text{-}T$
r_x	Biomass production rate	$m/L^3\text{-}T$
S	Substrate concentration	m/L^3
SCE	Solids capture efficiency (%)	none
SLR	Solids loading rate	$m/L^2\text{-}T$
SRT	Solids residence time	T
SS	Suspended solids concentration	m/L^3
T	Temperature	
t	Time	T
TKN	Total Kjeldahl Nitrogen concentration	m/L^3
t_{max}	Time to achieve the maximum deficit	T
m	Specific growth rate	T^{-1}
V	Reactor volume	L^3
VOL	Volume	L^3
w	Mass air flow rate	m/T
WLR	Weir loading rate	$L^3/L\text{-}T$
X	Biomass concentration	m/L^3
X	Suspended solids concentration	m/L^3
X_{O_2}	Mass fraction of oxygen in make-up air (decimal)	none
y_o	Organic loading	m/T
Y_{max}	Maximum yield	m/m
Y_{obs}	Observed yield	m/m

7.14 References

American Society of Civil Engineers and Water Pollution Control Federation. 1969. *Design and Construction of Sanitary and Storm Sewers*. New York and Washington, DC: ASCE/WPCF.

Bisogni, J.J., and A.W. Lawrence. 1971. Relationships Between Biological Solids Retention Time and Settling Characteristics of Activated Sludge. *Water Research*. 5:753.

Grady, C.P.L. 1990. Biodegradation of Toxic Organics: Status and Potential. *J Envir Engr Div ASCE*. 116:805.

Grady, C.P.L. and H.C. Lim. 1980. *Biological Wastewater Treatment*. New York: Marcel Dekker, Inc.

Keinath, T.M. et al. 1977. Activated Sludge - Unified System Design and Operation. *J Env Engr Div ASCE*. 103:829.

Lawrence, A.W., and P.L. McCarty. 1970. A Unified Basis for Biological Treatment Design and Operation. *J Sanit Engr Div ASCE*. 96:757.

McCarty, P.L. 1964. Anaerobic Waste Treatment Fundamentals. *Public Works*. 95:(9)107, 95:(10)123. 95:(11)91, 95:(12)95.

Metcalf & Eddy. 1979. *Wastewater Engineering*. 2nd ed. New York: McGraw-Hill.

Metcalf & Eddy. 1991. *Wastewater Engineering*. 3rd ed.. New York: McGraw-Hill.

Montgomery, James M., Inc. 1985. *Water Treatment Principles & Design*. New York: John Wiley & Sons.

Nemerow, N.L., and A. Dasgupta. 1991. *Industrial and Hazardous Waste Treatment*. New York: Van Nostrand Reinhold.

Reynolds, T.R. 1982. *Unit Operations and Processes in Environmental Engineering*. Monterey, CA: Brooks/Cole Engineering Division.

Rich, L.G. 1961. *Unit Operations of Sanitary Engineering*. New York: John Wiley & Sons.

Sawyer, C.N., P.L. McCarty, and G.F. Parkin. 1994. *Chemistry for Environmental Engineering*. 4th ed. New York: McGraw-Hill.

Ten States Standards (1978). Recommended Standards for Sewage Works. 1978 Edition. Great Lakes-Upper Mississippi River Board of State Sanitary Engineers. Albany, NY.

U.S. Enviromental Protection Agency. 1979. *Process Design Manual for Sludge Treatment and Disposal*. Washington, DC: EPA 625/1-79/011.

Chapter 8

AIR POLLUTION

by W. Christopher King, P.E., DEE
and
Jon M. Rueck, P.E., DEE

8.1 Introduction

Air pollution work is customarily divided into categories including ambient air quality, source emissions, and indoor air quality. The ambient air quality category encompasses the concentration of atmospheric constituents and the affect of these constituents on human health and the environment; e.g. acid rain. Also considered are nuisance concerns such as soot deposits on buildings, industrial fallout that damages cars and buildings, and the adverse aesthetic impacts of smog, a recurring blight in some of the nation's cities. Source emissions are usually divided into stationary sources such as power plants, incinerators, and industrial stacks and mobile sources dominated by cars and trucks. While ambient air quality concerns and source emissions have evolved along with the development of our cities and the industrial revolution, indoor air pollution is a more modern concern. It was triggered by the energy conservation efforts beginning in the mid 70s. Deprived of ample ventilation as building envelopes were tightened, the relatively small emissions from numerous sources and recirculated constituents were discovered to impact human health.

TIP FOR THE P.E. EXAMINATION CANDIDATE

The P.E. examination concentrates on stationary source emission control and to a much less extent on ambient air quality.

The major focus of this chapter is the design of air pollution control technology for stationary emission sources. Ambient air quality issues are also addressed, but in less detail. Indoor air quality is discussed in Chapter 10, as this subject has been historically connected to the specialty of industrial hygiene.

8.1.1 Historical Perspective Air pollution control and air quality have long been a matter of concern with much of the recorded history involving pollution episodes in London, England. The earliest efforts to control air pollution began there in the 1200s. Burning what we now know as low-energy, high-ash coal produced heavy smoke and haze in the city. Laws (with real teeth) authorized torturing and hanging anyone selling this dirty coal. These "nuisance" conditions continued as a common occurance during London's annual heating season, aggravated by inversions (a meteorological condition involving cold air overlying warmer air thus reducing the mixing and dispersion of the air at the ground surface). An inversion that lasted for several days in the 1950s trapped the emissions from coal and wood burning furnaces and these eventually rose to acutely hazardous particulate and sulfur dioxide ambient air levels; 4,000 "excess deaths" were blamed on this event. This dramatic proof of the acute effects of poor air quality led to the development of English air pollution regulations.

In the United States, air quality concerns began to receive attention shortly after the turn of the twentieth century. The major atmospheric concern stemmed from the presence of smoke; up to the late 1940's, most American urban centers had smoke abatement agencies. The transfer of interest from solely smoke to more comprehensive air quality issues came about as a result of developments in the field of Industrial Hygiene. Early atmospheric pollution studies were an extension of the science of industrial hygiene and included:

- a US Public Health Service (PHS) and US Bureau of Mines study of silicosis (1910)

- PHS organizes a Division of Industrial Hygiene (1912)

- studies of air in industrial workshops were well underway, e.g., carbon monoxide from the use of gas-heated eqipment (1916); and

- inustrial hygiene activities of the PHS included studies of municipal dust, the radium dial painting industry, and a comparative study of air pollution in fourteen of the largest cities (1931) (Dworsky, 1990).

Not withstanding the foregoing, the seminal event in the specialty of air pollution control was triggered by an air pollution episode in Donora, Pennsylvania, in 1944; twenty persons died and 5,910 became ill. It triggered the first major comprehensive study of air pollution by the PHS. This study confirmed the dearth of fundamental knowledge regarding the effects of atmospheric pollution on health. Additional studies followed and in 1963 the first federal Clean Air Act was adopted.

8.1.2 Elementary Definitions Air can be characterized chemically, meteorologically, and in other ways. For the purposes of this book, air's chemical composition, as presented in Table 8-1, is of primary concern. Additionally, water vapor,

Table 8-1
Composition of air

Gas	Volume percent	ppm $_v$	Weight percent
Nitrogen	78.084	780,840	76.85
Oxygen	20.946	209,460	23.15
Argon	0.934	9,340	
Carbon dioxide	0.033	330	
Neon		18	
Helium		5	
Methane		2	
Total	100		100

The molecular weight for air is 28.94 g/g-mole.

ozone, sulfur dioxide, nitrous oxide, nitrogen dioxide, nitric oxide, carbon monoxide, and other gases and vapors are present in air, in variable amounts.

Air pollution exists if the constituents differ significantly from the proportions provided in Table 8-1. Pollution varies in concentrations, most generally presented in (ppm$_v$) for gases, and in grams/cubic meter (or grains/dry standard cubic foot) for particulate matter.

In water related calculations, parts per million (ppm) is often used interchangeably with concentrations expressed in mg/L (milligrams per liter). As such, the ratio is a weight ratio, comparing the weight or mass of the contaminant to the mass of fluid which contains it. In air pollution, however, ppm generally is a *volume* fraction, which compares directly with a *mole* fraction in gases. The difference in the basis of units as seen Table 8-1, can create errors in calculating the amount of oxygen required to oxidize something, by as much as 10%.

The physical properties of air vary with temperature as shown in Table 8-2.

8.2 Air Pollution Regulations

Air pollution regulations are divided into air quality or ambient air regulations, and the regulation of air pollution sources, both stationary and mobile. Mobile source standards are not covered in this text.

8.2.1 Ambient Air Quality
Standards for ambient air quality are established pursuant to the Clean Air Act of 1970 as last amended in 1990. Table 8-3

TIP FOR THE P.E. EXAMINATION CANDIDATE

An error of 10% in air-related calculations would make it impossible to choose between a correct response and a distractor in a multiple choice problem. Therefore, while it doesn't matter which approach (mass basis or volume basis) is used, it is important to be consistent. Chapter 2 presents a method to convert from ppm_v to mg/L for gaseous pollutants.

Table 8-2
Properties of air

Temperature		Absolute viscosity (μ)		Density(P_g)	
(°F)	(°C)	(lb/hr-ft^2)	(kg/hr-m)	(lb/-ft^3)	(kg/m^3)
20	-6.67	0.040	0.060	0.0863	1.386
40	4.44	0.041	0.061	0.0827	1.328
60	15.6	0.043	0.064	0.0794	1.275
80	26.7	0.045	0.067	0.0763	1.225
100	37.8	0.047	0.070	0.0734	1.178
120	48.9	0.047	0.070	0.0684	1.098
140	60	0.048	0.071	0.0662	1.063
160	71.1	0.050	0.075	0.0639	1.026
180	82.2	0.051	0.076	0.0619	0.994
200	93.3	0.052	0.078	0.0600	0.963
250	121.1	0.055	0.082	0.0558	0.896
300	148.9	0.058	0.086	0.0521	0.836
350	176.7	0.060	0.089	0.0489	0.785
400	204.4	0.063	0.094	0.0460	0.739
500	260	0.067	0.100	0.0412	0.661
600	315.6	0.072	0.107	0.0373	0.599
700	371.1	0.076	0.113	0.0341	0.547
800	426.7	0.080	0.199	0.0314	0.504
900	482.2	0.085	0.127	0.0295	0.474
1000	537.8	0.089	0.133	0.0275	0.441

AP-40, Air Pollution Engineering Manual, USEPA, 1973.

lists the National Ambient Air Quality Standards (NAAQS) for the currently regulated pollutants. These standards embrace different averaging periods, a key difference which significantly affects engineering calculations. As an example, there are two standards for particulate matter with diameter of less than 10 microns (PM_{10}) — the average annual concentration must be less than 50 mm/M^3, while no

Table 8-3
National ambient air quality standards

Pollutant	NAAQS/Averaging Period
Sulfur dioxide	80 µg/m^3/yr
	365 µg/m^3/day
Ozone	235 µg/m^3/hr
Nitrogen dioxide	100 µg/m^3/yr
Carbon monoxide	10 µg/m^3/8 hr
	40 µg/m^3/hr
Lead	1.5 µg/m^3/3 mo
Total suspended particulate	75 µg/m^3/yr
	260 µg/m^3/day
Respirable particulate (PM_{10})	50 µg/m^3/yr
	150 µg/m^3/day

24-hour period can exceed 150 mm/M^3. Each area of the United States has been assigned to an Air Quality Control Region (AQCR) for the purpose of controlling compliance with the NAAQS. Initially, there were 235 air quality control regions established, but that number has grown slightly since 1971. The air quality of each AQCR is measured against the NAAQS. If the air quality in the region fail to meet any of the NAAQS, the region is defined as a *nonattainment* area which initiates a series of controls based on the level of nonattainment and the pollutant.

Table 8-4 presents 1991 data on the estimated emissions of air pollutants based on the general source categories and compares these values to the same data reported in 1985. The significant reduction in lead in a relatively short period of time, shown in Table 8-4, was accomplished by the systematic elimination of lead by regulations, first from fuels by eliminating leaded gasoline, and by tightly controlling other uses of lead. However, many of the other mass emissions have not improved significantly in spite of reductions in individual source emission rates because the number of sources have increased. Automobiles, as an example, are significantly cleaner than in 1985, but the unit savings is offset by the increased number of cars on the road today.

Table 8-4
National emissions estimate for 1991 (values in millions of metric tons/yr)

Source category	PM_{10}	SO_2	CO	NO_2	VOCs	Pb
Transportation	1.51	0.99	43.49	7.26	5.08	0.00162
Fuel combustion	1.10	16.55	4.67	10.59	0.67	0.00045
Industrial processes	1.84	3.16	4.69	0.60	7.86	0.00221
Solid waste disposal	0.26	0.02	2.06	0.10	0.69	0.00069
Miscellaneous	0.73	0.01	7.18	0.21	2.59	0.00000
Total	5.44	20.73	62.09	18.79	16.89	0.00497
Percentage of 1982 total	97%	101%	69%	92%	87%	9.5%

Source: deNevers, 1995.

8.2.2 Stationary Air Pollution Regulations

Fuel combustion and industry are the major stationary sources targeted by air pollution regulations. The air pollution regulations for stationary sources are constructed into a set of new source performance standards (NSPSs) and National Emission Standards for Hazardous Air Pollutants (NESHAPs). Many industries, 70 plus as of this publication, have specific emission standards based on the type of pollutants expected from the specific process. For example, coal-fired power plants are regulated for particulate, sulfur dioxide, and nitrogen dioxide emissions under the NSPSs. NESHAPS are also industry specific, and include many chemicals such as mercury, asbestos, and vinyl chloride and other volatile organics. A detailed discussion of the air pollution standards is beyond the scope of this text, nor is it necessary for purpose of this book.

TIP FOR THE P.E. EXAMINATION CANDIDATE

Specifics on the laws and regulations are generally not addressed in the P.E. examination because of the frequent changes in the regulations and the differences that exist between states.

8.3 Ambient Air Quality

8.3.1 Ambient Air Quality Quantification

A primary engineering analysis associated with air pollution management is quantification of ambient air quality. The first of these calculations applies the ideal gas law to samples taken at different temperatures and pressures. It is a straightforward conversion to develop Equation 8-1 from the ideal gas law.

$$V_2 = V_1 \cdot \frac{P_1}{P_2} \cdot \frac{T_2}{T_1} \qquad [8\text{-}1]$$

where:
- V_1 = volume or flowrate at initial condition
- V_2 = volume or flowrate at final condition
- P_1 = initial Pressure in any units
- P_2 = final pressure in same units as above
- T_1 = initial absolute temperature in °C or °K
- T_2 = final absolute temperature in same units as above

Example 8-1 — Calculating an air concentration, dry and humid conditions

A filter collects 0.035 g during 2.5 hr of operation. The flow rate of the air pump is set at 100 L/hr at 70°F and 760 mm Hg. The conditions under which the sample was collected were 95°F and 785 mm Hg.

a. Determine the particulate concentration in mg/m³ at 70°F and 760 mm Hg.

Solution:

$$V_2 = V_1 \cdot \frac{P_1}{P_2} \cdot \frac{T_2}{T_1}$$

$$V_1 = 100 \text{ L/hr} \cdot 25 \text{ hr} = 250 \text{ L}$$

$$V_2 = 250 \cdot \frac{785}{760} \left(\frac{460 + 70}{460 + 95} \right)$$

$$V_2 = 253 \text{ L}$$

$$C_{mass} = \frac{0.035 \text{ g}}{253 \text{ L}} = 0.14 \text{ mg/L} = 138 \text{ mg/m}^3$$

b. If the sample described had moisture that was removed from the gas stream prior to flow measurement (this is done to report concentrations as dry standard ft³), determine the particulate concentration at 70°F and 760 mm Hg. The moisture was collected in silica gel impingers and measured gravimetrically. The weight of moisture in the sample was 50 g.

Solution:

$$\frac{50 \text{ g of water}}{18 \text{ g/gmol}} = 2.78 \text{ gmol of water}$$

At standard conditions, 1 gmol of a substance equals approximately 22.4 L of gas.

$$22.4 \text{ L/gmol} \cdot 2.78 \text{ gmol} = 62.2 \text{ L of water}$$

$$C_{mass} = \frac{0.035 \text{ g}}{253.4 \text{ L} + 62.2 \text{ L}} = 0.11 \text{ mg/L} = 110 \text{ mg/m}^3$$

8.3.2 Opacity   Visible emissions have long been one air quality regulation measurement parameter for stack emissions. An early plume color (or opacity) sensing method was the *Ringleman* system. This system employed screens which passed various percentages of visible light, enabling printing a set of charts in numbers from 0 to 5 (20% increments). These Ringleman charts were used by trained observers to compare the "density" of a plume (or exhaust rising from a chimney or stack) to match the plume with a particular chart. The Ringleman system was the standard for observing stack emissions for many sources and serves as an indirect measure of particulate concentration..

Opacity, as it is used as a plume assessment criteria today, involves a measurement made with the human eye. It's not only possible, but common to train observers to assess the opacity of a plume within 5% increments. And, the method is not limited to only black plumes; plumes of any color can be assessed. Most anyone can be trained to make accurate opacity judgments, although not all states allow anyone other than regulatory officials to become "certified."

8.4 Stationary Source Air Pollution

Air pollution control devices for stationary sources typically include cyclones, electrostatic precipitators, and baghouses for particulate controls; incinerators for organic vapors; and scrubbers to control various gaseous and solid air pollutants.

8.4.1 Particle Size Distribution   Particulate controls are based on fundamental characteristics governing particles in air. The NAAQS in Table 8-3 indicates that size is an important characteristic of a pollutant particle. Total suspended particles are differentiated from PM_{10} because it has been determined that only particles less than 10 microns are respirable and, therefore, can be expected to contribute significantly to human health impacts. From an engineering standpoint, particle size is important because it, together with particle density, are defining properties governing design of particulate air pollution control devices.

The particles in most air pollution sources are produced over a wide distribution of sizes. Figure 8-1 shows the range of particle sizes for many common particles. Typically, there is a higher total number of small particles and a higher total mass in the large particles. For most applications, where design is based on mass removal efficiencies, Gaussian models have been found to effectively depict particle size distributions. Also, most sources produce irregularly shaped particles with different aerodynamic properties. To simplify the complexity posed by different sizes, shapes, and densities, the term *aerodynamic diameter*, which is defined as the diameter of a sphere with the density of water that will settle in air at the same rate as the actual particle, is employed. This lumps density, diameter, and aerodynamic drag into one variable that can be measured and is readily applicable in the design of air pollution control systems.

Figure 8-1
Characteristics of particles and particle dispersoids

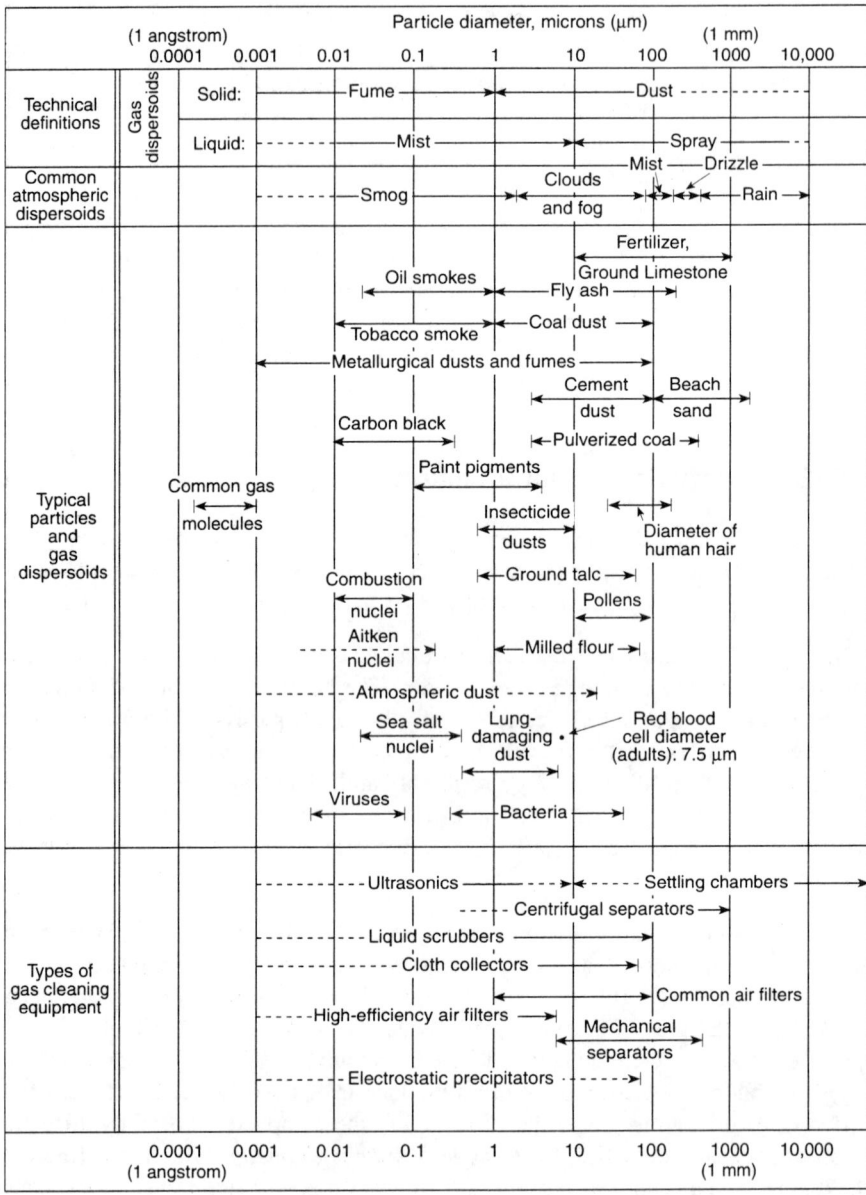

Source: Lapple, 1961.

Careful consideration of the particle size distribution is essential in the design of particulate collectors. Cascade impactors are used to measure the aerodynamic diameter distribution of a particulate laden gas. If the mass distribution is lognormally distributed, which is usually assumed, the distribution can be characterized by the geometric mean diameter and the geometric standard deviation. Example 8-2 shows how particle distribution data is reported, and how to calculate the geometric mean and standard deviation.

Example 8-2 — Particle size distribution

Given the cascade impactor data below, determine the geometric mean diameter and the geometric standard deviations for the distribution.

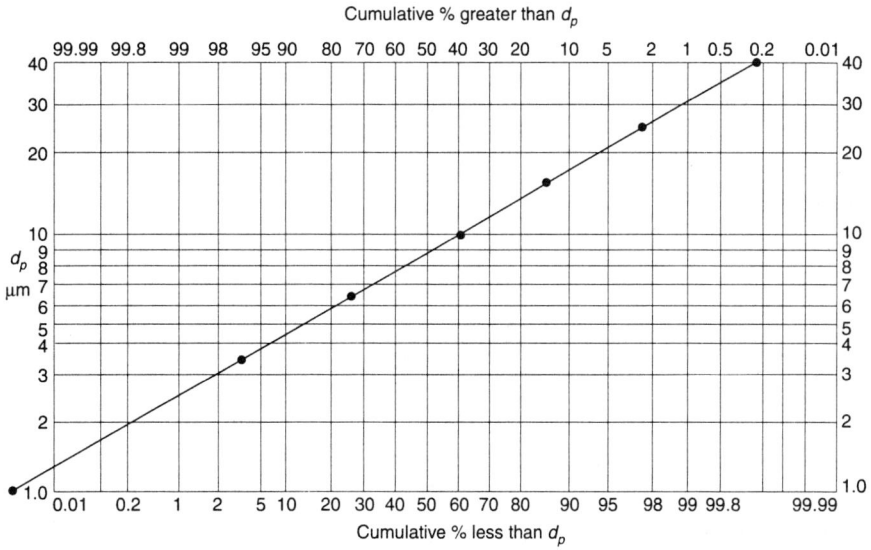

Aerodynamic, μm	Mass in particular range, mg
0-2	5
2-5	25
5-8	220
8-12	350
12-20	250
20-30	120
30-50	35
750	5

Solution: (Method of Coooper, 1990)

Step 1. Develop the expanded data Table shown.

dp, μm	dp	Mass, mg	Cumulative Mass, mg	Cumulative % less
0-2	1	5	5	0.5
2-5	3.5	25	30	3
5-8	6.5	220	250	25
8-12	10	350	600	60
12-20	16	250	850	85
20-30	25	120	970	97
30-50	40	25	995	99.5
750	50	5	1,000	99.9

Step 2. Plot dp versus cumulative mass percent less time values on a Log Probabilty graph.

Step 3. The geometric mean diameter, d_{50}, is read directly from the plot.

Step 4. The geometric standard deviation, Gg, is determined using the following formula:

$$Gg = \frac{d_{84.1}}{d_{50}} = \frac{d_{50}}{d_{15.9}}$$

where: $d_{15.9} \approx 5.3\ \mu m$

$d_{84.1} \approx 15.7\ \mu m$ (from Log Probabilty Plot)

$$Gg = \frac{15.7}{8.9} = 1.76$$

$$Gg = \frac{8.9}{5.3} = 1.68$$

$$\overline{Gg} = 1.72$$

8.4.2 General Design Considerations Several factors must be considered in the choice of an air pollution control device. First, it is important to characterize the source to be controlled. Next, opportunities to minimize pollution at its source need to be examined, including process changes, product modification, etc. Once the extent of emissions have been minimized by feasible process changes, it is necessary to characterize the emission using parameters such as exhaust temperature and volume.

Often, several control mechanisms are used in series to affect the most effective and economical control. For example, in the case of processes which emit hot, volatile organic compounds and particulate-laden gases, the process train could begin with a "drop-out box" to slow the flow enough to allow extremely large particles to be collected and not needlessly contaminate the subsequent processes. A drop-out box may be nothing more than a large chamber with a small inlet and outlet, but a large volume, sufficient to allow particles with high settling velocities to "drop-out" before the exhaust stream exits the box. Such a system might be followed by a cyclone collector because these can easily accommodate high temperatures. A regenerative incinerator could be used next to oxidize the VOCs. To this point in the process train, minimizing dilution air to the extent possible is desirable to limit the volumetric flow to the final control device and, accordingly, its size and cost. Finally, a venturi scrubber, a packed-tower scrubber, a bag filter, or electrostatic precipitator would be selected, depending upon the remaining characteristics of the exhaust stream and regulatory requirements. Figure 8-2 presents the overlapping nature of the particle size range of applicability for several classes of control devices.

Additionally, it is necessary to consider the power required in the selection of control devices. It is customary to express the power required as the pressure drop across the treatment unit when used in their most effective control ranges. Table 8-5 presents the "average" pressure drop (expressed in inches of water column) for several control technologies and their typical performance.

8.5 Particulate Control

8.5.1 Cyclone Collectors Cyclone separators are particulate treatment devices that remove solids from the air stream by centrifugal force. Figure 8-3 shows the dimensions for conventional and high efficiency cyclones with the fundamental dimensions ratios reported as a function of the cyclone diameter.

Operationally, the gas enters the top of the cyclone tangentially to the barrel and is forced to travel cyclonically inside the device. Because of the shape, the gas turns and forms a vortex in the center of the device as it moves upward to the exit duct. The particles are removed by centrifugal force driving them to the wall of the collector before the gas exits the cyclone and then by gravity to the bottom where particulates are discharged.

Cyclones are able to efficiently remove larger and/or dense particles at high efficiency, but are not as effective for pollutant streams with high mass concentrations of particles with diameters less than 5 mm. They are often employed to collect large particles that can be recycled into a manufacturing process or as a pre-cleaner to reduce the mass loading on other more efficient collectors.

Figure 8-2
Range of particle sizes, concentrations and collector performance

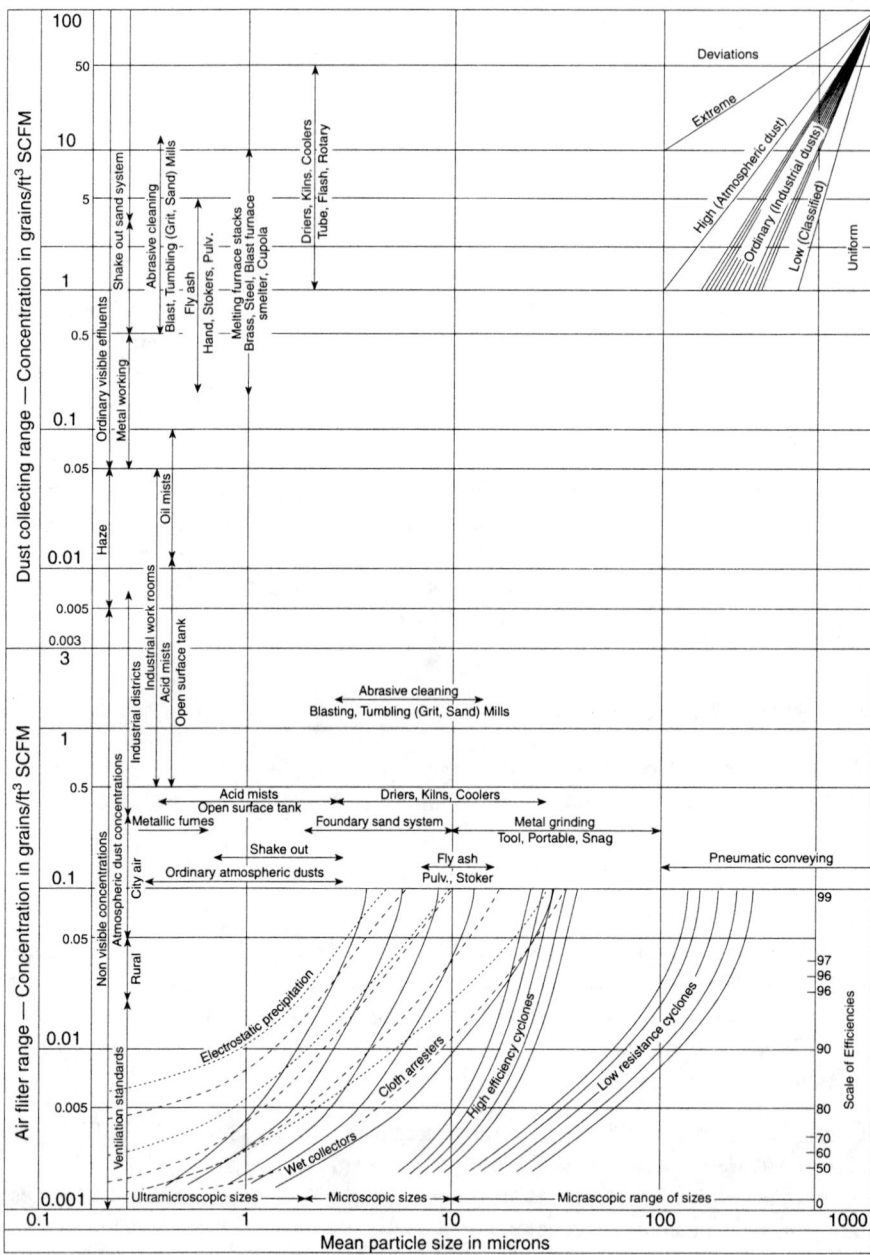

Source: American Air Filter Co., Inc., 1952.

Table 8-5
Air pollution control technologies performance

Equipment	Efficiency range (%)	Delta P (in H_2O)
Wet electrostatic precipitator	99+	0.6
Electrostatic precipitator	99+	0.9
Spray tower	94 - 95	0.4
Furnace filter	58 - 60	1.7
Fluidized bed scrubber	95	2.4
Shaker-type bag filter	99+	2.5
Reverse jet bag filter	99+	3.0
Medium efficiency cyclone	65 - 70	3.7
High efficiency cyclone	84 - 85	4.9
Jet-impingement scrubber	87 - 90	8.0
Medium energy venturi scrubber	99+	20.0
Hi energy venturi scrubber	99+	31.5

TIP FOR THE P.E. EXAMINATION CANDIDATE

The method of Lapple has stood the test of time and therefore would be considered acceptable if applied to a P.E. examination problem.

There are several methods available to predict the collection efficiency of a cyclone based on a given particle size distribution. All of the texts referenced at the end of this chapter contain these methods and provide example problems. The formulae required for the Lapple (1951) method are presented in Equations 8-2 through 8-5, and their use is illustrated in Example 8-3.

$$N_e = \frac{1}{H}[L_b + 0.5L_c] \quad [8\text{-}2]$$

where: N_e = number of air rotations inside the cyclone
H = height of inlet duct
L_b = length of cyclone body
L_c = length of cyclone cone

[8-3]

Figure 8-3
Standardized dimensions for standard and high efficiency cyclone collecters

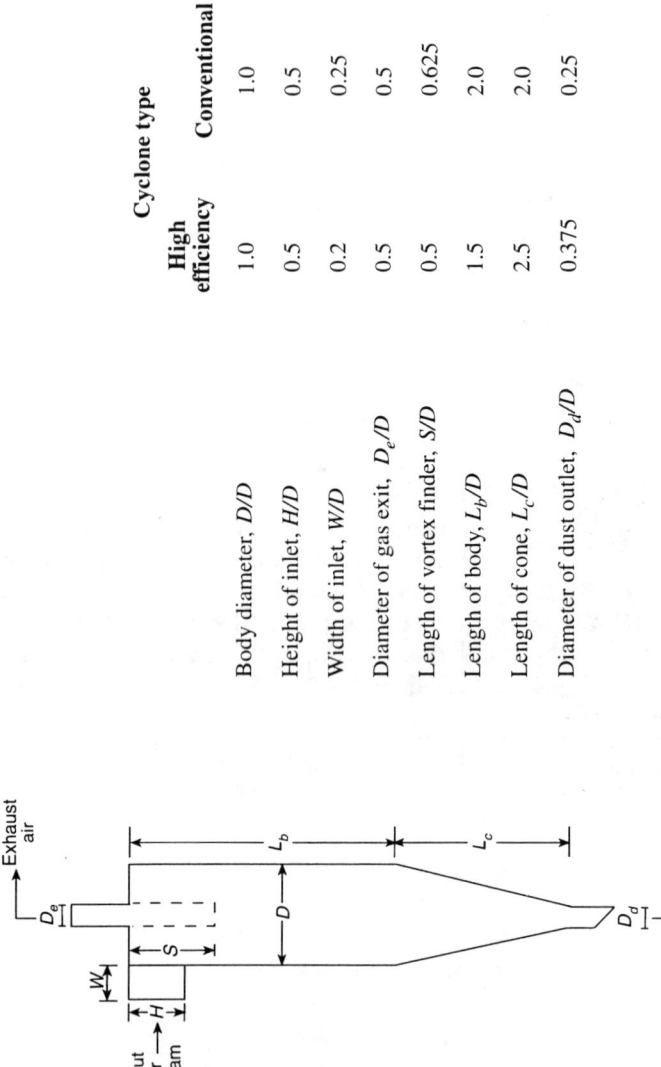

	Cyclone type	
	High efficiency	Conventional
Body diameter, D/D	1.0	1.0
Height of inlet, H/D	0.5	0.5
Width of inlet, W/D	0.2	0.25
Diameter of gas exit, D_e/D	0.5	0.5
Length of vortex finder, S/D	0.5	0.625
Length of body, L_b/D	1.5	2.0
Length of cone, L_c/D	2.5	2.0
Diameter of dust outlet, D_d/D	0.375	0.25

Source: Cooper, 1990.

$$d_{pc} = \left[\frac{9\mu W^2 H}{\pi \, Ne \, Q \, (\rho_p - \rho)} \right]^{1/2}$$

where: d_{pc} = cut diameter, that particle that will be removed at 50% efficiency
μ = viscosity
W = inlet width
H = inlet height
ρ_p = density of particle
ρ = density of gas
Q = volumetric flow rate

Note: Any consistent units work in Equation 8-3.

$$n_j = \frac{1}{1 + (d_{pc} / d_{pj})^2} \qquad [8\text{-}4]$$

where: d_{pj} = mean diameter for the size fraction

$$\eta_o = \Sigma \eta_j m_j \qquad [8\text{-}5]$$

where: η_o = overall mass removal efficiency
η_j = removal efficiency for each size fraction, from Equation 8-4
m_j = mass of each size fraction

Example 8-3 — Cyclone design

Determine the overall removal efficiency that can be achieved by a high efficiency cyclone with 1 m diameter treating 150 m³/min of sawdust laden gas at 120°F and with 1,000 mg/m³ dust concentration. This dust has the particle size distribution presented in Example 8-2. Assume the density of the particles is 800 kg/m³.

Solution:

Step 1. Calculate N_e using Equation 8-2.

From Figure 8-2, the dimension for a 1 m diameter high efficiency cyclone are:

H = 0.5 m
W = 0.2 m
L_b = 1.5 m
L_c = 2.5 m

Substituting these values in Equation 8-2

$$N_e = \frac{1}{H}\left[L_b + \frac{L_c}{2}\right] = \frac{1}{0.5}\left[1.5 + \frac{2.5}{2}\right]$$
$$= 5.5 \text{ air rotations}$$

Step 2. Calculate d_{pe} using Equation 8-3.

$$d_{pc} = \left[\frac{9\mu W^2 H}{\pi N_e Q(P_p - P_g)}\right]^{1/2}$$

from Table 8-2, @ 120°F, $\mu = 0.07$ kg/m-hr and $P_g = 1.098$ kg/m³

$$d_{pc} = \left[\frac{9(0.07) \text{ kg/m-hr} \bullet (.2)^2 \text{m}^2 \bullet (.5)\text{m}}{3.1416 \ (5.5)150 \text{ m}^3/\text{min} \bullet 60 \text{ min/hr} \bullet (800 - 1.098)\text{kg/m}^3}\right]^{1/2}$$

$$= (1.01 \bullet 10^{-10})^{1/2}$$
$$= 10.07 \ \mu\text{m}$$

Note: Be sure to check units carefully

Step 3. Calculate the removal efficiency for each mean particle diamter using Equation 8-4.

$$n_j = \frac{1}{1 + (d_{pc}/d_{pj})^2}$$

j	dp, μm	n_j	m, mg	$n_j m_j$
1	1	.01	5	0.05
2	3.5	.11	25	2.75
3	6.5	.29	220	63.8
4	10	.50	350	175
5	16	.72	250	180
6	25	.86	120	103.2
7	40	.94	25	23.5
8	50	.96	5	4.8
				Σ = 553.1

Step 4. Calculate overall removal efficiency with Equation 8-5.

The overall removal efficiency is determined by summing the removal efficiency for each particle. As indicated in the preceding tabulation the overall efficiency is 55.3%.

The efficiency of this cyclone could be improved by changing its dimensions. Since the particle parameters are fixed in this design, only the flow rate and the cyclone dimensions can be adjusted and, generally, only the diameter of the cyclone can be altered by the designer. As an example, if the diameter of this high efficiency cyclone was reduced to 0.5 m, its efficiency would increase to 87.8%.

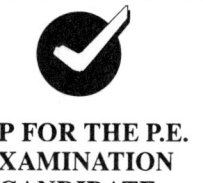

TIP FOR THE P.E. EXAMINATION CANDIDATE

Work this Example 8-3 using a 0.5 m high efficiency cyclone and confirm the 87.8% efficiency.

8.5.2 Electrostatic Precipitators (ESP)

An electrostatic precipitator (ESP) removes particles from an air stream by imparting an electrical charge to the particles, then passing them through a force field that causes them to migrate to an oppositely charged plate where they are collected. This process is depicted in the simplified diagram shown in Figure 8-4.

Figure 8-4 shows one duct with a single set of wires between two plates. In practice, the total area is a series of these ducts with dust being collected on both sides of the plates except for the two exterior plates. The particle charge is produced by free electrons released from the gas molecules in the electrostatic field attaching to the particle. The particles move to the plates at a *drift velocity*, w, where the charge is neutralized and the particles are collected in a cake. The drift velocity is primarily a function of charge, particle size and density, fluid viscous force, and a term called resistivity. The dust is collected from the plates by occasionally rapping the plates to loosen the cake allowing it to fall into a collection hopper at the bottom of the ESP.

Design entails providing an appropriate combination of controlling factors such that enough particles reach and are collected at the plate before the gas stream exits the ESP. This is analogous to gravitational settling and the relationship of settling

Figure 8-4
Operation of an electrostatic precipitator

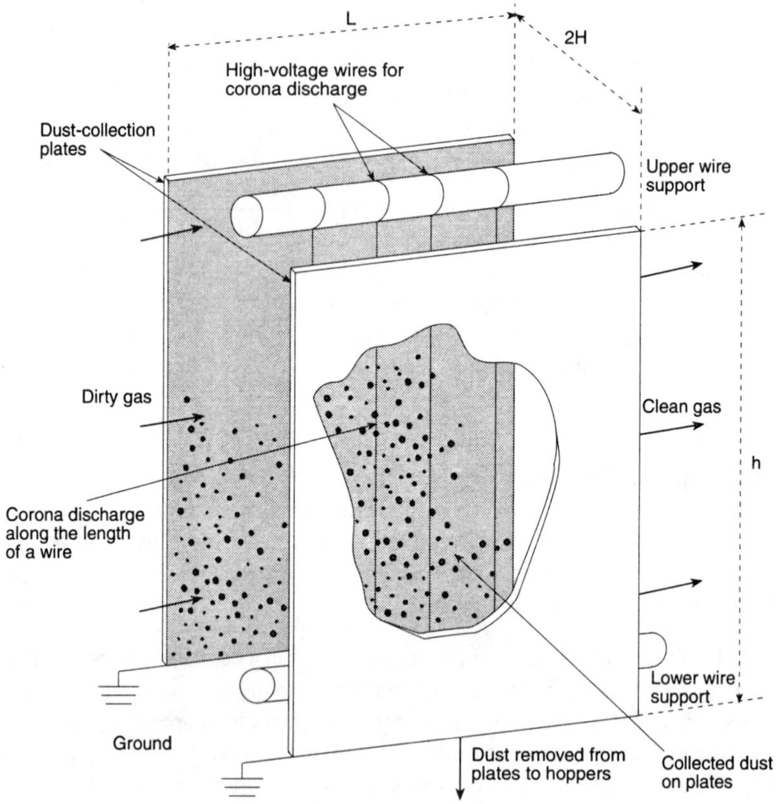

Source: deNevers, 1995.

velocity to surface loading rate. The most commonly accepted approach to design of ESPs is by the application of the Deutsch-Anderson equation:

$$\eta = 1 - \exp\left(\frac{-wA}{Q}\right) \qquad [8\text{-}6]$$

where: η = mass collection efficiency
  w = drift velocity
  A = plate Area
  Q = flow rate

A theoretical approach to the calculation of drift velocity is:

$$w = \frac{(6.64 \bullet 10^{-18}) \bullet E^2 \bullet d_p}{\mu} \quad [8\text{-}7]$$

where:
w = drift velocity in m/sec
E^2 = field strength in V/M
d_p = particle diameter in microns
μ = absolute viscosity in kg/m-sec

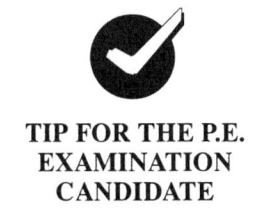

TIP FOR THE P.E. EXAMINATION CANDIDATE

Equation 8-7 is units specific. Failure to use the specified units will result in errors.

From a practical perspective, this equation has several variables that are difficult to analyze. Therefore, it is more common and reliable to use measured drift velocities in designing ESPs. Table 8-6 lists measured drift velocities for several common dust sources.

Example 8-4 presents a design approach based on the Deutsch-Anderson equation using both theoretical and measured values for drift velocity.

Table 8-6
Typical drift velocities encountered in industrial practice

Application	Drift velocity (ft/sec)
Pulverized coal (fly ash)	0.33-0.44
Paper mills	0.25
Open-hearth furnace	0.19
Secondary blast furnace (80% foundry iron)	0.41
Gypsum	0.52-0.64
Hot phosphorus	0.09
Acid mist (H_2SO_4)	0.19-0.25
Acid mist (TiO_2)	0.19-0.25
Flash roaster	0.25
Multiple-hearth roaster	0.26
Portland cement manufacturing (wet process)	0.33-0.37
Portland cement manufacturing (dry process)	0.19-0.23
Catalyst particles	0.25
Gray iron cupola (iron-coke ratio = 10)	0.10-0.12

Source: deNevers, 1995.

Example 8-4 — ESP design

An ESP is to be employed to control emissions from a coal-fired power plant. The gas stream is 100,000 ft³/min at 400°F. The design criteria provides for 98.5% overall removal with 100% removal of particles of 2 mm. Particle resistivity is 10" ohm-cm. The ESP is a 40 kV model with a plate to plate spacing of 8 in, a height of 16 ft, and a rated gas velocity of 4 ft/sec.

Determine the total area, plate length, and number of plates required for the ESP. Use standard drift velocity from Table 8-6 for the area calculation and calculated drift velocity for determining the plate length.

Solution:

Step 1. Calculate total plate area from Deutsch-Anderson Equation.

From Table 8-6, select $w = 0.4$ ft/s

$$\eta = 1 - \exp\left(\frac{-Aw}{Q}\right)$$

$$.985 - 1 = -\exp\left(\frac{-A(0.4 \text{ ft/sec})}{1,667 \text{ ft}^3/\text{sec}}\right)$$

$$\ln(.015) = -0.00024 \, A \, \text{ft}^2$$

$$A = 17,500 \, \text{ft}^2$$

Step 2. Calculate theoretical drift velocity using Equation 8-7.

$$w = \frac{(6.64 \bullet 10^{-12}) \bullet E^2 \bullet d_p}{\mu}$$

where: E = Charge/plate spacing in m
E = 40,000V/(4/12 ft • 0.3048 m/ft) = 393,700 V/m
d_p = 2 μm
μ = from Table 8-5 = 0.094 kg/m-hr = 2.6•10⁻⁵ kg/m-sec

$$w = \frac{(6.64 \bullet 10^{-12}) \bullet (393,700)^2 \bullet (2)}{2.5 \bullet 10^{-5}}$$

$$= 0.079 \, \text{m/sec} = 0.259 \, \text{ft/sec}$$

Step 3. Calculate length of each ESP plate.

To calculate the length of the plate it is necessary to note that the particle must migrate to the plate in the same time the gas moves from the front to the back of the ESP. Mathematically, detention time, t_d, is:

$$t_d = \frac{\text{distance to plate (1/2 duct width)}}{w} = \frac{\text{L plate}}{\text{V gas}}$$

or

$$\text{L plate} = \frac{1/2 \text{ width} \cdot \text{V gas}}{w} = \frac{\left(\frac{4}{12}\right) \text{ft} \cdot 4 \text{ ft/sec}}{0.259 \text{ ft/sec}} = 5.15 \text{ ft}$$

For this example, select a plate length of 6 ft to provide a safety factor.

Step 4. Determine plate configuration.

$$\text{Number of plate units} = \frac{\text{Total area}}{\text{L} \cdot \text{H}} = \frac{17{,}500 \text{ ft}^2}{(6 \cdot 16) \text{ ft}^2} = 183$$

Note: Both sides of all of the plates are used, except the two exterior plates. Therefore:

$$\text{Number of plates} = \frac{183}{2} + = 91 \text{ plates}$$

8.5.3 Baghouses A baghouse collects particles by filtering air through a fabric media, typically configured in long, vertically-suspended sock-like configurations. There are numerous variations of baghouses, but all have the dirty air entering from one side, often from the inside of the bag, passing through the filter media where a particulate cake forms on the dirty side of the bag. Many bags are enclosed in one housing, hence the term baghouse to describe this form of air treatment device. Baghouses are configured with several units installed so that units can be employed according to the applied flow rates, which may vary, and to provide other units as a backup for maintenance purposes.

Baghouses obtain high efficiencies over a wide range of particle size distributions. Selection of the correct type of fabric material is based on temperature of the gas, pH of the gas stream, abrasiveness of the particles being collected, and service life. Cooper (1990) compiled data on dynel, cotton, wool, nylon, polypropylene, orlon, dacron, nomex, teflon, and glass fabrics. Telfon, for example, is rated for temperatures up to 400 °F with excellent resistance to acids and bases.

The air cleaning process is stopped once the pressure drop across the filter reaches an economically unacceptable level. The trade-off to frequent cleaning and maintaining lower pressure drops is the wear and tear on the bags produced in the cleaning processes. Collection of the dust cake is accomplished by shaking or pulsing the fabric to loosen the cake from the filter, allowing it to fall into a collection bin at the bottom of the baghouse. In many industrial processes, this dust can be recycled because it is physically unchanged by the collection process. The most significant recent design changes to baghouses have been in the development of ways to effectively clean the filter bags with a minimum of wear and down time.

Baghouses are typically designed based on a standard loading factor chosen according to the type of process air stream being treated. These loading factors vary from a low of 1-2 (ft^3/min air flow)/(ft^2 of filter area) to a high of 12-14 ft^3/min-ft^2 (Cooper, 1990). The factors used are referred to as the air-to-cloth ratio and evolve to the approach velocity or the superficial velocity for the filter over the total cloth surface area. Keep in mind, efficiency is NOT a design a parameter for fabric filters. Efficiencies are 99% to 99.9% for most all applications. Design heavily emphasizes proper configuration and economic considerations over the life of the equipment. Example 8-5 shows the simple calculations required to size a baghouse.

TIP FOR THE P.E. EXAMINATION CANDIDATE

For P.E. examination purposes, the air-to-cloth ratio for the type of baghouse and specific application being designed would be supplied. Engineers experienced with water filtration will recognize that this concept is completely analogous to designing a sand filter.

Example 8-5 — Baghouse

A baghouse is installed to clean air stream from a gypsum processing plant. The process flow is 40,000 ft^3/min at 300°F and has a dust loading of 25 grains/ft^3. For a pulse-jet baghouse with nomex filter media with a rated air-to-cloth ratio of 10 ft/min, size and configure the units. Each bag is 8 inches in diameter and has an effective length of 12 feet while each baghouse has 80 bags per unit.

Solution:

    Step 1. Select an air-to-cloth ratio and type of material for the filter.

        This information is provided in the problem statement.

Step 2. Determine the area of a bag.

$$\text{Area} = \pi \bullet d_{bag} \bullet L_{bag} = 3.14 \bullet (8/12) \text{ ft} \bullet 12 \text{ ft} = 25.12 \text{ ft}^2$$

Step 3. Calculate the total fabric area and number of bags

$$\text{Total area} = \frac{40,000 \text{ ft}^3/\text{min}}{10 \text{ ft}^3 \text{ air}/\text{min} - \text{ft}^2 \text{ fabric}} = 4,000 \text{ ft}^2$$

Number of bags = $4,000 \text{ ft}^2 / 25 \text{ ft}^2 = 160$ bags

Step 4. Configure the number of units.

Select three units, two online and one for backup/maintenance.

8.6 Scrubbers

Air scrubbers are routinely employed for particulates gaseous SO_x and HCl and to effectively remove other soluble pollutants. However, the primary application is for SO_x control and to a lesser extent for particulates.

8.6.1 Scrubbers for SO_x Control Scrubbers are frequently employed to remove SO_x, primarily SO_2, from stationary source emissions. From a mass emissions perspective, coal-fired power plants are the major source of concern. The sulfur content of coal ranges from less than 1% by weight to over 3% for some high sulfur coals mined in the United States. This sulfur is in a form that is substantially transformed into a gas (mostly as sulfur dioxide) during coal burning. Other mineral fuels, oil and natural gas, also contain naturally-occurring sulfur. Crude oils are referred to as sour when they contain high concentrations of sulfur, typically 1 to 3 percent. For crude oil, this is not an insurmountable problem because modifications in the refining process can recover/remove the sulfur. Danielson (1973) reports that sulfur is present in some natural gases, but indicates no concentrations high enough to be of air pollution significance, particularly in comparison to the mass emission rates for coal.

There are numerous permutations to the basic scrubber system illustrated in Figure 8-5 which shows a typical countercurrent scrubber system. Hardware configurations include cross-flow, parallel flow, packed tower, bubbler systems, and dry injection of reaction chemicals. There is an equal variety in the types of chemicals added to the scrubber water to enhance the capture of the gaseous sulfur in the liquid phase. deNevers (1995) and Cooper (1990) both list limestone, lime, sodium carbonate, sodium bicarbonate, and combinations of these as routinely employed in gas scrubbing. Each of these references also list many, primarily

Figure 8-5
General SO_x scrubber system

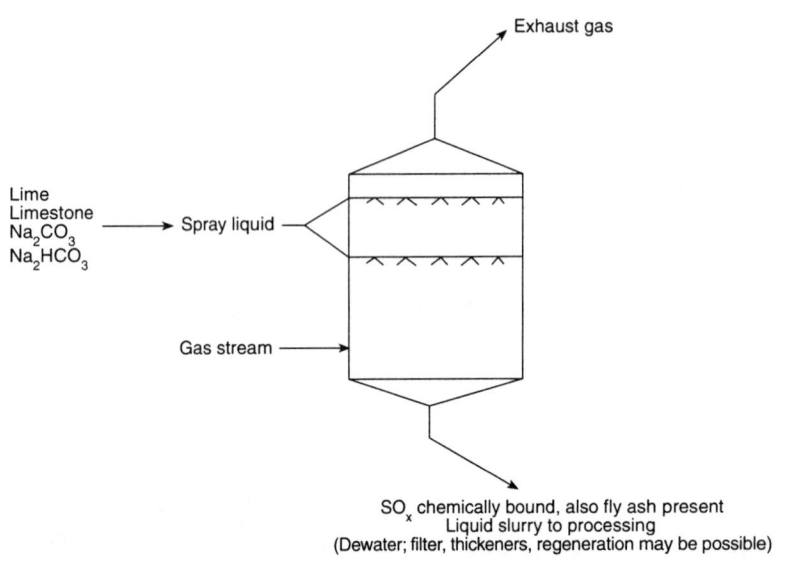

proprietary processes (Well-Lord process is an example of a sodium carbonate system that regenerates the reactant chemicals) which are employed to scrub sulfur dioxide from flue gas.

The important concepts for the design of scrubber systems are: 1) quantification of the raw chemicals required and waste volumes generated, 2) volumetric flow rates in the scrubber, and 3) operational considerations such as controlling scaling in the systems as precipitates are formed in the reactions. The reactions for the most common scrubbing chemicals are as follows:

Limestone scrubbing

$$CaCO_3 + H_2O + 2SO_2 \rightarrow 2CaSO_3 + 2CO_2 + H_2O \qquad [8\text{-}8]$$

Lime Scrubbing (as $Ca(OH)_2$)

$$Ca(OH)_2 + SO_2 + H_2O + \frac{1}{2}O_2 \rightarrow CaSO_4 \bullet 2H_2O \qquad [8\text{-}9]$$

Trona (natural sodium carbonate)

$$Na_2CO_3 + SO_2 \rightarrow Na_2SO_3 + CO_2 \qquad [8\text{-}10]$$

Double Alkali (lime or limestone plus sodium carbonate)

1) $Na_2CO_3 + SO_2 \rightarrow Na_2SO_3 + CO_2$ [8-10]

2) $Na_2SO_3 + CaCO_3 + \frac{1}{2}O_2 \rightarrow CaSO_4 + Na_2CO_3$ [8-11]

Example 8-6 — Wet scrubber for SO_2 removal

A counter-current wet scrubber with lime treatment, as shown, is used to contol SO_2.

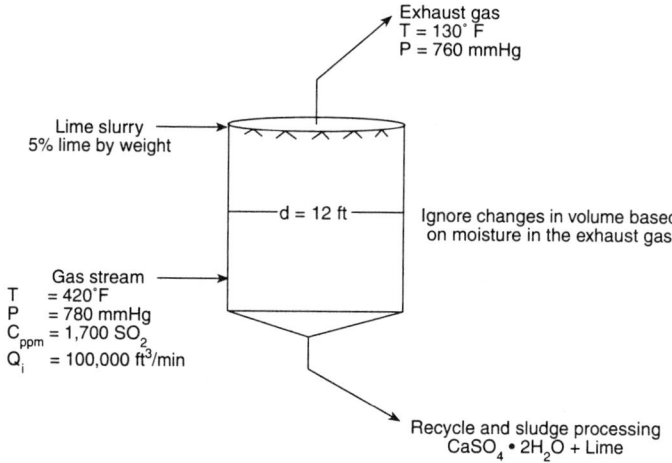

a) Determine if the gas velocity in the srubber is within design standards (7-20 ft/sec at entrance).

Gas velocity

$$V = \frac{Q_i}{Area}$$

$$= \frac{100,00 \, ft^3 / min}{3.14(6)^2 \, ft^2}$$

$$= 884.6 \, ft / min$$

$$= 14.74 \, ft / sec$$

b) Determine the concentration of SO_2 in the exhaust gas in both mg/m^3 and ppm_v, if the scrubber achieves a mass removal efficiency of 92%.

Mass of SO_2 in the gas stream (See Equation 2-3)

$$C(mg/m^3) = 4.09 \cdot 10^{-2}(Cppm_v)MW_s$$
$$= 4.09 \cdot 10^{-2}(1,700)64$$
$$= 4,450 \, mg/m^3$$
$$= 4.45 \, g/m^3$$

Mass $SO_2 = 4.45 \, g/m^3 \cdot 100,000 \, ft^3/min \cdot 0.0283 \, m^3/ft^3$
$$= 12,593 \, g/min$$

Effluent $SO_2 = 8\%$ of $12,593 \, g/min$
$$= 1,007 \, g/min$$

$$V_2 = V_1 \cdot \frac{P_1}{P_2} \cdot \frac{T_2}{T_1} \quad \text{(Equation 8-1)}$$
$$= 100,000 \cdot \frac{780}{760} \cdot \frac{(130+460)}{(420+460)}$$
$$= 68,810 \, ft^3/min \cdot 0.0283 \, m^3/ft^3$$
$$= 1,947 \, m^3/min$$

$$C_{mass} = \frac{1,007 \, g/min}{1,947 \, m^3/min} = 0.517 \, g/m^3$$

$$C_{ppm_v}\left(\frac{mg}{m^3}\right) = \frac{C_{mass}}{4.09 \cdot 10^{-2} \cdot MW_s}$$
$$= \frac{517 \, mg/m^3}{4.09 \cdot 10^{-2} \cdot 64}$$
$$C_{ppm} = 198 \, ppm$$

c) Determine the stoichiometric lime consumption rate and calcium sulfate generation rate.

$$CaO + SO_2 + H_2O + \frac{1}{2}O_2 \rightarrow CaSO_4 \bullet 2H_2O$$

$$56 + 64 + 36 + 16 \rightarrow 172$$

$$.875 + 1 + .56 + .25 \rightarrow 2.69$$

SO_2 removed = 92% of 12,593 g / min

$$= 11,586 \text{ g / min} \bullet \frac{1 \text{ lb}}{453.6 \text{ g}}$$

$$= 25.54 \text{ lbs / min}$$

$$\text{Lime} = 25.54 \text{ lbs / min} \bullet .875 \frac{\text{lbs lime}}{\text{lbs SO}_2} = 22.35 \frac{\text{lbs lime}}{\text{min}}$$

$$CaSO_4 = 25.54 \text{ lbs / min} \bullet 2.69 \frac{\text{lbs CaSO}_4}{\text{lbs SO}_2} = 68.65 \text{ lbs / min}$$

d) Calculate the lime slurry feed rate based on either, five times the stoichiometric requirement or a liquid-gas ratio of 6.0 lbs liquid/lbs gas. Specify the largest required liquid feed rate.

Lime slurry is 5% by weight CaO and 22.35 lbs lime/min is required.

$$\frac{22.35 \text{ lbs / min}}{.05} = 447 \text{ lbs / min slurry weight}$$

Assuming specific gravity of water

$$Q_{liq} = 447 \text{ lbs / min} \bullet \frac{1 \text{ gal}}{8.34 \text{ lbs}}$$

$$= 53.6 \text{ gal / min} \bullet 5 \text{ safety factor}$$

$$= 268 \text{ gal / min}$$

The other option is to consider the liquid-gas ratio by weight

weight of air = 100,000 ft^3 / min • 0.045 lbs / ft^3
= 4,500 lbs air / min

liquid = 4,500 lbs air / min • 6 lbs liquid / lbs air
= 27,000 lb / min slurry • 0.12 gal / lb
= 3,237 gal / min

Use liquid-gas ratio as the controlling flow rate.

8.6.2 Scrubbers for Particulate Control As previously noted, there are many different configurations for scrubbers, each with different capabilities and limitations. This section discusses wet scrubbing for the single purpose of particulate control. Most of the typical wet scrubbing systems have proven incapable of obtaining high efficiencies in particulate removal. However, a special modification, the venturi scrubber, can achieve good removals under the appropriate operating conditions. Figure 8-6 depicts a simplified venturi scrubber.

Dust-laden gases enter at the top and flow down at increasing velocities as the duct narrows to the throat width of the venturi. Water is injected at the throat and becomes aspirated into fine droplets by the force of the gas. Since the liquid is initially at zero velocity relative to the gas stream, particles in the gas stream collide with the water droplets and are entrained in the droplet. The liquids accelerate as they join the gas stream and quickly reach the velocity of the gas downstream of the throat. The gas/liquid mixture is then taken into a collection device such as a cyclone where the liquid plus the particles are separated from the gas stream.

High gas velocities exist in the throat area which produce, a pressure drop. This pressure drop controls the removal efficiency of a venturi scrubber. Since the key parameter controlling a venturi scrubber is pressure drop, removal efficiency is directly related to cost in the form of fan motor energy cost.

Several different ways are available to estimate venturi scrubber efficiency based on the variables of gas flow rate, liquid flow rate, gas velocity, and particle size. Most texts indicate that, under nominal conditions (20-40 inches of water pressure drop), venturi scrubbers will remove particles greater than 3 to 5 microns at nearly 100 percent. For pressure drops at the high end of this range, 99 percent removals are possible for particles with aerodynamic diameters in the range of 0.5 microns.

Figure 8-6
Venturi scrubber for particulate collection

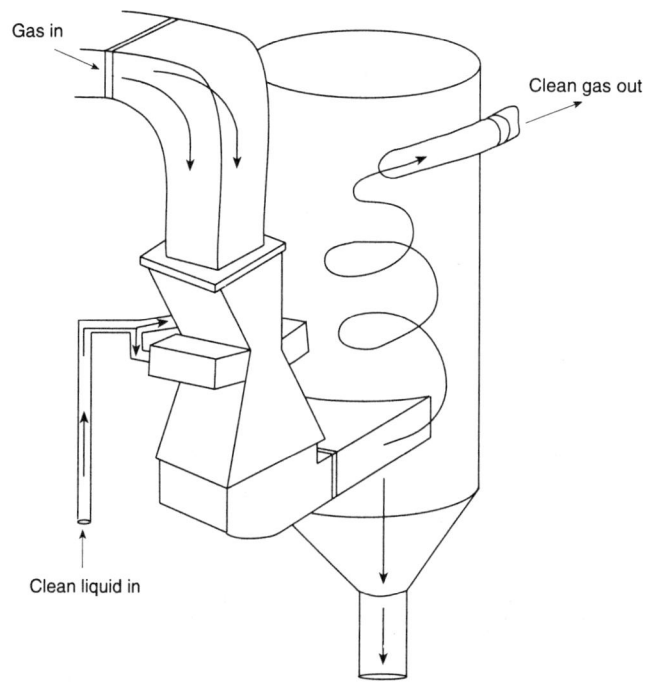

Hesketh (1974) developed a simple method of estimating particle collection efficiency;

$$P_t = C_o / C_i = 3.47(\Delta P)^{-1.43} \qquad [8\text{-}12]$$

where: P_t = mass fraction penetration = 1 - fractional efficiency
C_o = mass concentration of particles, less than 5 μm out
C_i = mass concentration of particles, less than 5 μm in
DP = pressure drop in inches of water

A second important calculation for venturi scrubbers is to estimate the pressure drop based on liquid and gas flow rates and then to determine the power costs associated with that pressure drop. There is no general agreement on 'the method' for determining pressure drop; many exist. Presented below is a method developed by Calvert (1972) and reported in Wark and Warner (1981) that is based on fluid fundamentals and can be determined from the basic operating parameters of the venturi;

$$\Delta P = -\rho_l v^2 (Q_l / Q_g) \qquad [8\text{-}13]$$

where: ΔP = pressure drop
ρ_l = liquid density
v = gas velocity in the venturi throat
Q_l = volumetric flow rate of liquid
Q_g = volumetric flow rate of gas

note that $Q_g = v \cdot$ Area of throat

Power consumption is determined by:

$$P = Q_g \bullet \Delta P \qquad [8\text{-}14]$$

8.7 Control of Nitrogen from Stationary Sources

Nitrogen oxides as air pollutants are quite different in that they are primarily an unwanted by-product of processes that combust hydrocarbons, such as automobiles and power plants. In the presence of a carbon source and with sufficient temperature and oxygen, atmospheric nitrogen is converted to NO and NO_2. These chemicals are of concern in the atmosphere because they combine with water vapor to form nitric acid, one of the primary components of acid rain. Automobiles and the like are, by far, the major emitters of NO_x, but any facility that burns carbon-based fuels at high temperatures will create NO_x emissions. In automobiles, NO is controlled by catalytic converters following the reaction below;

$$2NO + 2CO \xrightarrow{c} N_2 + 2CO_2 \qquad [8\text{-}15]$$

c= platinum- rhodium

For stationary combustion sources the NO_x must be reduced; ammonia is the most common chemical used. The reactions are shown below;

$$6NO + 4NH_3 \rightarrow 5N_2 + 6H_2O \qquad [8\text{-}16]$$

or

$$2NO_2 + 4NH_3 \rightarrow 3N_2 + 6H_2O \qquad [8\text{-}17]$$

8.8 Treatment of Volatile Organic Gases

A discussion of air pollution control of volatile organic chemicals could and often does require entire texts. First, there are many sources for volatile organic compounds (VOCs) and each has its own control technology. The data in Table 8-4 lists transportation and industrial processes as the major contributors of VOCs. In transportation, VOCs are unburned fuels emitted from vehicles and fueling operations. Fueling operations provide a variety of sources, e.g., tank vapors displaced during refueling and the breathing losses as storage tanks change gas volumes during the daily warming and cooling cycle. *AP 42, Compilation of Air Pollution Emission Factors*, provides data to estimate the mass emissions from fuel handling operations.

TIP FOR THE P.E. EXAMINATION CANDIDATE

It is certainly beyond the scope of this text to try to cover this subject in detail, nor is great attention to this area normally included in the P.E. examination.

Industrial sources provide a wide variety of VOC emissions, each with its unique control problems. Incineration of exhaust gases containing VOCs is discussed.

From Cooper(1990), the formula for the combustion of any simple hydrocarbon (C_xH_y) can be related to the oxygen supplied by air as:

$$C_xH_y + (b)O_2 + 3.76(b)N_2 \rightarrow xCO_2 + (y/2)H_2O + 3.76(b)N_2 \qquad [8\text{-}18]$$

where: $b = x + (y/4)$

Control of organics must also account for any other ions in the organic molecule that may prove of concern if released to the environment. Chlorine is certainly the most common of the ions bound in organics, polyvinyl chloride (PVC) is one example. The chlorine liberated in the combustion process will react to form HCl which has it own emission standards and treatment technology (typically HCl is controlled with scrubbers). Metals are another concern that may exist in the processing of certain organic compounds.

The keys to successfully destroying VOCs in thermal processes are the three T's — time, temperature and turbulence. The three T's imply that in a well-mixed combustion chamber with sufficiently high temperature and adequate residence time, the reaction above will substantially go to completion. For toxic materials regulated under the RCRA (see chapter 9 for details), substantially complete means 99.99% percent destruction of the organic and 99.9999% for a few extremely hazardous organic substances. The stoichiometric relationship shown in Equation 8-18 is not able to accurately describe what accurs in an real incinerator because the kinetics of the reaction are not precise and the degree of contact of the reactants is variable. This is normally overcome by providing air in excess of the stoichiometric requirements. In combustion, this is referred to as *excess air*. Typically, excess air in the 50 % range provides for the best burning conditions while maintaining the highest possible temperatures.

Design of VOC incinerators involves selecting the appropriate combustion temperature, conducting a heat balance to determine what supplemental fuels are required, calculating air flows, sizing the chamber for a selected detention time, and determining if any other pollution controls are needed based on the presence of contaminants such as HCl or SO_x. In the incineration of dilute VOC gas streams, the heat value of the VOC is often neglected in the calculation of fuel required to achieve required combustion temperatures. For the most common VOCs, there is considerable data compiled concerning the relationship between residence time, combustion temperature, and destruction efficiency.

The important properties of the chemical contaminants for air pollution control are the combustion temperature, the heat value, and the reaction rate constant. However, for many applications, design is greatly simplified to using standardized minimum operating temperatures and residence times that provide substantial safety factors. For example, a 2 second residence time with a temperature of 1,600 °F is a common specification for a VOC incineration process. This residence time allows for the autoignition (the temperature at which a gas will combustion without a flame) of most hydrocarbons and more than sufficient reaction times for even the slower-reacting compounds. Design calculations determine the volume of gas based on the VOC gas stream flow rate and the fuel feed rate based on an energy balance. Example 8-7 provides the easiest way to describe the design of VOC incinerator.

Example 8-7 — Incineration of VOCs

Design an incinerator to dispose of chlorobenzene (C_6H_5Cl) at a feed rate of 500 lbs/hr. Data describing chlorobenzene is listed below. Assume chlorobenzene follows a first-order kinetic model for destruction with the rate constant in the data listed. A destruction efficiency of 99.99% must be acheived. Combustion is to be accomplished at a minimum of 1,400°F with 50% excess air. Natural gas is available as a supplemental fuel, if required.

Chapter 8: Air Pollution   295

Chlorobenzene

> Molecular Weight = 112.6
> Vapor pressure = 12 mm Hg
> Autoignition temperature = 1,245°F
> Heat of combustion = 14,000 Btu/lb
> Heat of vaporization = 170 Btu/lb
> 8-hr TWA = 350 mg/m^3
> k (1st order at 1,400°F) = 8.4 sec^{-1}

Natural Gas

> Available heat at 1,400°F = 615 Btu/ft^3

a) Determine the size of the incinerator.

Step 1. Find the residence time based on first order kinetics (Equation 3-9).

$$k \cdot t_d = \ln(C_o / C_t)$$

$$t_d = \frac{1}{k} \cdot \ln(C_o / C_t)$$

Note: C can be in concentration or mass.

$$t_d = \frac{1}{8.4} \sec^{-1} \cdot \ln(500 / (500) \cdot 0.0001)$$

$$= 1.1 \text{ sec}$$

Choose t_d = 2.0 sec to provide safety factor.

Step 2. Detemine the gas volume.

Begin with a calculation of combustion gases. It is necessary to modify Equation 8-18 to account for the one hydrogen atom of benzene being replaced by one chlorine atom. Recognizing that chlorine forms HCl in the product stream, it is then possible to use the combustion formula as follows:

$$C_6H_5Cl + 7O_2 + 7(3.76)N_2 \rightarrow 6CO_2 + 2H_2O + HCl + 26.32N_2$$

$$\text{Chlorobenzene} = 500 \frac{\text{lbs}}{\text{hr}} \text{ or } \frac{500 \text{ lbs / hr}}{112 \text{ lbs / lb} - \text{mole}}$$

$$= 4.44 \text{ lb} - \text{moles / hr}$$

The products of combustion are:

$$CO_2 = 6\frac{\text{lbs} - \text{moles } CO_2}{\text{lbs} - \text{moles } C_6H_5Cl} \bullet 4.44 \text{ lbs} - \text{moles } C_6H_5Cl / \text{hr}$$
$$= 26.64 \text{ lb} - \text{molesCO}_2 / \text{hr}$$

Similarily:

$H_2O = 2(4.44) = 8.88 \text{ lb} - \text{moles} / \text{hr}$
$HCl = 1(4.44) = 4.44 \text{ lb} - \text{moles} / \text{hr}$
$N_2 = 26.32(4.44) = 117 \text{ lb} - \text{moles} / \text{hr}$

The foregoing did not include the *excess air* necessary to acheive the desired destruction. Therefore, to include excess air:

$N_2 = 1.5(117) = 175 \text{ lb} - \text{moles} / \text{hr}$

$$O_2 = \frac{1}{2}\left(7\frac{\text{lb} - \text{moles } O_2}{\text{lb} - \text{moles } C_6H_5Cl}\right) \bullet 4.44 \frac{\text{lb} - \text{moles } C_6H_5Cl}{\text{hr}}$$
$$= 15.54 \text{ lb} - \text{moles} / \text{hr}$$

Total gases = 26.64 + 8.88 + 4.44 + 175 + 15.54 lb - moles / hr
= 230.5 lb - moles / hr

Using the Ideal Gas Law, (PV = nRT), the volume of gas is calculated by first rearranging that equation to solve for V.

V = nRT/p

where: P = 1 atm
T = 1,400°F (1,860°R)
n = 230.5 lb-moles/hr
R = 0.7302 atm-ft³/lb-mole °R

V =(230.5 lb-moles/hr)(0.7302 atm-ft³/lb-mole °R)(1860°R/1 atm)
V =313,060 ft³/hr or 86.9 ft³/sec

Chapter 8: Air Pollution

Step 3. Using a residence time in the incinerator of 2 sec, the size of the incinerator is equal to 2 sec • 86.9 ft³/sec or 174 ft³.

b) Determine the requirements for supplemental fuel, if any, and estimate the temperature in the incinerator.

Step 1. Conduct a heat balance to determine the need for supplemental fuel to achieve 1,400°F.

$Heat_{in}$ = $Heat_{out}$ + Losses

$Heat_{in}$ = C_6H_5Cl Heat of Combustion + Air Heat of Combustion where the Air for combustion = $(O_2 + N_2)1.5$. From part a, it was determined that there are 175 lb-moles of N_2 and 15.54 lb-moles of O_2 with 50% excess air. Therefore, the pounds of air introduced to the incinerator is:

N_2 = 175 lb - moles / hr • 28 lb / lb - mole = 4,900 lb / hr

$$O_2 = 1.5\left(7\frac{\text{lb - moles }O_2}{\text{lb - moles }C_6H_5Cl}\right) \cdot 4.44\frac{\text{lb - moles }C_6H_5Cl}{\text{hr}}$$

= 1,492 lb / hr

Air = 1,492 + 4,900

= 6,392 lb / hr (Air has specific enthalpy of 33.6 Btu / lb @ 200°F.)

$Heat_{in}$ = 500 lb / hr • 14,000 Btu / hr + 6,392 lbs / hr(33.6 Btu / lb)

= $(7 \cdot 10^6 + 0.215 \cdot 10^6)$ Btu / h +

= $7.215 \cdot 10^6$ Btu / hr

$Heat_{out}$ = Heat in products at 1,400°F + Losses

Mass out must equal mass in and losses are estimated at 25% of input, therefore:

$Heat_{out}$ = 341 Btu / lb • (6,392 lbs / hr + 500 lbs / hr) + 0.25(7.215 • 10^6 Btu / hr)

= $2.35 \cdot 10^6$ Btu / hr + 1.80 Btu / hr

= $4.15 \cdot 10^6$ Btu / hr

The preceding calculation reveals that burning chlorobenzene will produce more than enough energy to acheive the minimum temperature, 1,400°F. In fact, the incinerator will be hotter than 1,400°F which enables a reduction in the required residence time, and therefore the incinerator size. However, this excess energy could damage the incinerator if the temperatures in it are too high. The final temperature can be estimated as follows:

Mass (specific enthalpy @ $t_d = 2$) = Available heat

6,892 lbs / hr (enthalpy) = $(7.215 \cdot 10^6 - 4.15 \cdot 10^6)$ Btu / hr

enthalpy = 445 Btu / lb

By interpreting an enthalpy chart for air at various temperatures it is found that an enthalpy of 445 Btu/hr corresponds to a temperature of approximately 1,800°F. This temperature provides a safety factor to ensure destruction of the C_6H_5Cl, but it is not so high as to damage a well-designed incinerator.

c) Identify any air pollution concern failures incenerator and describe the best approach to acheive compliance.

The air pollution concern presented by the destruction of C_6H_5Cl is the emission of HCl. The rate of emission is:

$$500 \frac{\text{lbs } C_6H_5Cl}{\text{hr}} \cdot \frac{36.5 \text{ lbs HCl}}{112.6 \text{ lbs } C_6H_5Cl} = 162 \text{ lbs / hr}$$

Typically, RCRA regulations would require treatment to reduce HCl emissions to 4 lbs/hr or 99% removal. Since HCl is very soluble, removal could be easily accomplished using a wet scrubber system.

8.9 References

Brunner, Calvin R. *Handbook of Incineration Systems.*

Calvert, S. 1968. "Source Control by Liquid Scrubbers 112.6," *Air Pollution.* Academic Press.

Cooper, C. D., and Alley, F.C. 1990 (reissue) *Air Pollution Control: A Design Approach.* Prospect Heights, Illinois: Waveland Press.

Cross, Frank L. 1973. *Handbook on Air Pollution Control.* Lancaster, Pennsylvania: Technomic Publishing Company.

Danielson, John A. May 1973. *Air Pollution Engineering Manual.* USEPA.

deNevers, Noel. 1995. *Air Pollution Control Engineering*. New York: McGraw-Hill.

Dworsky, L. B. 1990. "The United States Public Health Sciences," *The Diplomate*. vol. 26, no. 4. Annapolis, Maryland: American Academy of Environmental Engineers.

Lapple, C.E. May 1961. "Process Use-Many Collector Types," Chemical Engineering. vol. 58, no. 5.

Rossano, August T. *Air Pollution Control; Guidebook for Management*.

Hesketh, H.E. 1974. "Fine Particle Collection Efficiency Related to Pressure Drop, Scrubbant and Particle Properties, and Contact Mechanism," *Journal of Air Pollution Control Association*.

—. USEPA. 1991. AP-42, Compilation of Air Pollutant Emission Factors, Supplements.

Wark, Kenneth, and Warner, Cecil. 1981. *Air Pollution: Its Origin and Control*. 2nd Ed. New York: Harper & Row.

Chapter 9

SOLID AND HAZARDOUS WASTE MANAGEMENT

by James Braithwaite, P.E., DEE
and
Katherine R. King, P.E., DEE

9.1 Introduction

Solid and hazardous waste engineering and management require a mixture of technical skills including civil, chemical, and mechanical engineering; the sciences of biology, chemistry, geology, and hydrology; and a thorough knowledge of federal, state and local regulations. This chapter includes an overview of current federal regulations and solid waste management systems — land disposal, incineration and energy recovery, waste minimization and recycling, hazardous waste treatment and disposal, and contaminated site remediation. A list of publications is provided at the end of this chapter for reference and further study.

9.2 Regulations

The modern history of solid and hazardous waste federal regulation began with enactment of the Solid Waste Disposal Act of 1965. That act provided for the promulgation of technical guidelines to promote effective solid waste management

>
>
> **TIP FOR THE P.E. EXAMINATION CANDIDATE**
>
> The problems on the P.E. examination are intended to test a candidate's *general* understanding and application of the fundamental federal regulations. Details regarding these regulations and state or local regulations are not addressed.

practices to protect environmental quality. Additional legislation enacted in the years since has further defined the measures necessary to ensure safe handling, treatment and disposal of solid and hazardous waste.

In 1976, Congress enacted the most comprehensive environmental protection measure ever with the passage of the Resource Conservation and Recovery Act (RCRA) which amended and replaced the Solid Waste Disposal Act of 1965. This law set the federal standards for all solid waste management and, for the first time, identified certain solid wastes as hazardous wastes, requiring special management from their generation until finally disposed; this became known as "cradle to grave" management. Pursuant to the law, regulations were adopted which established standards that govern the engineering of solid waste management systems. For RCRA, these regulations are published in 40 CFR 238-282 and contain standards for recycling and recovery of solid wastes; thermal processing of solid wastes; storage and collection of solid wastes; criteria for municipal solid waste landfills; regulation of underground storage tanks; rules for all aspects of hazardous waste management; and numerous waste specific standards for different solid wastes such as, regulation of used oil.

The most detailed criteria promulgated under the RCRA are the hazardous waste management standards, encoded in 40 CFR 260-272. Table 9-1 delineates the coverage RCRA provides and clearly demonstrates its comprehensiveness. These standards define generators of hazardous wastes and which solid wastes are hazardous wastes, regulate transporters of hazardous wastes, and establish criteria for disposal of all hazardous wastes including regulating and permitting of all facilities that treat, store, or dispose of hazardous wastes, known as TSD facilities. The most substantial amendment of the RCRA, the Hazardous and Solid Waste Amendments (HSWA) of 1984, rules banning the land disposal of hazardous wastes which possess a high potential to harm public health (see 40 CFR 268 for the specifics of the "land ban" rules).

In 1980, Congress enacted the Comprehensive Environmental Response, Compensation and Liability Act (CERCLA), also known as the Superfund law, to address response activity and funding of remediation of hazardous waste disposal sites not covered by RCRA. CERCLA affects many active and closed municipal solid waste landfills where hazardous wastes were commingled with municipal solid waste. The Superfund Amendments and Reauthorization Act (SARA) of 1986 included as Title III the Emergency Planning and Community Right-to-Know Act. This Act

Table 9-1
40 CFR parts applicable to hazardous waste

40 CFR part 260	Hazardous waste management system-general (definitions of terms, procedures for petitions, delisting)
40 CFR part 261	Identification and listing of hazardous wastes (determining if a waste is hazardous, exemptions)
40 CFR part 262	Standards applicable to generators of hazardous wastes
40 CFR part 263	Standards applicable to transporters of hazardous waste
40 CFR part 264	Standards for owners and operators of hazardous waste treatment, storage, and disposal facilities
40 CFR part 265	Interim status standards for owners and operators of hazardous waste treatment, storage, and disposal facilities
40 CFR part 266	Standards for materials being recycled/reused
40 CFR part 268	Land disposal restrictions
40 CFR part 270	EPA-administered permit of state hazardous waste programs
40 CFR part 271	Requirements for authorization of state hazardous waste programs
40 CFR part 272	Approved state hazardous waste management programs
40 CFR part 279	Used oil management standards
40 CFR part 280	Underground storage tanks
40 CFR part 281	State programs for administering the underground storage tank program
40 CFR part 148	Restrictions on the underground injection of hazardous waste

required industry to develop emergency preparedness plans and coordinate these plans with their communities in the event of a release or industrial accident.

Other regulations that affect solid and hazardous waste management include:

- The National Environmental Policy Act (NEPA) of 1969. It requires preparation of environmental impact statements (EIS) for public review and comment prior to development of a federally-permitted solid or hazardous waste treatment or disposal facility to determine the primary and secondary impacts the proposed facility may have on the natural and sociological environment of the site and its surrounding area and to present mitigating measures to address those impacts. Environmental assessments are also made for evaluating new construction or facility modifications and their potential impact on the surrounding environment.

- The Public Utility Regulation and Policy Act (PURPA) of 1981. It requires public and private utilities to purchase power from waste-to-energy facilities at competitive rates.

- The Clean Air Act Amendments (CAAA) of 1990. It regulates the emission of criteria pollutants and hazardous air pollutants. For example, landfills must control fugitive dust emissions and methane. Incineration facilities must comply with exhaust gas emissions standards. During remediation of hazardous waste sites, volatile organic carbons (VOCs), process emissions, and fugitive dusts must be controlled.

- The Clean Water Act (CWA) of 1977. As amended, it requires permitting and treatment of contaminated surface water and groundwater as well as leachate and non-contact storm water collected from a land disposal, incineration, or hazardous waste treatment facility. The CWA also governs wastewater treatment at federally- or publicly-owned treatment works.

- The Toxic Substances Control Act (TSCA) of 1988. A provision of the Act governs the treatment and disposal of wastes containing polychlorinated biphenyls (PCBs).

9.2.1 Hazardous Waste Management Determining whether a particular material is a hazardous waste can be a very complex process requiring detailed review of regulations found in 40 CFR 261. In general, a material must first meet the definition of a solid waste and then be either a material on one of the hazardous waste lists in 40 CFR 261 or meet one or more of the criteria of a characteristic hazardous waste. Table 9-2 summarizes the standards for evaluating a solid waste as a possible hazardous waste. Listed wastes are the easiest to identify; the waste is compared by its source and chemical composition to the lists in 40 CFR 261. If the substance is on any of the lists, it is a *Listed* hazardous waste with the appropriate waste code. If a waste is not listed, it may be a *Characteristic* hazardous waste based on the properties of corrosivity, ignitability, reactivity, or toxicity. A corrosive hazardous waste is a material with a pH of less than 2.0 or greater than 12.5. An ignitable waste has a flash point of less than 140°F or is a flammable compressed gas. Reactivity is more difficult to define; it includes materials that are unstable and react violently in air or water, generate toxic fumes, or are capable of explosive reactions including Class A or B explosives. Toxicity, the potential to leach toxic substances into the groundwater in a landfill environment, is measured by specified test protocols which determine if any of 40 substances listed in Table 1, 40 CFR 261.24, will leach from a solid into the liquid phase of a digested sample of the waste.

Once a material is determined to be a hazardous waste, the next step is to quantify the amount of hazardous waste generated per month. The Federal standards prescribe three different classes of hazardous waste generators based on mass. If the total quantity of all hazardous wastes generated in any month is less than 100 kg, the entity is a conditionally-exempt, small quantity generator (CESQG). This classification eliminates most reporting and management requirements under RCRA.

Chapter 9: Solid and Hazardous Waste Management

Table 9-2
Basic classifications of hazardous waste

Waste	Hazard code	Hazardous waste number
Ignitable	I	D001
Corrosive	C	D002
Reactive	R	D003
Toxicity Characteristic, TCLP	E	D004-D043
Acute Hazardous	H	
Toxic	T	
Listed Wastes		
From non-specific sources	Code varies by listed waste number	F001-F039
From specific sources	Code varies by listed wate number	K001-K151
Acute hazardous *(from discarded commercial chemical products, off specification species, container residues, and spill residues)*	H (may also have additional primary hazardous properties)	P001-P123
Toxic *(from discharded commercial chemical products, off-specification species, container residues, and spill residues)*	T, unless otherwise designated	U001-U359

A small quantity generator (SQG) generates 100 to 1,000 kg/mo SQGs are required to meet numerous criteria for manifesting wastes, storage restrictions, and the full disposal standards. A large quantity generator (LQG) generates more than 1,000 kg/mo and must comply with the RCRA requirements. One significant exception to the SQG cirteria is that there is a group of extremely hazardous chemicals listed in 40 CFR 261 which, if generated, require full RCRA compliance if over 1 kg/mo is generated. Table 9-3 summarizes the criteria for determining the generator status for a facility and then highlights the management requirements imposed by RCRA for each type generator.

Any person who moves a RCRA hazardous waste from a generator site to a TSD facility must be a licensed hazardous waste transporter. This person completes the transporter portion of the waste manifest and assures that the wastes are delivered only to the specified TSD facility.

TIP FOR THE P.E. EXAMINATION CANDIDATE

The full details regarding RCRA management responsibilities for each of the generator classifications is beyond the scope of this text and is normally not covered by the P.E. examination. Wagner (1991) is an excellent reference for RCRA management rules.

A TSD facility is a permitted facility that may accept hazardous waste from any generator for the purpose for processing and/or disposal. The standards for TSD facilities are provided in 40 CFR 264 or 265 depending on whether the facility existed at the time the RCRA came into effect, or was built or significantly modified since 1980. RCRA TSD facilities may include treatment in tank units, landfills, incinerators, waste piles, land treatment units, and storage in containers.

Table 9-3
Hazardous waste generator status standards

	Generator Status		
Standard	Conditionally exempt small quantity generator kg/mo	Small quantity generator kg/mo	Large quantity generator kg/mo
Generation rate			
[Regular HW]	<100	>100 and <1,000	>1,000
[Acutely HW]	<1	<1	>1
Storage [regular HW]	<999	<6,000	any amount for <90 days
Requirements			
EPA ID number	NR	Yes	Yes
Manifest waste	NR	Yes	Yes
TSD facility	State approve	RCRA permitted HW	RCRA permitted HW
Reporting	NR	Partial	Full
Record keeping	NR	Partial	Full
Contingency planning	NR	Partial	Full

NR — Not required

Example 9-1 — Hazardous waste classification and management

A steel company generates 750 gal of waste pickling liquor per month. The liquor's specific gravity is 1.258 (Hc s/b HCl pure). It has a pH of 2.0.

a) Why would this material be considered a hazardous waste and how would it be classified?

Solution:

If the pickling liquor is used continuously in the process, i.e., regenerated, it is not a waste. RCRA does not apply to hazardous materials until they are classified by the generator as waste. However, the company indicates that pickling liquor is *waste*.

Process knowledge and testing can be used to characterize the waste. Based on its pH of 2.0, the stream is corrosive, or D002. A Toxic Characteristic Leaching Procedure (TCLP) analysis for RCRA metals such as lead (D008) and chromium (D007) is also advisable. Pickling liquor does not meet any listed waste classifications (F#, K#, P#, U#). Therefore, it is a D002, with possible characteristic code(s) for toxic characteristic metal(s).

b) The waste pickling liquor is placed in 50 gal drums for shipment to a disposal facility. What requirements must the generator fulfill before releasing the waste containers from the plant?

Solution:

According to the provisions of 40 CFR 262, containers of 110-gal or less require a hazardous waste marking including the generator's name and address, the date of accumulation of waste in the container, and the hazardous waste number(s). Labeling must also comply with U.S. Department of Transportation (US DOT) regulations found in 49 CFR Part 172.

Additionally, the generator must prepare a hazardous waste manifest prior to shipment offsite. Both the form and content of this manifest are prescribed by 40 CFR 262.20 and must include:

- Generator's name, address and EPA ID number
- Transporter's name and EPA ID number
- Designated Facility's name, address and EPA ID number
- Manifest document number
- US DOT description (including proper shipping name, hazard class, and ID number)
- Quantity of containers, container type any special handling instructions
- Generator's certification

c) Would the steel company be considered a small- or large-quantity generator of waste and why?

The steel company is a large quantity generator because 750 gal/mo far exceeds the 1,000 kg quantity limit to be classified as a large quantity generator.

d) How long could the generator store the material at the site?

Since the problem statement does not indicate that the steel company has a RCRA permitted storage facility, the company could store the waste up to 90 days before it must be transported to a permitted TSD. The company may petition for another 30 days if additional time is necessary and all good faith efforts to remove the waste have been taken (see 40 CFR Part 262.34(b)).

9.2.2 Hazardous and Solid Waste Amendments (HWSA) The 1984 HSWA and subsequent amendments restrict the land disposal of hazardous wastes unless the waste can be demonstrated to fall below the EPA-promulgated treatment standards for the particular classification of hazardous waste. Hazardous wastes exceeding the land-ban treatment standard concentrations must be treated so that the concentrations in the treated waste, and any subsequent wastes generated from the treatment process, fall below the standard concentrations. Waste material containing one or more restricted hazardous wastes requires careful assessment of the source(s) to determine the applicable treatment standard. The wastes currently subject to land disposal restriction and the respective treatment standards are listed in 40 CFR 268.

9.2.3 Solid Waste Management Criteria for classification of solid waste disposal facilities and practices are provided in 40 CFR 257, while 40 CFR 258 presents minimum standards for the siting, design, and construction of sanitary landfills for non-hazardous waste. Table 9-4 summarizes these standards. Enforcement of these standards has been delegated to many individual states. These states have established siting, design, and construction standards which are often more restrictive; permitting programs; operating standards; and community planning requirements.

Table 9-4
Landfill siting and design criteria summary[a]

Location

Airports —	FAA notification if within 5-mile radius of end of any airport runway used by turbocraft or piston-type aircraft; precluded within 10,000 ft of end of a runway used by jetcraft or 5,000 ft of a runway used by piston-type aircraft.
Floodplains —	It must be demonstrated that the design and operation of a landfill will not restrict the flow of a 100-yr flood, reduce water storage capacity of the floodplain, or cause washout of solid waste.
Wetlands —	Precluded unless it can be demonstrated that there will be: no net loss of wetlands; no degradation to quality of wetlands; no threat to environmental quality; no harm to endangered species; and no other alternative available.

Table 9-4 (cont.)
Landfill siting and design criteria summary[a]

Fault areas —	Precluded within 200 ft of a fault.
Seismic —	Precluded from seismic impact zones.
Unstable areas —	No harm to structural integrity of the landfill must be demonstrated.
Existing landfills —	Must close by October 9, 1996, if compliance with airport, floodplain, or unstable-area restrictions cannot be demonstrated.

Operating Criteria

Exclusion of hazardous waste —	Implement program to exclude hazardous waste.
Daily cover —	Provide a minimum 6 in. of earthen cover or regulatory approved alternative.
Disease and vector control —	Provide prevention and control measures for vectors.
Explosive gas control —	Control concentration of methane gas to 25% or less of lower explosive limit in on-site structures such as scale houses; concentration of methane to the lower explosive limit at the facility boundary; and comply with minimum requirements for gas monitoring.
Air criteria —	Comply with State Implementation Program and prohibit open burning.
Access —	Control public access.
Surface water runon/ runoff —	Comply with minimum design requirements for runoff discharge from 24-hr, 25-yr storm and for runon control from peak discharge from 25-yr storm.
Surface water —	Comply with Clean Water Act for all liquid discharges to surface water.
Liquid restrictions —	Prevent disposal of bulk or noncontainerized liquids.
Recordkeeping —	Maintain written record demonstrating compliance with design and operating requirements.

Design Criteria

Liner —	Provide a composite liner consisting of minimum 30-ml flexible membrane liner (FML) overlying a minimum 2-ft layer of compacted soil with a hydraulic conductivity less than $1 \cdot 10^{-7}$ cm/sec; high-density polyethylene FML must be 60-ml thick or more.
Leachate collection system —	Provide a system to maintain less than 30-cm depth of leachate over liner.
Groundwater protection —	Limit concentrations of contaminants listed in 40 CFR 258.40 Table 1 in the uppermost aquifer, at the relevant point of compliance as defined by federal or state regulatory authority, to less than regulated levels.

Table 9-4 (cont.)
Landfill siting and design criteria summary[a]

Groundwater Monitoring and Corrective Action

Groundwater monitoring systems —	Install a groundwater monitoring system that enables collection of samples representative of water quality of the uppermost aquifer at locations upgradient (background) and downgradient of the landfill; number, spacing, and depth of wells shall be determined based on thorough evaluation of site-specific information on aquifer thickness, groundwater flow rate and direction, saturated and unsaturated conditions, and materials comprising the lower limits of the uppermost aquifer.
Groundwater sampling and analysis —	a) Document procedures and techniques used for sample collection, preservation and shipment; analytical procedures; chain of custody control; and quality assurance and quality control; b) use appropriate sampling and analytical procedures to accurately measure hazardous constituents; c) evaluate field-filtering of samples prior to laboratory analysis; d) use sampling procedures and frequency which are protective of human health, welfare and safety; e) measure groundwater elevations prior to well purging at each sampling event; f) establish background water quality; g) statistically evaluate analytical data.
Corrective action —	If groundwater contamination has been detected based on statistical evaluation, must select a remedy that protects human health and the environment; attain groundwater protection standard pursuant to 40 CFR 258.55; and control sources causing contamination.

Closure and Post-Closure

Final cover —	Provide a final cover system to minimize infiltration and erosion. Minimum design consists of an 18-in. infiltration layer with permeability less than or equal to bottom liner permeability, natural subsoils present must have hydraulic conductivity no greater than $1 \cdot 10^{-5}$ cm/sec and a 6-in. erosion layer capable of supporting vegetation.
Closure plan —	Provide a description of the final cover system and methods used to construct it; an estimate of the largest area of landfill requiring final cover at any time during active life; an estimate of the maximum inventory of wastes ever on site over the landfill's active life; and a schedule for completing closure activities.
Schedule —	Begin closure within 30 days of final receipt of waste and complete closure within 180 days of commencing closure activities.
Closure certification —	Provide certification by an independent licensed professional engineer that closure has been completed in accordance with the Closure Plan.
Deed restriction —	Record notation on deed to landfill property indicating land used for landfill facility and that use is restricted under 40 CFR 258.60 following closure.

Table 9-4 (cont.)
Landfill siting and design criteria summary[a]

Post-closure care —	Provide 30 years post-closure care including maintenance of final cover integrity and effectiveness; maintenance of leachate collection system; monitoring of groundwater; and maintenance and operation of gas monitoring system.
Post-closure certification —	Provide certification by an independent licensed professional engineer that post-closure care has been completed in accordance with the post-closure plan.
Financial Assurance	
Closure —	Provide written cost estimate (must account for inflation) for third-party closure of largest area of landfill and establish financial assurance for closure.
Post-closure —	Provide written cost estimate (must account for inflation) for third party to conduct post-closure activities and establish financial assurance for post-closure.
Financial assurance —	Approved financial assurance mechanisms include trust fund, surety bond guaranteeing payment or performance, letter of credit, or insurance.

[a]Refer to 40 CFR 258.10 thru 258.40 for comprehensive list

Example 9-2 — Landfill siting constraints

A national solid waste management firm desires to locate a landfill at one of three sites.

a) Assuming the state and local standards are equal to RCRA Subtitle D, what minimum siting criteria must be considered?

Solution:

Siting must be in an area that is not in seismic or fault areas, nor in unstable areas that would compromise the structural integrity of the landfill. Wetlands should be avoided, unless it can be demonstrated no other practicable alternative exists and provisions made for no net loss of wetlands. Finally, the landfill must not restrict the flow of a 100-yr flood, reduce water holding capacity of a floodplain or cause a washout of waste.

b) What minimum protection for groundwater is needed?

Solution:

A composite liner (minimum 30-ml flexible membrane liner overlying a minimum of 2 ft of compacted soil with hydraulic conductivity less than $1 \cdot 10^{-7}$ cm/sec), a leachate collection system, and other techniques as appropriate, e.g., engineered barrier, must be provided to limit contaminant concentrations listed in 40 CFR 258.4 Table 1 in the uppermost aquifer's relevant point of compliance.

c) Would a 40-ml, high density polyethylene liner placed over native sandy silt having a hydraulic conductivity of $1 \cdot 10^{-6}$ cm/sec meet the RCRA minimum standards? Defend the answer.

Solution:

No. First, the high density polyethylene liner does not meet the minimum 60-ml thickness criteria listed in 40 CFR 258.40. Second, native sandy silt does not have a hydraulic conductivity less than $1 \cdot 10^{-7}$ cm/sec.

9.3 Basic Solid Waste Management

9.3.1 Characterization of Solid Waste

Solid waste, as defined by RCRA, encompasses hazardous and nonhazardous wastes, including liquid, solid and semi-solid materials. As discussed in Section 9.2, hazardous waste includes listed wastes, characteristic hazardous wastes, materials and mixtures of materials containing hazardous wastes, and residues from the treatment of hazardous waste. Although municipal solid waste (MSW) is not considered a hazardous waste, there are often small quantities of household and commercially-generated hazardous materials present in MSW.

The typical composition of MSW and how it varies by season is important information necessary for design and economic analysis of landfills, collection programs, volume-reduction alternatives, incineration, and recycling and material recovery programs. The following tables present the typical characteristics of MSW.

- Table 9-5 presents the typical composition of MSW in the United States in 1990, including materials that may be recycled.

Table 9-5
Typical physical composition of residential MSW in the United States in 1990

Component	Percent by weight Solid Waste as collected[a]
Organic	
Food wastes	9.0
Paper	34.0
Cardboard	6.0
Plastics	7.0
Textiles	2.0
Rubber	0.5
Leather	0.5
Yard Wastes	18.5
Wood	2.0
Misc. organics	–
Inorganic	
Glass	8.0
Tin Cans	6.0
Aluminum	0.5
Other metal	3.0
Dirt, ash, etc.	3.0
Total	100.0

a) Excludes waste components now recycled and food waste that is ground up and discharged with wastewater
b) Twenty percent (20%) of the households in the United States are assumed to have food waste grinders. Additionally, it is assumed that the percentage of food waste ground up and discharged with wastewater is 25%.
c) Current (1990) recycling rate for the United States is assumed to be 11%.

Chapter 9: Solid and Hazardous Waste Management

- Table 9-6 presents an example of the seasonal variation of the composition of MSW.

- Table 9-7 presents the specific weight and moisture content for typical MSW constituents.

- Table 9-8 provides the percent residue and energy content for typical MSW constituents.

- Table 9-9 presents typical generation-rate data for MSW in the United States.

The text by Tchnobanoglous, Theisen, and Vigil (1993) provides additional information on solid waste composition.

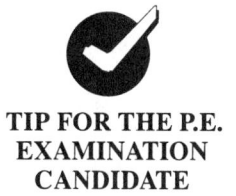

TIP FOR THE P.E. EXAMINATION CANDIDATE

For the P.E. examination, the candidate needs to know the physical, chemical and biological properties of the diverse materials contained in MSW. MSW may contain waste materials from residences, commercial operations, institutions, construction and demolition sites, as well as water and wastewater treatment plant sludge, and yard waste.

Table 9-6
Typical seasonal variation observed in the composition of residential MSW as collected

Waste	Winter season	Summer season	% by weight decrease	% variation increase
Food waste	11.1	13.5		21.6
Paper	45.2	40.0	11.5	
Plastics	9.1	8.2	9.9	
Other organics	4.0	4.6		15.0
Yard wastes	18.7	24.0		28.3
Glass	3.5	2.5	28.6	
Metals	4.1	3.1	24.4	
Inert and other waste	4.3	4.1	4.7	
Total	100.0	100.0		

a) Based on winter season.
Source: Tchnobanoglous, Theisen, and Vigil, Integrated Solid Waste Management (1993). Reprinted with permission of the McGraw-Hill Companies

Table 9-7
Typical specific weight and moisture content data for residential, commercial, industrial, and agricultural wastes

Type of waste	Specific weight lb/yd³		Moisture content % by weight	
	Range	Typical	Range	Typical
RESIDENTIAL (UNCOMPACTED)				
Food wastes (mixed)	220-810	490	50-80	70
Paper	70-220	150	4-10	6
Cardboard	70-135	85	4-8	5
Plastics	70-220	110	1-4	2
Textiles	70-170	110	6-15	10
Rubber	170-340	220	1-4	2
Leather	170-440	270	8-12	10
Yard wastes	100-380	170	30-80	60
Wood	220-540	400	15-40	20
Glass	270-810	330	1-4	2
Tin Cans	85-270	150	2-4	3
Aluminum	110-405	270	2-4	2
Other metals	220-1940	540	2-4	3
Dirt, ashes, etc.	540-1685	810	6-12	8
Ashes	1095-1400	1255	6-12	6
Rubbish	150-305	220	5-20	15
RESIDENTIAL YARD WASTES				
Leaves (loose and dry)	50-250	100	20-40	30
Green grass (loose and moist)	350-500	400	40-80	60
Green grass (wet and compacted)	1000-1400	1000	50-90	80
Yard waste (shredded)	450-600	500	20-70	50
Yard waste (composted)	450-650	550	40-60	50
MUNICIPAL				
In compactor truck	300-760	500	15-40	20
In landfill				
Normally compacted	610-840	760	15-40	25
Well compacted	995-1250	1010	15-40	25
COMMERCIAL				
Food wastes (wet)	800-1600	910	50-80	70
Appliances	250-340	305	0-2	1
Wooden crates	185-270	185	10-30	20
Tree trimmings	170-305	250	20-80	5
Rubbish (combustible)	85-305	200	10-30	15
Rubbish (noncombustible)	305-610	505	5-15	10
Rubbish (mixed)	235-305	270	10-25	15

Table 9-7 (cont.)
Typical specific weight and moisture content data for residential, commercial, industrial, and agricultural wastes

CONSTRUCTION AND DEMOLITION				
Mixed demolition (noncombustible)	1685-2695	2395	2-10	4
Mixed demolition (combustible)	505-675	605	4-15	8
Mixed construction (combustible)	305-605	440	4-15	8
Broken concrete	2020-3035	2595	0-5	–
INDUSTIAL				
Chemical sludges (wet)	1350-1855	1685	75-99	80
Fly ash	1180-1515	1350	2-10	4
Leather scraps	170-420	270	6-15	10
Metal scrap (heavy)	2530-3370	3000	0-5	–
Metal scrap (light)	840-1515	1245	0-5	–
Metal scrap (mixed)	1180-2530	1515	0-5	–
Oils, tars, asphalts	1350-1685	1600	0-5	2
Sawdust	170-590	490	10-40	20
Textile wastes	170-370	305	6-15	10
Wood (mixed)	675-1140	840	30-60	25
AGRICULTURAL				
Agricultural (mixed)	675-1265	945	40-80	50
Dead animals	340-840	605	–	–
Fruit waste (mixed)	420-1265	605	60-90	75
Manure (wet)	1515-1770	1685	75-96	94
Vegetable wastes (mixed)	340-1180	605	60-90	75

Source: Tchnobanoglous, Theisen, and Vigil Integrated Solid Waste Management (1993). Reprinted with permission of the McGraw-Hill Companies.

The organic constituents of MSW—predominately paper, food waste and yard waste — are highly biodegradable and generate gases during decomposition. Gas generation is an important consideration in the design and operation of landfills, including how the gases are formed, their composition and characteristics, the means by which they move, and how they can be collected and controlled. Table 9-10 presents the typical constituents of MSW landfill gas. Trace concentrations of toxic constituents can also be present in landfill gas, which can present significant environmental and health hazards.

Because the decomposition of the readily degradable organic constituents is primarily anaerobic, odors result. Sulfate in the waste reduces to form sulfide, which can, in turn, form hydrogen sulfide gas or biochemically react with other organic compounds in the waste to form malodorous compounds such as methyl mercaptan and aminobutyric acid.

Table 9-8
Typical values for inert residue and energy content of residential MSW

Component	Inert residue[a], percent Range	Typical	Energy[b], Btu/lb Range	Typical
Organic				
Food Wastes	2-8	5.0	1,500 -3,000	2,000
Paper	4-8	6.0	5,000 -8,000	7,200
Cardboard	3-6	5.0	6,000 -7,500	7,000
Plastics	6-20	10.0	12,000 -16,000	14,000
Textiles	2-4	2.5	6,500 -8,000	7,500
Rubber	8-20	10.0	9,000 -12,000	10,000
Leather	8-20	10.0	6,500 -8,500	7,500
Yard Wastes	2-6	4.5	1,000 -8,000	2,800
Wood	0.6 -2	1.5	7,500 -8,500	8,000
Misc. organics	–	–	–	
Inorganic				
Glass	96-99+	98.0	50-100[c]	60
Tin cans	96-99+	98.0	100-500[c]	300
Aluminum	90-99+	96.0	–	–
Other metal	94-99+	98.0	100-500[c]	300
Dirt, ashes, etc.	60-80	70.00	1,000 -5,000	3,000
Municipal solid wastes			4,000 -6,000	5,000

a) After complete combustion.
b) As discarded basis.
c) Energy content is from coatings, labels, and attached materials.
Source: Tchnobanoglous, Theisen, and Vigil, Integrated Solid Waste Management (1993). Reprinted with permission of the McGraw-Hill Companies.

Table 9-9
Estimated quantities for the waste categories comprising MSW generated per capita in the United States for the year 1990

Component	Solid wastes, lb	Energy, Btu/lb	Total energy, Btu
Organic			
Food Wastes	9.0	2,000	18,000
Paper	34.0	7,200	244,800
Cardboard	6.0	7,000	42,000
Plastics	7.0	14,000	98,000
Textiles	2.0	7,500	15,000
Rubber	0.5	10,000	5,000
Leather	0.5	7,500	3,750
Yard Wastes	18.5	2,800	51,800
Wood	2.0	8,000	16,000
Inorganic			
Glass	8.0	60	480
Tin cans	6.0	300	1,800
Aluminum	0.5	–	–
Other metal	3.0	300	900
Dirt, ashes, etc.	3.0	3,000	9,000
Total	100.0		506,530

Source: Tchnobanoglous, Theisen, and Vigil, Integrated Solid Waste Management (1993). Reprinted with permission of the McGraw-Hill Companies.

Table 9-10
Typical constituents of MSW landfill gas

Component	% (dry volume basis) [a]
Methane	45-60
Carbon dioxide	40-60
Nitrogen	2-5
Oxygen	0.1-1.0
Sulfides, disulfides, mercaptans, etc.	0-1.0
Ammonia	0.1-1.0
Hydrogen	0-0.2
Carbon monoxide	0-0.2
Trace constituents	0.01-0.6
Characteristic	Value
Temperature, °F	100-120
Specific gravity	1.02-1.06
Moisture content	Saturated
High heating value, Btu/ft^3	400-550

a) Exact percentage distribution will vary with the age of the landfill.

Source: Tchnobanoglous, Theisen, and Vigil, Integrated Solid Waste Management (1993). Reprinted with permission of the McGraw-Hill Companies.

9.3.2 MSW Collection The volume and density of MSW varies from community to community, depending on the availability of and participation in recycling, material recovery and yard waste composting programs. Generation rates alone do not provide enough information to quantify the amount of MSW that requires processing or disposal; it is also necessary to know the density, moisture content, and energy content of MSW. MSW can arrive at a processing or disposal facility at different densities, depending on the compaction achieved by the truck or transfer station that handles the waste prior to delivery. Table 9-11 provides typical ranges of compaction rates achieved by various equipment.

TIP FOR THE P.E. EXAMINATION CANDIDATE

Knowing how to interpret and use waste generation rates is valuable in solving waste collection and planning problems on the P.E. examination.

Technical and planning issues that must be considered in designing a cost-effective collection program include the volume and compaction capacity of the collection vehicles; source separation and additional handling of recyclables; the type of refuse receptacles; the number of crew members required to handle the waste and operate collection vehicles (whether mechanically or manually loaded); the collection area layout; the round-trip

TIP FOR THE P.E. EXAMINATION CANDIDATE

The candidate needs to know how to use generation rates, average densities and material balance diagrams to design properly-sized transfer stations, incinerators and landfills.

distance between the collection area and the transfer station, processing facility or landfill; and the number of trips required to serve a particular area. Economic analysis is vital in designing an MSW collection system. Tchnobanoglous, Thiesen, and Vigil (1993) provide several formulas for calculating time per collection trip using a variety of different types of collection vehicles, the truck volume needed, the number of pick-up locations and number of trips required to serve an area, labor requirements, and haul-route layout for designing a cost-effective collection system.

Table 9-11
Typical compaction ranges for municipal solid waste

	lb/yd^3 [a]
Uncompacted MSW, as generated	100 - 300
In compactor truck	400 - 800
At transfer station after compaction	600 - 800
In stationary bailer	1,000 - 1,200
Landfilled with dozer	600 - 1,000
Landfilled with compactor	1,000 - 1,200

a) Densities are highly variable depending on types of refuse and equipment

---

Example 9-3 — The impact of recycling on waste disposal volume

A municipal solid waste department plans to separate a portion of the ferrous metal, newsprint and cardboard from its MSW waste stream. The department operates 50 collection trucks, each having a volume capacity of 16 yd^3 and a compaction capability of 600 lb/yd^3. Each truck collects an average of 8 loads per week over the year, with allowances for partial loads and downtime.

a) Using the data provided in Table 9-11, calculate the amount of waste in yd^3 to be landfilled per year assuming the current generation rate remains the same and no recycling occurs.

Solution:

Step 1. Waste Volume Collected
50 trucks • 16yd^3/load • 600 lb/yd^3 • 8 loads/wk • 5 wk/yr
= 199,680,000 lb/yr

Step 2. Waste to be Landfilled
199,680,000 lb - 49,760,000 lb
= 149,760,000 lb/yr

b) Estimate the savings in the volume of waste to be landfilled per year if 25% (by weight) of the waste stream is recycled.

Chapter 9: Solid and Hazardous Waste Management

Solution:

Step 1. Recycling Volume
199,680,000 lb • 0.25 = 49,920,000 lb/yr

Step 2. Landfilling Volume without Recycling
$$\frac{199{,}680{,}000 \text{ lb}}{1{,}000 \text{ lb}/\text{yd}^3} = 199{,}680 \text{ yd}^3/\text{yr}$$

Step 3. Landfilling Volume with Recycling
$$\frac{149{,}760{,}000}{1{,}000 \text{ lb}/\text{yd}^3} = 149{,}760 \text{ yd}^3/\text{yr}$$

Step 4. Volume Savings
149,680 - 149,760 = 49,920 yd³/yr

Example 9-4 — Solid waste collection

A hauling firm is preparing a bid to provide solid waste collection services to a city. The city has a population of 4,734 people and an average per capita waste generation rate of 2.5 lb/day. An additional 25% of the city's waste comes from commercial establishments in the city that will be included in the collection contract. The waste has an average density of 300 lb/yd³ at the curb. The hauling company has 25-yd³ trucks with an average compaction rate of 550 lb/yd³. Each truck can make, on average, 8 trips per week to the processing facility to off-load the waste, with allowances for downtime. The terms of the contract restrict collection to a five day work week. How many trucks will the hauling company need to commit to this job if they win the bid?

Step 1. Residential Waste Generation
4,734 people • 2.5 lb/capita • 365 day/yr = 4,319,775 lb/yr

Step 2. Commercial Waste Generation
4,319,775 lb/yr • 25% = 1,079,943 lb/yr

Step 3. Total Waste Generation
4,319,775 + 1,079,943 = 5,399,718 lb/yr

Step 4. Truck Capacity
25 yd³ truck load • 550 lb/yd³ = 13,750 lb/truck load
13,750 lb/truck load • 8 loads/wk • 52 wk/yr = 5,720,000 lb/yr

Step 5. $\dfrac{5{,}379{,}718.75 \text{ lb / yr of waste}}{5{,}720{,}000 \text{ lb / yr truck capacity}} = 0.94$ or 1 truck

---

9.3.3 Hazard Management Both MSW and hazardous waste pose potential hazards that can affect human health and safety. Environmental engineers must design solid waste management systems to prevent unsafe situations from occurring.

The possibility of loss to the environment exists whenever wastes are transferred from one container to another. All forms of wastes must be handled in areas with good ventilation and proper emission controls to protect against inhalation and dermal absorption of toxic or hazardous vapors. When liquid materials are handled, secondary containment is necessary in the event of spillage or failure in mechanical transfer systems.

Many materials are incompatible and, if commingled, may generate severe spontaneous chemical reactions. When designing a storage system for solid and hazardous materials, physical barriers between incompatible materials are necessary to prevent accidental mixing during transfer operations or in case of a spill. Hazardous and incompatible materials may exist in a commingled state in MSW and can cause a spontaneous reaction during collection or at the transfer station or landfill. For example, the friction caused during compaction at a landfill can result in the explosion of a gas tank. Sound waste management programs provide planned contingency measures to rapidly control random chemical events.

Although regulatory measures have been or are being taken to eliminate medical waste from MSW, significant biological hazards still exist in MSW from other sources such as baby diapers, food waste, and uncontrolled sources. Operational procedures are required to minimize the spread of disease and viruses during waste handling and storage activities and to protect solid waste workers from such hazards.

Other worker safety issues in solid waste management include confined-space entry, mechanical-equipment operation, and slip-and-fall hazards. Engineered waste management systems must incorporate appropriate features to prevent accidents or deaths.

Many wastes are corrosive and require the incorporation of corrosion-resistant materials in liquid-control equipment and air-ventilation equipment. Chemical compatibility of the materials that will be in contact with MSW or hazardous waste is vital to the long-term integrity of engineered waste handling systems.

9.4 Landfilling

9.4.1 Basic Design Considerations The basic function of a modern landfill is to provide for economical placement of solid waste material into a controlled environmental setting which protects the surrounding environment. The design, operation, closure, and monitoring of a modern landfill depend upon the type of solid waste the facility is intended to accept. A modern landfill can vary from a simple facility capable of handling only relatively inert materials such as demolition debris to the more complex, multiple-liner, totally-enclosed vault-type landfill used for RCRA Subtitle C hazardous waste. Figure 9-1 provides a cross-section of both a landfill capable of disposing hazardous waste and a landfill for MSW. These are several variations of each design based on site conditions, materials of construction, and applicable safety factors.

At the beginning of design, the type and volume of solid wastes to be disposed must be determined. Once the waste type and generation rate have been established (ideally through precursor studies) design can begin. The first step is to establish the useful life span and internal disposal volume, also known as the "air volume of the landfill." Typically, landfills are designed to have sufficient capacity to contain at least 20 years of waste from their intended service area. Except for special circumstances, landfills with less than a 20 year life are generally uneconomical due to the cost and time required to site, permit, design, and construct the landfill and then maintain it during its post-closure period. At this early stage, landfill capacity is often measured in acre-ft.

Equation 9-1 presents one method to estimate the total volume required for a municipal solid waste landfill.

$$V = 1.2 \cdot t \cdot \left(\frac{W}{\rho_w}\right) \qquad [9\text{-}1]$$

where:
V = total volume for the landfill, (L³)
t = design operating life for the landfill, typically 20 to 40 yr
W = waste generation rate, mass/yr
ρ_w = density of the waste, see Table 9-11

The factor of 1.2 adjusts the total volume for the loss in available volume for the soil added to the fill as daily cover. Bagchi (1994) reports the volume of the 6-in. daily cover on the active face equals one-fifth to one-sixth of the volume of the waste. The density of the waste depends primarily of the methods of compaction both during collection and during construction of the landfill, with the latter being most important; Table 9-11 reports a range of from 600 to 1,200 lb/ft³. Corbitt (1990) reports 1,100 lb/ft³ as a reasonable estimate for landfill operations. Waste

Figure 9-1
Generalized cross sections of landfill (a) hazardous waste landfill; (b) solid waste landfill

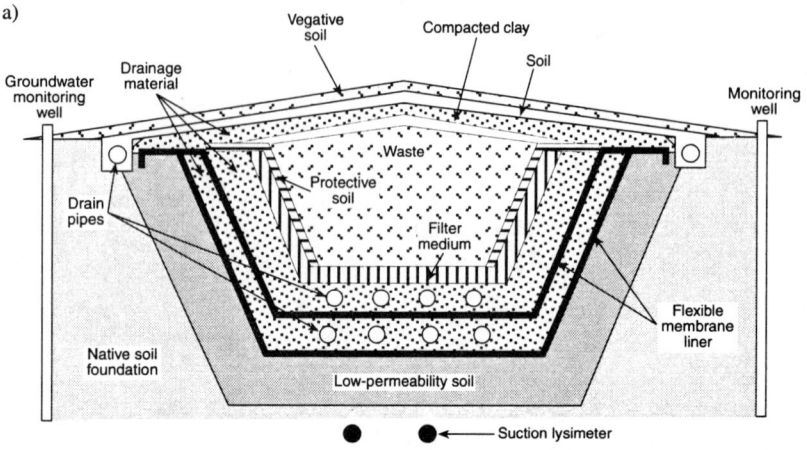

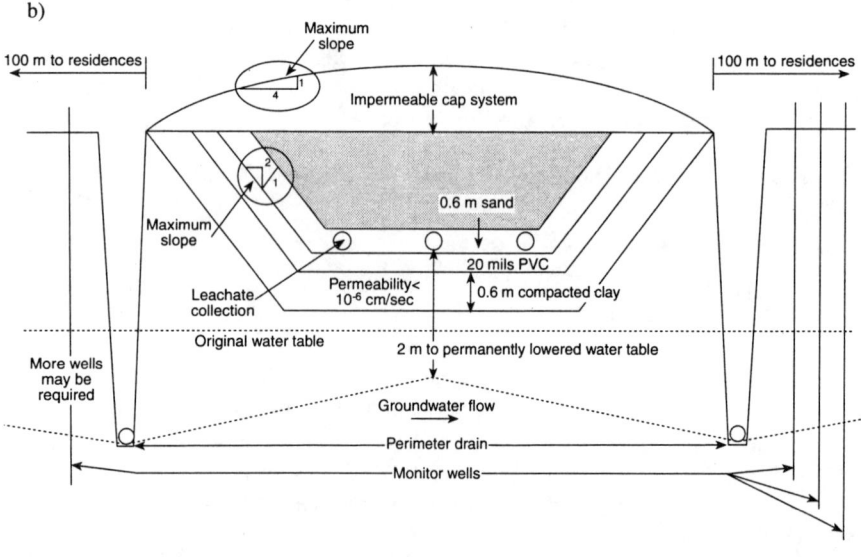

Source: Sincero, 1996

generation rates can be estimated in many ways including using historical rates when data is available, conducting waste generation rates studies, and using typical values such as 5 lb/person-day, a value seen in several references. In actual design, the commercial and industrial generation rates are the most difficult to estimate over the life time of a landfill.

TIP FOR THE P.E. EXAMINATION CANDIDATE

For the P.E. examination, waste generation rates will normally be given in multiple choice equations and the method of determining a value will be specified for essay problems.

Once the necessary landfill volume has been established, a rough sizing of the landfill footprint and depth can be established based on practical limitations and applicable (federal, state, and local) regulations. Typically, the external above-grade surface of the landfill is designed to have 4:1 (horizontal:vertical) sideslopes to facilitate construction, revegetation, erosion control, and maintenance, and to optimize internal volume.

Landfill sideslopes and terrain are often varied to integrate the site into the natural environmental features of the surrounding property and to facilitate the ultimate long-term planned use of the property. Filled areas typically have two to four percent minimum slopes to promote runoff and minimize standing water on the completed landfill surface. Systems of terraces, dikes, access roads, sideslope drains, channels and ditches are used to collect and direct stormwater from the top of the landfill so that the runoff does not create erosive velocity or flooding conditions.

The below-grade sideslopes of a modern landfill usually do not exceed 3:1, and often have flatter slopes to ensure sideslope stability and ease of construction. Slopes in excess of 3:1 can create liner stress and seaming problems when synthetic liners are used and construction can be very difficult, requiring workers to be lowered on ropes down the often slippery sideslopes of the synthetic liners. Slope stability analyses are generally performed to ensure that the design slopes of the landfill are consistent with soil conditions and provide an adequate safety factor.

Landfill depths below surrounding grade are generally constrained by the hydrogeological conditions encountered at the site or other naturally-occurring features such as bedrock. On an ideal site, the excavation depth below natural ground surface is established to provide sufficient daily and final cover such that no net import of soils is required.

Horizontal isolation distances around the perimeter of a landfill vary with local regulations and surrounding land use. At a minimum, there should be at least a 100-ft setback from the property line to the edge of the refuse boundary to accommodate the final cap, perimeter drainage systems, access roads and monitoring systems. Hazardous waste landfills typically have a minimum perimeter isolation distance of 300 to 500 ft from the waste boundary to the property line. For MSW landfills, protecting sensitive adjacent land uses such as residences, institutions, or commercial activities often requires isolation of a quarter mile or more to reasonably minimize the landfill's impact on these uses. The Federal Aviation Administration requires that landfills near airports receiving federal improvement funds be sufficiently isolated to minimize bird-strike hazards. Depending on the character of the airport, between 5,000 and 10,000 ft of separation is required, with the larger isolation distance required for runways that serve jet-propelled aircraft.

9.4.2 Siting The landfill siting process involves a very challenging balance of science, engineering, politics and public relations. The ideal landfill site is a geologic setting that will provide most, if not all, of the basic landfill soil components required for construction, i.e., liner and final cover clays and suitable drainage-layer sands. The site would be underlain with thick clay deposits having little or no groundwater, or with usable groundwater only at great depth below low-permeability clays. It would be reasonably flat to permit easy construction staging and integration into the surrounding landscape, and would be isolated from any sensitive surrounding land use, yet located close to the centroid of waste generation and have good access on a class A road to a well-constructed interchange on an interstate transportation route.

TIP FOR THE P.E. EXAMINATION CANDIDATE

The candidate should be able to develop and/or employ a simple decision matrix to numerically score, weight, and rank alternative landfill sites.

In reality, the ideal site does not exist. Siting a landfill requires a weighted compromise of many factors — politics, local opposition, hydrogeology, land use, soil type, proximity to waste generation areas, access, wetlands, flood plains, proximity to airports, and other sensitive environmental issues, such as archeology, historical preservation, and endangered species. Complex matrices are often employed using all available maps and record documents during an initial feasibility study to screen, compare and rank alternative sites. After the initial study and ranking, typically one to three sites are selected for preliminary site walkovers, environmental assessments, and reconnaissance borings. Based on the results of these examinations one site is selected for detailed investigation.

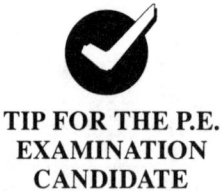

TIP FOR THE P.E. EXAMINATION CANDIDATE

The candidate must be able to design a simple hydrologic study and understand the basic use and interpretation of borings, wells, soil testing, and groundwater monitoring results.

A thorough hydrogeologic study is conducted to fully evaluate the environmental setting of the selected site. Sufficient borings and wells are installed to define the site's soil types, geologic stratification, depth to groundwater, and groundwater-flow direction.

Example 9-5 — Landfill evaluation

During an investigation to determine the suitability of a property for landfill development the following information was obtained:

The property is on a topographic high having an average surface elevation of 780 ft above mean sea level (MSL). Surface elevations decline toward a river, which is located southwest of the property and has an average elevation of 700 ft above MSL. A small pond exists at the northwest corner of the property at an elevation of 775 ft MSL.

A low-permeability clay extends from the ground surface to a depth of 50 ft. Beneath the clay is a poorly-graded sand, which has a permeability greater than the clay above. The uppermost aquifer is encountered at a depth of 60 ft (720 ft MSL) and is within the sand unit.

Using this information and the figure below, determine the following:

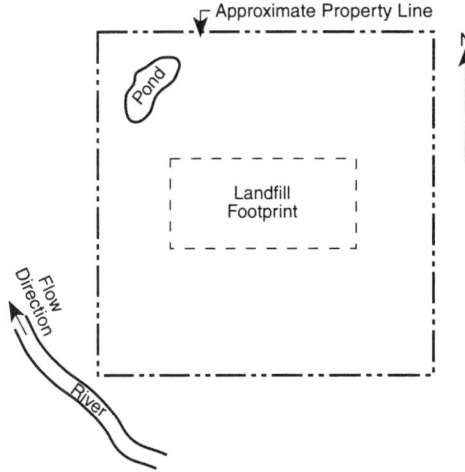

a) In which direction is groundwater most likely flowing? and why?

Solution:

Groundwater is most likely flowing toward the river and not to the pond, because the groundwater elevation (720 ft above MSL) is similar to that of the river (700 ft above MSL). In addition, the surface topography declines in elevation towards the river and water table surfaces typically mimic the land surface. The fact that the surficial soils are clay and the elevation of the pond surface (775 ft above MSL) is much higher than that of the groundwater surface (720 ft above MSL) at the site suggests that the pond is perched and its water level is sustained by surface water runoff and not groundwater.

b) Using a minimum number of wells to keep costs down, how would the groundwater flow direction at the property be determined? Wells should be strategically placed to be useful for future landfill monitoring needs

Solution:

The minimum number of wells to determine the direction of groundwater flow is three. The three points should be placed such that they form a triangle enabling solution as a standard three-point problem. Because it is judged that groundwater flow is to the southwest, one well should be placed to the northeast of the landfill footprint. This will provide an upgradient monitoring location for background water quality. One of the remaining wells should be placed on the assumed downgradient side of the landfill footprint to be used to detect potential releases from the landfill into the groundwater. The third well should be placed such that it completes the triangle, and preferably, also downgradient of the landfill.

c) In which geological unit should the screen of each well be placed to determine the groundwater flow direction? How far into the water table should the screen extend, and why?

Solution:

The geologic unit in which the screen should be placed is the sand because it is the upper-most aquifer and would be the unit of concern if the landfill were to leak. The screen should be placed such that the water table surface intercepts the screen.

TIP FOR THE P.E. EXAMINATION CANDIDATE

RCRA subtitle D regulations outline the basic design elements required in landfill liner and final cover design, as well as a variety of other ancillary requirements. The candidate should be generally familiar with these regulations and have them available during the examination for reference.

After all the siting studies have been completed, design of the landfill begins. Design is usually divided into two stages. The first stage, generally referred to as the "permit-level design," provides the necessary information for the regulatory review and permitting process. The second phase, "construction-level design," provides the details required for construction of the landfill if the site receives a construction permit. The permit-level design involves the conceptual development of the engineering features needed to control runoff and runon, to control or avoid groundwater at the base of the landfill, to protect area groundwater, to control gas emissions, and to meet the basic volume and regulatory requirements for the site.

9.4.3 Soil Liner Systems

Modern landfill technology requires the installation of a liner system to collect leachate at the base of a landfill and prevent its escape into the surrounding environment. Leachate, which results from the downward migration of liquid through the waste, has site-specific characteristics related to the type of waste and cover material, waste placement, local climatic conditions, and waste age which must be accomodated in liner design.

Constituents in the leachate may cause the degradation of liner materials. It is essential that the engineer determine the potential constituents in a particular waste stream and evaluate the impact of those constituents on the various liner materials. The most reliable method of demonstrating that the leachate will not cause degradation of synthetic liner materials is to conduct a leachate compatibility study. This study employs an EPA-prescribed protocol to expose the liner material to a representative sample of the landfill's leachate under controlled conditions for a specified duration ranging from 30 to 120 days. The degradation of the material is evaluated at the conclusion of the test period(s) using a series of mechanical properties tests. If the specific leachate characteristics are not known in advance, published information is used to assess compatibility. Table 9-12 reports the typical composition of leachate from MSW landfills.

Table 9-12
Composition of leachate from landfills

Consituent*	Range	Typical
BOD_5 (5-day biochemical oxygen demand)	2,000-30,000	10,000
TOC (total organic carbon)	500-20,000	6,000
COD (chemical oxygen demand)	3,000-45,000	18,000
Total suspended solids	200-1,000	500
Organic nitrogen	10-600	200
Ammonio nitrogen	10-800	200
Nitrate	5-40	25
Total phosphorus	1-70	30
Ortho phosphorus	1-50	20
Alkalinity as $CaCO_3$	1,000-10,000	3,000
pH	5.3-8.5	6
Total hardness as $CaCO_3$	300-10,000	3,500
Calcium	200-3,000	1,000
Magnesium	50-1,500	250
Potassium	200-2,000	300
Sodium	200-2,000	500
Chloride	100-3,000	500
Sulfate	100-1,500	300
Total iron	50-600	60

*All units in mg/L except pH
Source: Corbitt (1990)

The liner configuration required at a specific landfill depends on the type of waste disposed. For instance, several industries generate high-volume, low-hazard waste, which will have significantly lower potential for impact on the environment than will MSW. The liner requirements for these two types of conditions would be different. The regulatory requirements in many states allow the installation of a less sophisticated liner for many industrial applications than those required at MSW sites.

Soils are often used to line landfills. Historically clay-rich soils have been the most common material used for the construction of landfill liners because relatively low hydraulic conductivity can be achieved with soils high in clay content. The selection of the liner soil material depends on several factors. The major factor, according to EPA recommendations (40 CFR 258.40), is that the soil liner have an in-place hydraulic conductivity less than $1 \cdot 10^{-7}$ cm/sec. Other characteristics such as the extent of clay content, Atterburg limits, and the moisture-density relationships of the soil, are important considerations which affect the constructability and integrity of the liner. Soil characteristics are dictated by local availability of soils (on-site or within an economical haul distance) and their use is governed by state and/or local review agency requirements. A properly-constructed clay liner will be uniform and homogeneous throughout its cross-section. The ability to achieve this goal is determined by several variables including: the soil's water content, type of compaction equipment used, size of clods in the soil, and the "knitting" of material between adjacent lifts. The most important of these variables are the relationships between moisture, density, and permeability in the selected material and the soil water content during the course of construction.

Figure 9-2 shows the relationship between the moisture content and the dry density of a typical clay soil. The moisture content corresponding to the *maximum dry*

Figure 9-2
Relationship between the moisture content and the dry density of a typical clay soil

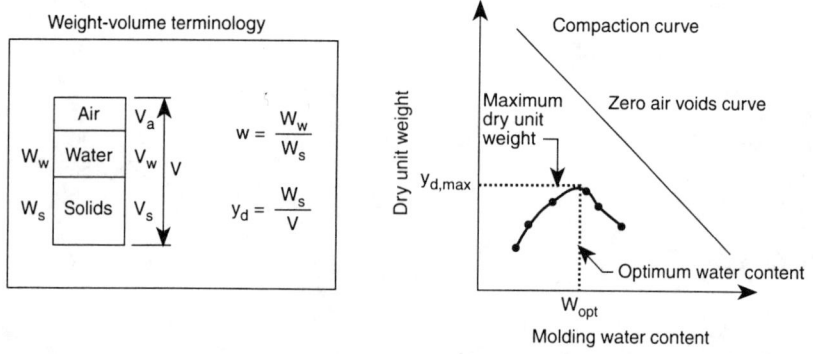

unit weight (gd) is referred to as the optimum water content. The zero air voids curve, also known as the 100% saturation curve is a curve that relates dry unit weight to water content for a saturated soil that contains no air.

It is common practice to install a clay liner slightly on the wet side of the "optimum water content" for maximum density in order to take advantage of the relatively low hydraulic conductivity that can be achieved at "wet of optimum" and avoid the rapid increase in permeability that occurs when soils are compacted "dry of optimum." Installing clay soil marginally on the wet side of optimum (1% to 4%) also provides a construction advantage. Under these moisture conditions, the clay will be more workable than a dryer clay, clods will break-down more easily, and the material can be "kneaded" into a relatively consistent, homogeneous matrix. Conversely, relatively dry clay will hamper breaking of any clods present and be difficult to compact, creating voids which increase the overall permeability of the liner. Moisture conditioning is often required during construction. If the material is too dry, water is added to increase the moisture content, usually by spraying water from a tanker over the loosely-placed material before it is worked. If the water content of the clay is too high, the material can be bladed and/or disked to expose more of the material to the atmosphere and accelerate drying before compaction.

It is important during liner design to ensure that the liner will be compacted to a water content and dry unit weight that will lead to low hydraulic conductivity and adequate engineering performance. Specifications of water content/dry unit weight criteria should be based upon test data specific to the particular soil to be used in construction. Given that compactive effort will vary in the field, a rational approach is to select several compactive efforts in the laboratory that span the range of compactive effort anticipated in the field. The water content/dry unit weight criterion that evolves would then be expected to apply to any reasonable compactive effort.

The Modified Proctor laboratory test represents a reasonable upper limit on the compactive effort likely to be delivered to the soil in the field. The Standard Proctor compaction test (ASTM D-698) represents a medium compactive effort. A reasonable estimate of the lower limit of compactive energy can be obtained using the "reduced compaction" modification in which standard compaction procedures (ASTM D-698) are followed except that only 15 drops of the hammer per lift are used instead of the usual 25 drops. The key is to span the range of compactive effort anticipated in the field with laboratory procedures. (USEPA, 1993).

One satisfactory approach, adapted from USEPA, (1993), is as follows:

1. Compact soil in the laboratory with modified, standard, and reduced compaction procedures to develop compaction curves as shown in Figure 9-3a. As few as two compaction procedures can be used if field construction procedures make either the lowest or highest compactive energy irrelevant.

2. The compacted specimens should be permeated, e.g., per ASTM D-5084. The measured hydraulic conductivity is then plotted as a function of molding water content as shown in Figure 9-3b.

3. Figure 9-3c shows the dry unit weight/water content points replotted with different symbols used to represent compacted specimens with hydraulic conductivities greater than the maximum acceptable value and specimens with hydraulic conductivities less than or equal to the maximum acceptable value. An "Acceptable Zone" should be drawn to encompass the data points representing results meeting or exceeding the design criteria.

4. The Acceptable Zone should be as shown in Figure 9-3d based on other engineering considerations such as shear strength and/or shrink/swell potential. Additional tests are usually necessary in order to define the acceptable range of water content and dry unit weight that satisfies both hydraulic conductivity and shear strength criteria. Figure 9-4 illustrates how Acceptable Zones derived from hydraulic conductivity and shear strength can be overlapped to define a single Acceptable Zone.

Figure 9-3
Clay soil moisture content determination

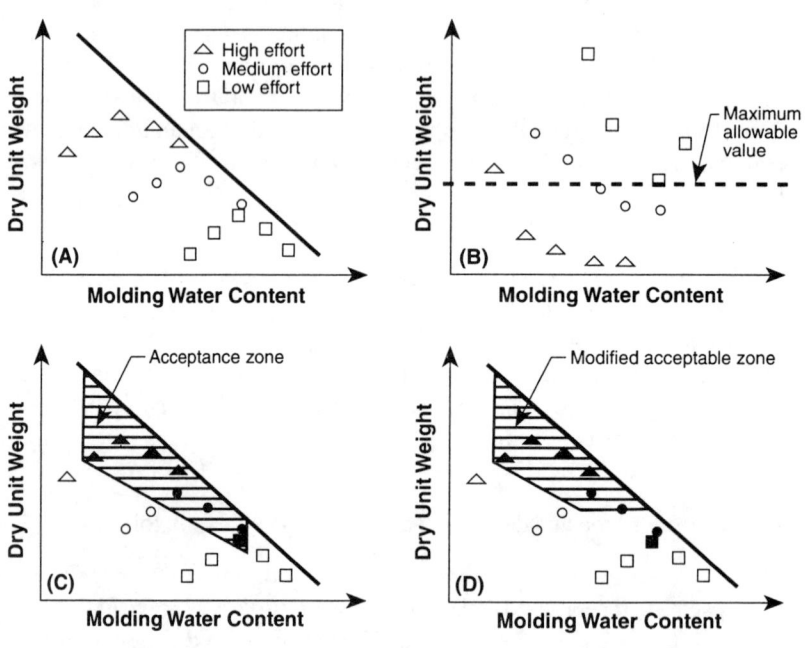

Figure 9-4
Clay soil moisture content derived from hydraulic conductivity and shear strength

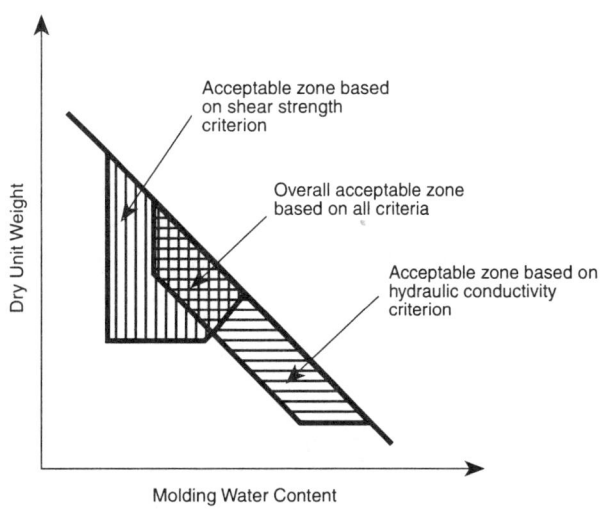

The selection of proper compaction equipment is critical to the construction of a liner with an acceptable hydraulic conductivity. The equipment should provide both a clod-breaking and a knitting/kneading action. A sheepsfoot roller, with feet long enough to fully penetrate the loose soil lift and partially penetrate the underlying lift, is commonly used. The feet provide the knitting action necessary to break up clods in the loose lift and at the same time penetrate the underlying lift to blend the two lifts into one homogeneous mass. Another technique which is often used to increase the blending between adjacent lifts is to scarify the surface of the underlying lift with a disc prior to placing the succeeding lift. Discs can also be used to break clods prior to compaction. Equipment weight is another important consideration. Generally, heavier equipment will provide the necessary compactive effort with fewer passes than lighter equipment. The tracking of loaded scrapers over previously-placed soil during construction can be used to supplement the compactive effort of roller compactors.

If a cost-effective source of acceptable clay is not available locally, onsite soils can be blended with an imported conditioning material to achieve the required hydraulic conductivity. The most common conditioning material is bentonite, a smectite mineral clay mined and manufactured to produce a commercial product with consistent characteristics. Mixing is accomplished in a pugmill or by spreading the material and working it into the in-place soils in thin lifts with construction or rototilling equipment. After the composited material is adequately blended, it is spread in uniform lifts, combined with an appropriate amount of water to achieve the desired water content as determined by laboratory tests, and then compacted.

Another increasingly popular way to reap the advantages of bentonite in a soil liner is to use bentonite attached to or sandwiched between geotextiles or geomembranes. These bentonite mats are much easier to use than soil admixtures and are considered especially effective when placed immediately beneath a flexible membrane liner because of bentonite's swelling and sealing capabilities.

Alternative soil materials that do not provide a maximum hydraulic conductivity less than $1 \cdot 10^{-7}$ cm/sec may be considered by design engineers and regulatory agencies for specific cases. The primary factors, in these instances, include the type of waste and its leaching characteristics, local hydrogeologic and climatic conditions, and applicable regulations.

A compacted clay liner is subject to degradation once installed. Desiccation-cracking caused by drying and/or the impacts of freezing/thawing during and after construction will increase the liner permeability and compromise its ability to control leachate. The degree of desiccation can be reduced significantly by covering the liner with a drainage layer or some other type of protection. The depth of the initial waste layer should be adequate to protect the liner from freezing. During warm weather, the liner should be kept well-watered prior to placement of the overlying protective layers to prevent dessication. If dessication cracks occur, they must be scarified and recompacted to the original liner specifications.

Although soil liners are generally constructed to achieve a maximum hydraulic conductivity less than $1 \cdot 10^{-7}$ cm/sec, the soil liner is considered a porous material. The fundamental law governing groundwater flow and velocity in porous media flow evaluations was developed in 1856 by Henry Darcy, a French hydraulic engineer. Darcy's Law (see also Chapter 5), developed through experiments where water was made to flow through cylinders filled with sand, is expressed as follows:

$$Q = -KiA \qquad [9\text{-}1]$$

where: Q = the flow rate
K = the saturated hydraulic conductivity (typically expressed in cm/sec)
i = the nondimensional hydraulic gradient of groundwater flow, (equal to the change in height divided by the change in length)
A = the cross-sectional area of flow

Darcy's Law illustrates that the flow of liquids through a soil liner is directly proportional to the hydraulic conductivity of the soil and the hydraulic gradient. The hydraulic gradient is directly proportional to the depth of leachate above the liner.

Under equal hydraulic gradients, migration through a saturated soil with a hydraulic conductivity of $1 \cdot 10^{-6}$ cm/sec will be 10 times greater than soil with a hydraulic conductivity of $1 \cdot 10^{-7}$ cm/sec. It is, therefore, important to achieve minimum hydraulic conductivity during the construction of a soil liner and to minimize leachate accumulation during operation.

Darcy's law can be rearranged to define specific discharge (μ), also known as Darcy velocity, which is:

$$\mu = \frac{Q}{A} = -Ki \quad\quad [9\text{-}2]$$

The actual velocity of groundwater through porous media must also account for the fact that the cross-sectional area of flow includes both granular solids and the voids through which the groundwater actually flows. The following equation applies:

$$v = \frac{-Ki}{n} \quad\quad [9\text{-}3]$$

where: v = average linear groundwater velocity (pore velocity)
n = effective porosity of the soil media

The uses and limitations of these Equations in defining groundwater flow and velocity are described in a number of references, including those by Freeze and Cherry (1979), Todd (1959), and Fetter (1965).

9.4.4 Flexible Membrane Liners
The development of synthetic materials since the 1970s has provided environmental engineers with several options for landfill liners, in addition to clay and other soils, which can be tailored to satisfy a variety of landfill conditions and needs. Flexible membrane liners (FMLs) are constructed of continuous sheets of polymeric materials supplied in rolls with widths ranging from 10 to 20 ft. Seams between adjacent panels are field constructed using chemical-bonding or heat-welding techniques. Some of the many material options include the following:

TIP FOR THE P.E. EXAMINATION CANDIDATE

The candidate should understand how to calculate groundwater gradient, understand the measurement of hydraulic conductivity in porous media, and be able to calculate velocity using the above equations. In addition, a basic understanding of aquifer and well analysis, as described in the above references, will be useful.

- Polyvinyl Chloride (PVC)
- High-Density Polyethylene (HDPE)

- Linear Low-Density Polyethylene (LLDPE)
- Chlorinated Polyethylene (CPE)
- Ethylene Propylene Diene Monomer (EPDM)

Each of these materials has unique characteristics described in the manufacturers' literature that make it suitable for specific applications. Additionally, these polymeric materials may be manufactured with an integral grid of fiberglass reinforcing strands. These fiber-reinforced materials have increased strength and puncture resistance.

A major design consideration in material selection is chemical compatibility. Leachate compatibility testing, as prescribed by EPA Method 9090, must be conducted on the material proposed for a specific installation to verify that the material will not degrade when exposed to the site-specific leachate.

A second major design consideration for a FML is physical stress. This is particularly true on the sideslopes of a landfill. The FML, which is anchored at the top of the sideslope, must be capable of supporting its own weight, as well as stresses induced by waste settlement. If the FML cannot overcome these loadings, the material will fail by tearing. In addition, the anchor trench must be designed to prevent the FML from pulling out. FML design requires that the material's specific gravity, friction angle, thickness and yield stress be known. In addition, the engineer must know the friction angle between the FML and the underlying supporting materials.

An FML provides a relatively impermeable barrier. The major factor that influences the amount of leakage through the FML is the number of penetrations in the completed installation. Penetrations are generally caused by seaming defects and pinholes that may be the result of improper manufacturing, subgrade preparation and/or construction techniques. The flow through FML pinholes or defects can be approximated with orifice hydraulics.

A rigorous construction quality assurance (CQA) program is essential in installing FML liners. The program, often dictated by state regulations, must include material conformance tests prior to placement and both non-destructive seam tests and destructive seam tests during construction. Material conformance testing is generally completed in a materials laboratory, prior to construction, on samples taken from the actual rolls to be installed. Tests are conducted to verify that the material characteristics of the FML meet the project specifications. Non-destructive seam tests are conducted on the FML following the completion of seaming. The specific test to be conducted depends on the type of material and type of seam. In general, the non-destructive seam test will indicate whether the seam has a pinhole or other defect. Destructive seam tests are conducted on a specimen cut out of the completed seam. A specimen is subjected to various types of load testing until it fails.

The results of the destructive tests are compared to the values included in the CQA plan to determine whether the seam is acceptable.

An FML must be installed over a carefully prepared and approved subgrade. The subgrade should be properly compacted and free from rocks, clods, sticks, and other significant irregularities. The surface should be rolled smooth with a steel-wheel roller just prior to installation. Prior to the deployment of the FML panels, the rolls must be inspected for defects and the manufacturer's testing results evaluated. If no defects are present, the material should be unrolled, taking care to minimize sliding of the material over the subgrade. The deployment techniques must result in the prescribed overlap between adjacent panels to allow for proper seaming. Ballast, most often sand bags (without wire ties), should be placed on the panels concurrently with the deployment process to prevent the panels from being lifted by the wind. Seaming should be conducted as soon as practical by qualified personnel on clean, dry material.

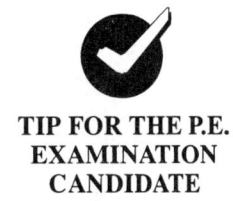

TIP FOR THE P.E. EXAMINATION CANDIDATE

The candidate should be generally familiar with seaming and testing techniques for PVC and HDPE, which are the most commonly employed FMLs.

The FML material may be subject to degradation from exposure to sunlight. This is generally not a problem because the FML is covered by granular material of the leachate collection and recovery system shortly after the FML is installed. Both the FML and compacted soil components should be protected from frost. Freezing/thawing cause displacement of the soil component and can cause significant stress in the FML material. The installation of a lift of waste thick enough to protect the composite liner from freezing will control damage from freezing and thawing.

9.4.5 Composite Liner Systems Current design practices for MSW and hazardous waste landfills incorporate composite and double-composite liner systems (see Figure 9-1). A composite liner is comprised of an FML overlying a low-permeability soil liner. A composite liner will (theoretically) outperform both FML and clay liners alone. In the composite liner, the FML is placed directly over the clay liner. Because the FML is in direct, firm contact with the clay material, leakage through penetrations in the FML will not spread out between the two layers. Since the saturated area under the opening in the FML is relatively small, the clay liner assures that the liner system will perform as intended.

Many liner designs incorporate a second composite liner below the upper (primary) composite liner separated by a leachate collection and recovery layer/system (described in detail in section 9.4.7). This configuration can be used to detect and measure leachate movement through the primary liner, adding an important component to the environmental monitoring capability and performance assurance at the site.

9.4.6 Final Cover System   The final cover system on a landfill serves the important function of controlling the amount of precipitation from rainfall or snow melt that percolates into the waste. It is important to limit precipitation entering the landfill to reduce the amount of leachate generated. Current standards require that the permeability through the landfill cover be less than or equal to the permeability of the installed bottom liner system. In addition, the final cover must be designed to minimize erosion from wind and water, accommodate settlement and subsidence of the underlying waste, minimize maintenance requirements, and be consistent with the ultimate land use planned for the site.

The final cover is a multi-layer system with each layer serving a specific function. A properly-designed and constructed cover system will facilitate runoff rather than infiltration, sustain a good vegetative cover that minimizes erosion damage to the cover, provide a barrier layer that restricts percolation into the waste, include drainage layers and protection layers above the barrier layer and, in many cases, include a gas management layer(s) below the barrier layer.

The uppermost layer is the vegetative layer, which normally consists of two parts. The top part provides the medium, usually topsoil, to support the final cover vegetation. The lower part of the vegetative layer protects the underlying layers from damage caused by root penetration and also provides frost protection in northern climates. A good vegetative cover will include a mixture of grasses with minimum root penetration. The final cover is subject to the erosive forces of wind and water. The major source of erosion is stormwater runoff. The potential for erosion is influenced by the slope of the final cover, type of soil, and integrity of the vegetation. During facility design, it is important to estimate the amount of erosion that will occur using the Universal Soil Loss Equation to ensure that the estimated soil loss does not exceed design standards.

The lower part of the root-zone/frost-protection layer is often designed to have high permeability to promote drainage away from the top of the underlying barrier layer. Provisions must be included in the design to collect the water from this drainage layer at the landfill perimeter and discharge it to adjacent waterways. Common materials used for the drainage layer are granular soils (sand) or synthetic materials such as geonets.

A barrier layer is provided underneath the vegetative and drainage layers. Its purpose is to limit the amount of stormwater infiltration that reaches the waste. The material(s) used for the barrier layer will be a compacted soil, a FML, or a soil/FML composite similar to those described in the preceeding sections 9.4.3 to 9.4.5.

9.4.7 Leachate Management   Most modern landfills are designed with sophisticated leachate collection and recovery systems (LCRS) to control and collect leachate above the underlying liner. The LCRS is placed directly above the completed liner system and removes leachate for treatment and disposal. The current performance standard for a properly-designed LCRS is to limit the leachate head above the liner to a maximum of 12 in. LCRS components include a drainage blanket, collection piping system and leachate removal facilities.

The drainage blanket is placed above the liner on both the sideslopes and landfill base. The function of the drainage blanket is to provide relatively unimpeded flow of leachate to the leachate collection piping. Common materials for this purpose are granular soils, such as sand and gravel, and synthetic geomembranes, geonets and geocomposites. The environmental engineer must consider several factors when selecting the drainage blanket material, including the effective hydraulic conductivity of the material, stability (particularly on the sideslopes), material strength, and the potential for clogging. Clogging can result from biological sources and from the intrusion of fines. A proper understanding of potential types of clogging and methods of control is essential to the design of a proper drainage blanket. The landfill base is generally designed to slope toward the leachate collection piping interceptors.

The leachate piping system is typically constructed of perforated drainage pipe and collects the leachate from the drainage blanket and transports it to the leachate removal facilities. PVC and HDPE are the most common materials used for collection piping because of their chemical compatibility with most landfill environments. The pipes must be designed to withstand the loads induced by the overlying waste, provide adequate flow capacity for the leachate, and provide cleanout and inspection capabilities.

Leachate withdrawal in early landfill designs was commonly accomplished by the extension of solid wall piping through the landfill's sideslope. A penetration of the landfill liner was required wherever this occurred creating a potential pathway for leachate to leak from the landfill at each penetration. Current design practice favors the installation of leachate collection sumps along the landfill sidewall as part of the landfill liner and LCRS. Access to the sumps is provided by large-diameter riser pipes above the sidewall liners. Withdrawal pumps, related discharge piping or hose, and power conductors are lowered into the collection sump through the riser pipes to pump the collected leachate to a treatment/disposal facility.

Leachate must be properly treated prior to disposal. Several options exist for managing leachate after it is pumped from the landfill. Storage and loadout facilities can be installed to accommodate a quantity of leachate until it can be transported to a wastewater treatment plant. If sanitary sewer service is available, the facility may be allowed to discharge leachate directly to the sewer system under controlled conditions or after prescribed pretreatment. Or, a wastewater treatment plant can be constructed to treat the leachate to the degree required for a permitted discharge to adjoining streams or lands.

The volume of leachate generated at a landfill and the efficiency of the liner system can be estimated using EPA's "Hydrologic Evaluation of Landfill Performance" (HELP) computer model. In absence of specific data, EPA has defined the minimum leachate generation to be 10% of the annual rainfall falling on the surface of the landfill according to Corbitt (1990). After the leachate generation and liner efficiency are determined, the leachate head on the landfill can be calculated for a given collection pipe spacing, base liner slope and thickness and hydraulic conductivity of the drainage blanket.

Example 9-6 — Leachate collection pipe and drainage layer sizing

A 10-acre lined landfill is constructed with a sand drainage layer tributary to a system of leachate collection pipes. The leachate collection pipes will be perforated, 6 in. nominal diameter, SDR-9.33, high density polyethylene (HDPE). HDPE has a Manning's "n" value of 0.009. The pipes will be set at minimum 1.0% flow grades (post-consolidation) towards the leachate collection sumps. The hydraulic conductivity of the sand drainage layer is 0.1 cm/sec.

The Hydraulic Evaluation of Landfill Performance (HELP) Model estimates the peak infiltration rate into the drainage area to be equivalent to 0.00002 cm/sec. Leachate flow rates are also derived from HELP Model which calculates the peak daily lateral drainage rate in the leachate collection layer to be approximately 2 in. For the 10-acre site, this is equivalent to 72,600 ft^3/day, or 0.84 ft^3/sec. The maximum drainage length along the bottom of the landfill cell is 200 ft. The slope of the base grades is 1% (0.57 deg) to the leachate collection pipeline.

a) Determine the maximum flow capacity of leachate collection pipes and their adequacy to handle anticipated leachate flow rates.

Solution:

a) Assume a peak gravity flow depth of 95%. Since open channel flow in pipes (gravity flow) occurs up to a flow depth equal to 95% of the pipe diameter, the pipe's maximum flow area, wetted perimeter, and hydraulic radius can be determined using established ratios for partially-filled pipes.

The pipe flow velocity can be determined using Manning's formula for open channel flow:

$$V = (1.49 / n) \bullet Rh^{2/3} \bullet Sp^{1/2} \qquad [9\text{-}4]$$

where:
V = pipe flow velocity (ft/sec)
n = Manning's coefficient (dimensionless)
Rh = hydraulic radius (ft)
Sp = pipe slope (ft/ft)

Chapter 9: Solid and Hazardous Waste Management

thus,

$$V = (1.49 / 0.009) \bullet (0.2864 \bullet 0.5 \text{ ft})^{2/3} \bullet (0.01 \text{ ft} / \text{ft})^{1/2}$$

where: $V = 4.53$ ft per second

The flow capacity of the pipe is equivalent to the product of the pipe flow velocity and the flow area.

$$Q = V \bullet A \qquad [9\text{-}5]$$

where: Q = pipe flow capacity (cubic ft per second)
A = Pipe flow cross sectional area (ft^2)

thus,

$$Q = (4.53 \text{ ft} / \text{sec}) \bullet (0.7707 \bullet (0.5 \text{ ft})^2)$$

$$Q = 0.87 \text{ ft}^3/\text{sec}$$

This is greater than the peak leachate drainage rate of 0.84 ft^3/sec estimated by the HELP Model. The leachate collection pipe is adequately sized.

b) Determine the maximum potential head buildup on the liner resulting from leachate drainage from the farthest point of the landfill cell to the leachate collection pipeline.

Solution:

Leachate head can be estimated using Moore's (1980, 1989) methodology:

$$Y = L \bullet [(r / K + \tan^2 \alpha)^{1/2} - \tan \alpha] \qquad [9\text{-}6]$$

where: Y = maximum saturated depth on the landfill liner (ft)
L = maximum horizontal drainage distance (ft)
r = rate of infiltration into the drainage area (cm/sec)
K = saturated hydraulic conductivity of the drainage layer (cm/sec)
a = slope angle, measured from the horizontal (deg)

$$Y = (200 \text{ ft}) \bullet [(0.00002 / 0.1 + \tan^2(0.57\text{deg}))^{1/2} - \tan(0.57\text{deg})]$$
$$Y = (200 \text{ ft}) \bullet [(0.0002 + 0.0001)^{1/2} - 0.01]$$
$$Y = 1.46 \text{ ft}$$

This value exceeds the maximum allowable head of 1 ft (30 cm). Either the configuration of the drainage layer or the material characteristics of its components needs to be revised.

9.4.8 Landfill Gas Management The generation and fate of anaerobic decomposition gases in landfills has been extensively researched using both field and laboratory studies. More recently, studies have correlated exterior influences such as atmospheric pressure to subsurface movement of landfill gases. These studies have quantified the extent of lateral gas migration, vegetation damage, and enabled the design of landfill gas control and recovery systems.

Landfill gas is generated during the anaerobic decomposition of organic matter. It is composed of approximately 40 to 60% methane (CH_4), 40 to 60% carbon dioxide (CO_2), and trace-levels of non-methane organic compounds (NMOCs) and volatile organic compounds (VOCs). These gases move under pressure and concentration gradients within the landfill. Unless they are managed, they can eventually leave the site, moving up into the atmosphere or migrating horizontally to adjacent property. Exterior influences such as atmospheric pressure and other climatic conditions influence the manifestation, longevity, and regularity of gas migration. Regulatory and health concerns related to these gases include:

- Explosive gas buildup
- Hazardous air emissions
- Groundwater contamination
- Odor control
- Greenhouse gas buildup

These concerns can be minimized by early detection, which can be accomplished by properly-trained site personnel without the need for specialized equipment.

Landfill gas migration results from convection, the movement of landfill gas in response to pressure gradients, and diffusion, the movement of gas from areas of higher to lower concentration. Anaerobic decomposition within a landfill produces gas pressures generally in the range of 1 to 12 in. of water column. Instances of landfill gas pressures exceeding 100 in. of water column have been documented. Methane concentrations within a landfill in excess of 50% by volume in air create the concentration gradients necessary for diffusion.

Monitoring techniques and equipment have been developed to monitor surface and subsoil gas concentrations, pressure variations, and other factors that may influence migration. The complexity and type of instrumentation depends largely upon the information required. Magnehelic-style pressure gauges can be used to monitor landfill pressures through subsurface monitoring probes. Methane and carbon dioxide concentrations can be measured by using direct-reading instruments designed to monitor explosive atmospheres. NMOC and VOC concentrations must be determined through laboratory analysis of discrete samples.

The typical methane concentration in landfill gas, at approximately 40 to 60% by volume in air, exceeds its explosive range, 5 to 15% by volume in air. Since oxygen is not present in the landfill after the early stages of decomposition, the risk of an explosion or fire is limited to those instances when the landfill gas escapes the landfill, is diluted with air, and accumulates in a structure. Accumulation is possible because methane has a specific gravity lower than air. Uncontrolled subsurface migration of landfill gas has caused explosions and loss of life and property. Methane's explosive potential is the single largest hazard associated with landfill gas and the reason it is closely regulated. Conversely, methane's energy content is the reason that collected landfill gas can be used as an energy resource.

The frequency and location of areas where landfill gas is detected can vary because of changes in atmospheric pressure and the degree of saturation of the soil surface. These influences are important in documenting the presence of landfill gas with appropriate meters after its presence is first recognized. It has been demonstrated that, during periods of increased atmospheric pressure, landfill gas migration can be reduced because higher pressures within the landfill are required to push the gas outward. Conversely, during periods of low atmospheric pressure, landfill gas migration is encouraged. The change in migration trends can occur in a relatively short time period. Similar fluctuation in migration patterns can be caused by temporary seasonal sealing of the surface soils over and around a landfill by frost penetration during the winter months and precipitation during other seasons. Because of these fluctuations, landfill owners must regularly monitor for gas in order to detect these fluctuations and avoid potentially hazardous situations.

Landfill gas migration into surface soils, whether it occurs on or off site, can often be visually detected by observing discolored or dying vegetation. Landfill gas, itself, is generally not toxic to vegetation; rather, it affects certain plants by excluding oxygen from the root zone. When vegetation stress occurs on sideslopes, surface soil erosion and gully formation can damage the cover systems. Similar damage can occur off the landfill property limits to woodlots, lawns, orchards, and ornamental shubbery and trees. Offsite gas migration can occur while the site is operational, or it may not become apparent until after the site has been closed. However, stressed or dead vegetation surrounding a landfill is not necessarily caused by gas migration. Differential settlement on the landfill site can result in ponding of surface water on portions of the landfill which, if continued for an extended time, can also damage vegetation in those areas. Altered offsite drainage patterns that cause siltation or erosion is another cause of vegetation damage.

Hazardous air emissions related to landfill gas consist of VOCs and NMOCs. While traditionally these emissions have been unregulated, state and federal regulations now exist to control these emissions. These compounds, while present only in trace concentrations, present threats to human health and the earth's atmosphere. Additionally, as landfill gas moves outward from a landfill through the subsoil, some constituents can become dissolved in groundwater, causing additional contaminant transport.

Gas can be controlled by passive or active systems. Passive systems rely on natural pressure, concentration gradients and barriers, while active systems create a positive or negative pressure gradient to control migration. Selection of a control system is driven by regulatory requirements and site characteristics. Passive venting systems can control convective gas flow, but do not affect diffusive flow unless an impermeable barrier underlies the landfill. Active systems are effective in controlling both convective and diffusive flow.

Conceptually, lateral gas movement can be controlled by providing a pathway that is consistently more permeable than the surrounding soil. *Passive* control systems consist of deep gravel-filled trenches, shallow vent trenches placed beneath the final cover, vent wells, and low-permeability barriers that direct collected gases to venting structures discharging to the atmosphere. To be effective, passive systems constructed around the perimeter of the landfill should extend to an impermeable geologic unit or the water table. Low-permeability gas barriers (such as synthetic liners, admixed materials and natural soil) can be very effective in controlling all types of migration. When low-permeability barriers are used without passive venting capability, they may create vegetation stress problems inside the barrier.

Active gas control systems control gas migration through development of a vacuum in the soil, which either collects soil gases within the zone of influence or by reducing the pressure to the extent that it is to overcome by the pressure generated within the landfill. Negative-pressure systems are an accepted and widely-used control strategy and are efficient in providing the degree of safety required. Depending on location, the extracted gas may be exhausted to the atmosphere, combusted to prevent migration of odors and emissions, or recovered for use. Specific site data will determine whether the extraction system should be placed within the refuse boundaries or on the site exterior. Positive-pressure gas control systems have not been as widely accepted as negative-pressure systems for migration control. The primary limitation is the lack of accurate design criteria to provide adequate soil gas pressure, yet limit air infiltration into the refuse.

The energy value of landfill gas is directly related to its methane content. A landfill gas with an average methane concentration of 50% has an energy value of approximately 500 Btu/ft^3. Landfill gas can be used as fuel for a variety of applications or upgraded and converted to compressed natural gas. Energy-recovery options that have been implemented successfully include direct burning in industrial boilers, direct fueling of engines to run electrical generating equipment, and upgrading of the gas to pipeline quality for direct sale to a utility. Many of these options can be implemented with minimal gas treatment, but care must be taken to minimize the impacts of potentially corrosive compounds that form as the gas is compressed and combusted.

9.5 Incineration/Energy Recovery

9.5.1 Introduction Combustion of MSW converts the organic constituents into their natural combustion end-products of carbon dioxide and water. The process involves four sequential phases — drying, ignition, volatilization, and, finally, combusion.

The chemical reactions in combustion can, in broad terms, be represented as follows:

$$2 C + O_2 = 2 CO \qquad [9\text{-}7]$$

$$2 CO + O_2 = 2 CO_2 \qquad [9\text{-}8]$$

$$C + O_2 = CO_2 \qquad [9\text{-}9]$$

$$C + H_2O = CO + H_2 \qquad [9\text{-}10]$$

$$2 H_2 + O_2 = 2 H_2O \qquad [9\text{-}11]$$

Drying and gasification usually occur in the first 40% of the burning grate surface in a mass-burn furnace. However, it is difficult to pinpoint specific reaction zones within the furnace because of the heterogeneous nature of MSW. In suspension burning, such as in a refuse-derived fuel (RDF) plant, the waste is injected into the combustion chamber and drying, ignition, and volatilization occur throughout the combusion chamber. Such furnaces are usually equipped with a grate on which combustion of the fixed carbon is completed.

9.5.2 Mass Burning Older mass-burning plants usually consisted of a scale, the tipping floor, a pit (sometimes called a bunker), cranes for feeding the waste, the furnace, a secondary combustion chamber, a cooling chamber, air pollution control equipment, an induced draft fan, and a stack. In newer plants, the secondary combustion chamber and cooling chamber are replaced by a boiler usually positioned directly over the grates.

Some smaller (less than 300 T/D capacity) mass-burning facilities do not have a pit, but use the tipping floor to receive and store the waste as do some of the RDF-fired plants. The pit or floor should have a capacity to store wastes equal to $3^1/_2$ days of furnace burning capacity at minimum. This storage is necessary to keep the furnaces on line during three-day holiday weekends when no waste is delivered to the facility.

Waste is fed into the incinerator furnace hopper by overhead cranes, or by front-end loaders onto a conveyor in the case of tipping floor storage. The crane or front-end loader/conveyor is a critical piece of equipment in the plant without which the plant would not operate. Accordingly, good design provides redundant units — one for use and the other for standby. Good operation practice is to alternate the use of feed equipment on a regular schedule, such as weekly. The hoppers are rectangular funnels connected to chutes that convey the MSW to a furnace feed mechanism. The chutes can be water-cooled or refractory-lined to protect the chutes from the heat of the igniting waste. An encased structural system supports the chamber refractory and insulation, the grate system, the hoppers, and the overfire and underfire ductwork.

In older plants, the primary combustion chamber is refractory-lined with special refractory block, backed by insulating block, hung or supported from the structural system. Some parts of the lining near the grate surfaces may be lined with silicon carbide air-cooled shapes to minimize slag buildup on the walls. Both the underfire air and overfire air are usually taken from the pit area to control dust and odors. The underfire air, as the name implies, is fed into hoppers under the grates to provide most of the combustion air. The overfire air is fed above the fire to provide additional oxygen and turbulence to promote better combustion of volatiles in the gas stream emanating from the combustion bed. Combustion gas temperatures vary from 1,800° to 2,200° F.

The grates support the firebed and provide the source of agitation, transportation, and combustion air. Grates come in many designs, sizes and shapes. The different types include: traveling grates, rocking grates, reciprocating grates, reverse reciprocating grates, and drum grates. Some units, such as rotary kilns, have no grates. There are also special designs such as the O'Connor combustor which is a rotary tube bundle in which the refuse is burned; and the tube bundle imparts the heat energy to water running through the tubes. Some attempts have been made to combust processed RDF using fluidized beds and pyrolysis, but the results have been mixed. Fluidized beds are widely used in industry with homogeneous feedstocks.

In older plants, a secondary combustion chamber may be included to provide a residence time of about one second, preferably at 1800° F, to allow for complete combustion of organic compounds in the gas stream. The secondary combustion chamber is followed by the cooling chamber (in cases where no energy is recovered). The cooling chamber cools the gases to approximately 500°F to protect the air pollution control devices which follow.

In newer plants, the combustion chamber over the grates is enclosed by water filled tubes or water walls. Underfire air is introduced through the grates, as in older plants, while overfire air is fed into the combustion chamber through nozzles projecting through the chamber water walls. The hot gases pass out of the combustion chamber and through boiler superheater and convection tube heat transfer surfaces. Economizer heat transfer surfaces may also be used to increase boiler heat extraction efficiency. However, this heat transfer surface tends to consist of smaller diameter, relatively closely-spaced tubes and, therefore, is more subject to plugging by deposits caused by dirty flue gases. Superheater surfaces may be located at the exit from the combustion chamber, which is economical and efficient, but also exposes the tubes to greater risk from high temperature metal wastage, or downstream following a portion of the convection heat transfer surface. After passing through the boiler heat transfer surface, the gases are discharged to the air pollution control devices. The temperature of the gases exiting boiler heat transfer surface would normally be controlled to approximately 500°F ± 100°F.

About 150 mass-burn units have been built in the United States since the early 1970s. Most were built with water-filled tubes enclosing the primary/secondary combustion chambers (replacing the older refractory construction), with the water circulating in these tubes integrated into the downstream superheater and convection tube boiler. The steam generated in these boilers can be used to power electricity-generating turbines or as a heat source for district heating or industrial process heat. These are known as waste-to-energy (WTE) plants.

9.5.3 Refuse Derived Fuel (RDF) Furnaces

RDF furnaces are very similar to furnaces that burn other solid homogenous fuels, such as coal, wood, wood and coal waste, etc. The energy produced is used in the same fashion as the energy produced by mass burning furnaces. The distinguishing element in RDF plants is the required waste processing prior to burning. The unit process descriptions which follow represent a typical arrangement of the various phases of RDF processing equipment in common use today. This arrangement is not the only combination of processes that can be used, but it is one that has achieved consistent success.

Refuse is typically discharged from the refuse collecting vehicle onto a receiving floor where it is temporarily stored and then pushed onto a conveyor with a front-end loader. This provides an opportunity to inspect the material and remove oversize objects that would interfere with handling and processing; explosive objects such as gas tanks; automobile batteries; and other items that would cause problems during processing, combustion, or ash disposal. Once the waste has been examined and pushed onto a conveyor, it is passed through a flail mill. The flail mill is intended to open bags and expose the refuse material without shredding it to any great extent. During processing, it is important to avoid breaking glass into such small sizes that it would be impregnated into the paper materials which would prevent separating the glass from the paper prior to combustion and thereby increase the ultimate ash content of the RDF.

The next step is to pass the refuse by a magnet on a conveyor to extract some, but not all, of the ferrous metal. The next piece of equipment in the process is frequently a trommel, which is a very large (10- to 12-ft diameter) drum with openings of 2 and 4 in. The trommel tumbles the material to separate the smallest pieces from the refuse by allowing them to fall through the 2-in. openings. This process removes broken glass, dirt, stones, and other non-combustible materials. When the recycling of ferrous materials is to be maximized, the material less than 2 in. in size that falls from the trommel is also directed by a magnet a second time to remove any metal in that material stream. The waste then passes by the 4-in. openings in the trommel through which fall wastes that can be transferred directly to the RDF storage area (alternatively, this relatively clean material may also be shredded). The remaining oversize material exits the outlet of the trommel and is directed to a shredding unit, typically a horizontal hammermill shredder. Removing glass and metals by the magnet and trommel prior to shredding processes reduces wear and tear on the hammers in the hammermill.

The hammers in a horizontal hammermill shredder are designed to shred the remaining waste material to a size that will pass through grate bars normally set less than 4 in. apart. To classify this material after shredding, the refuse is often be passed through another trommel, where 4-in. openings provide for the separation of the fuel materials less than 4 in. in size. The oversized materials larger than 4 in. are directed back to the hammermill shredder. The material separated by the second trommel can also be directed past a magnet again to extract ferrous metal. The RDF is then stored in reserve to provide a constant fuel supply for onsite RDF-burning facilities or for shipping to an offsite RDF combustion location.

9.5.4 Air Pollution Control In the past, air pollution control for MSW furnaces consisted of a baghouse filter or an electrostatic precipitator (ESP), primarily to remove particulates. However, the Clean Air Act Amendments of 1990 and related regulations have significantly tightened the air emission limits for municipal waste combustors. Table 9-13 presents a summary of the current federal air emission standards.

Table 9-13
Summary of air emission standards for municipal waste combustors

Applicability
The new source performance standards (NSPS) apply to municipal waste combusters (MWCs) with unit capacities above 225 mg/day (250 ton/day) that combust residential, commercial, and/or institutional wastes. Industrial discards are not covered by the NSPS.

Good Combustion Practices
- Maximum load level demonstrated during dioxin/furan performance test.
- Maximum particulate matter (PM) control device inlet temperature demonstrated during dioxin/furan performance test.
- Carbon monoxide (CO) level (averaging time) as follows:

Modular, starved and excess air, MWCs	50 ppmv(4 hr)
Mass burn waterwall and refractory MWCs	100 ppmv(4 hr)
MWCs using fluidized bed combustion	100 ppmv (4 hr)
Mass burn rotary waterwall MWCs	100 ppmv (24 hr)
Refuse-derived fuel (RDF) stokers	150 ppmv (24 hr)
Coal/RDF mixed fuel-fired MWCs	150 ppmv (4 hr)

- ASME or state certification for MWC supervisors. Operator training and training manual for other MWC personnel.

MWC Organic Emissions (Measured as total dioxins/furans)
- Dioxins/furans[a,b]   30 mg/dscm
- Basis   GCP (Good Compustion Practices), spray dryer, fabric filter

MWC Metal Emissions (Measured as PM)[a, b]
- PM (Particulate Matter)   34 mg/dscm
- Opacity   10% (6-min average)
- Basis   fabric filter

MWC Acid Gas Emissions (Measured as SO_2 and HCl)[a]
- SO_2 80% reduction or 30 ppmv (24 hr)
- HCl   95% reduction or 25 ppmv
- Basis   spray dryer and fabric filter

Nitrogen Oxide Emissions[a]
- NO_x 180 ppmv (24 hr)
- Basis   selective noncatalytic reduction

Monitoring Requirements
Continous Emission Monitoring System (CEMS) except as noted:

SO_2	CEMS, 24-hr geometric mean
NO_x	CEMS, 24-hr arithmetic average
Opacity	CEMS, 6-min average
CO, load, temperature	CEMS, 4 or 24-hr average
PM, dioxins/furans[b] and HCl	Annual stack test

a) All emission levels are at 7% O_2, dry basis.
b) Particulate master
c) Dioxins/furans measured as total tetra-through-octa-chlorinated dibenzo-p-dioxins and dibenzofurans, and not as toxic equivalents.

Most municipal waste combustors are being equipped with baghouses and acid gas scrubbers to remove particulates, acid gases, and most of the heavy metals. Additional control equipment may be required to meet NO_x, dioxin, and mercury permit limits. They must also be equipped with continuous emission monitoring systems (CEMS), which consist of sampling probes and analyzers to indicate concentrations in the flue gas of substances such as CO, SO_2, NO_x, and O_2, and to monitor temperature and opacity.

NO_x gas emission control has been demonstrated by the use of Selective Non-Catalytic Reduction (SNCR). The SNCR system consists of reagent injection nozzles that atomize the reagent into the gas stream in critical temperature zones in the boiler. Reagents that can be used are urea or ammonia. When properly installed and operated, this system can limit emissions to substantially below 180 ppm.

A spray dryer absorber system can be used to control both acid gas and trace organics such as dioxin. The acid gas control is effected by use of lime injection. It will control acid gases such as sulphur dioxide, hydrochloric acid, and fluoride and sulfuric acid mist. An activated carbon injection system using powdered activated carbon may also be used for control of mercury and trace organics. Some plants have been equipped with a carbon column filled with activated carbon through which the gases pass for control of mercury and trace organics.

Under the new EPA rules (in draft), a fabric filter is the preferred equipment to remove particulates from the gas stream. The coating of captured material on the filterbag surface provides a layer of the reactants introduced in preceding air pollution equipment, further reducing the pollutant. Baghouses can now control particulate emissions from the combustion gases to 0.007 grains per dry standard ft^3 (corrected to 7% O_2). The maximum inlet temperatures for a baghouse varies from 200°F to 350°F, depending on the bag material.

9.5.5 Ash Management Combustion produces two types of ash — bottom ash and fly ash. Bottom ash is the remaining non-combustible waste material which falls through or off the end of the furnace grates. Fly ash is particulates entrained in the combusion air stream.

One of two general design approaches may be used to handle the bottom ash from a MSW furnace. In one approach the ash is discharged through a chute, the end of which is submerged in a water-filled trough providing an air seal for the furnace. The ash is scraped along the bottom of the trough, and up a slope on which it is dewatered, by a drag conveyor. The dewatered ash is discharged by the conveyor into a hopper, and then into a truck for conveyance off-site. The second approach uses a ram discharge and a series of conveyors. In this approach, the bottom ash falls off the end of the grate into a water-filled tank on the bottom of which an oscillating ram operates. After the ash is quenched, the ram pushes the material onto a short transfer belt conveyor which can discharge the material to one of two parallel transverse belt conveyors. These transverse conveyors move the quenched waste to a point where the material can be loaded onto trucks for conveyance off-site. At some plants, the bottom ash is passed over a magnetic separator to remove additional metals.

Fly ash is collected by the air pollution control system. It may be combined with bottom ash or kept separate for disposal. Whether or not it is combined with bottom ash depends upon the heavy metal content of the fly ash which can affect its ultimate disposal.

Typically, MSW combuster ash is disposed by landfilling. However, there is considerable interest in recycling this material as there is for ash from coal-fueled power plants. Tests of most MSW incinerator ash prove that it is not a hazardous waste. Ash may be disposed in a monofill, i.e., a landfill specially designed to handle only one material, or a properly-lined landfill designed to accept municipal waste. Some jurisdictions require dedicated cells for disposal of incinerator residue. The moisture content of the ash as it leaves the incinerator facility generally varies between 40% and 60% by weight.

9.5.6 Energy Production The roughly 150 mass-burn units built in the United States since the early 1970s were all built with energy recovery in mind. The most common energy recovery system produces steam through radiant and convective heat transfer in the combustion chamber by use of waterwalls enclosing the combustion gas stream, and by convection tube bundles and superheater surfaces in the following boiler. The steam can be used to produce electricity using turbine generator sets and/or as a heat source for district heating or in industrial processes. RDF boilers are very similar to boilers that burn solid fuels such as pulverized coal, sawdust, and coal waste.

Older MSW mass-fired WTE plants with refractory-lined combustion chambers and flues could generally produce approximately 2 lb of steam/lb of MSW with a system efficiency of 50 to 60%. Substantial heat loss occured in these plants because of the need to use large amounts of combustion air to control furnace temperatures. In newer mass-fired water wall units, efficiency of heat recovery has been found to range from 65 to 70% with steam production of approximately 3 lb of steam/lb of MSW. RDF-fired plants have a heat extraction efficiency ranging from 60 to 80%, with a steam production rate somewhat greater than 3 lb of steam/lb of RDF. However, steam production will be somewhat less than 3 lb of steam/lb of MSW because some combustible material is lost in processing the MSW to produce RDF.

Revenue from sale of the energy value in the steam can result in an income of from $25 to $50/ton of MSW. With a total cost of amortization and operation of such plants ranging from $75 to $150/ton of rated MSW capacity, tipping fees at such plants generally range from $50 to $100/ton of rated capacity. Costs at such plants are found to vary substantially from site to site depending on labor costs, features included in the plant affecting plant construction and operation costs, power costs and energy sales agreements, and many other local and site-specific factors.

9.6 Resource Recovery Systems

9.6.1 Introduction Several factors have focused attention on the recovery of resources present in solid wastes including:

- the increasing difficulty in siting new landfills and the potential scarcity of landfill capacity;
- regulations mandating reduction in waste quanities requiring disposal;
- increasing stringency of air emissions limits compromising the viability of incineration/energy recovery as a management method; and
- recognition of the value of waste constituents and/or the economic loss accompanying their disposal.

The phrase "integrated solid waste management system" has been coined to refer to solid waste management programs that incorporate one or more resource recovery systems. USEPA has developed a hierarchy of resource recovery methods designed to limit the amount of waste requiring disposal in landfills. That hierarchy (in order of preference) follows:

- *Waste reduction* — reducing the quantity of waste produced by changing production processes, product designs, and consumer behavior.
- *Recycling* — taking a waste product and using it as a raw material in the production of a new product.
- *Composting* — reducing the volume and toxicity of organic waste streams by enhancing conditions for biological decomposition of the waste in a controlled manner.
- *Energy recovery* — burning waste as means of volume reduction and recovery of the energy contained in the waste materials.

9.6.2 Waste Reduction/Waste Minimization Waste reduction/waste minimization programs have been adopted by both the public and private sectors as a means of reducing waste generation rates. Industrial waste minimization programs involve fine tuning of industrial production processes to reduce the amount or toxicity of waste that is produced. Economic incentives are also provided to industry to reduce packaging. Public sector waste reduction programs have typically involved educational efforts, container deposits, product charges, and restrictions on materials placed in landfills to change the behavior of individual citizens.

Each year, industrial facilities in the United States generate large quantities of solid and hazardous wastes that have varying degrees of toxicity. The cost to dispose of the waste can be significant. In addition, there are potential long-term liabilities associated with waste disposal systems. These factors encourage industries to minimize the volume of waste they generate.

Examples of industrial waste minimization include: reducing water usage to reduce the amount of wastewater treatment plant sludge generated; recycling post-consumer-waste fiber as a source of cellulose fiber in paper mills; and collection, treatment, and re-use of off-specification materials in the production of cement. The effectiveness of waste minimization systems depends on the type of waste generated, age of facility, production process, and many other site-specific variables.

Industrial waste minimization programs are generally driven by economics and a desire to reduce production costs. There are a number of direct and hidden costs associated with waste management. Direct costs include the transportation and disposal of waste. These costs have increased substantially in recent years because of the development of strict new environmental regulations. Hidden costs include the cost to remediate facilities that have disposed of waste improperly, the cost of raw materials to replace material that is disposed in a landfill, and the cost of manufacturing processes that generate waste.

In the public sector, waste reduction programs have been largely driven by regulations or government incentives. Examples include: banning recyclables and yard waste from landfills; enacting bottle and can deposit laws; requiring developers to include space for recycling programs in building plans; encouraging bulk packaging of consumer goods and foods; and enacting laws requiring fast-food restaurants (major waste sources) to reduce their waste generation rates.

Example 9-6 — Consumer waste reduction

Identify eight methods to effect consumer waste reduction.

Solution:

1. Encourage shoppers to bring their own bags and containers to stores.

2. Rent or borrow appliances such as garden tillers, floor sanders, and rug shampooers rather than purchase them.

3. Seek extended warranties on automobiles, tires, furnaces, and TVs.

4. Repair items rather than throw them away.

5. Use reclosable and reusable containers, such as lunch boxes and garbage cans, rather than disposable bags.

6. Buy beer, soft drinks, and milk in refillable containers. If this is not possible, buy a biodegradable or recyclable container.

7. Buy items in larger, more economical packages.

8. Buy bulk or unpackaged goods when possible.

9.6.3 Recycling/Materials Recovery

Recycling (otherwise known as materials recovery) has been primarily associated with the public sector. However, industrial recycling programs also exist. Some industries have developed programs to capture and reuse resources that would otherwise end up in the waste stream. Industry Waste Exchanges have formed to facilitate recycling of industrial waste. Through these Exchanges, industries that generate specific types of waste are matched with industries that can use the waste as a feed material for a production process. Other options for recycling industrial waste include using high-volume, low-hazard wastes such as foundry sand and coal ash as general fill material or in road or parking lot construction projects. The ability to recycle industrial waste is very waste-specific. Physical and chemical testing must be performed on each waste stream to determine its properties and the suitability of the material for reuse.

Communities develop recycling programs for a variety of reasons including environmental and economic concerns, regulations, and the resistance to developing new landfills or expanding existing ones. Whatever the motive, the key factor for success with recycling programs is the ability of the recycler to develop stable material markets. The lack of established, stable markets was a significant problem in the early days of recycling (1970s). Existing markets for recyclables were inundated as new recycling programs came into existence, forcing down the price paid for recyclables. As a result, recycling programs were forced to dispose of excess materials in landfills and the credibility of the programs was damaged. This lack of stability continues; it is an inherent characteristic of the market for which provisions must be made in the design of consumer-waste recycling programs.

A wide variety of materials are routinely recycled through curbside collection and other source separation programs. The typical materials recycled are:

- Paper, such as newsprint, magazines, and mixed paper,
- Glass,

- Aluminum,
- Plastics, such as PETE, HDPE, and polystyrene,
- Ferrous metals, and
- Corrugated cardboard.

Table 9-5 provides a breakdown of the typical components of municipal solid waste. In using this information to design a recycling program, it is important to remember that there are seasonal and regional variations in the waste steam (see Table 9-6).

A recycling program consists of the following components:

- Collection and transportation
- Separation and materials handling
- Marketing and sales
- Public education

There are a variety of collection systems used in recycling programs. Some of the common types are:

- *Source-Separated Collection* collects recyclables at the curbside of each residence. The recyclables are either sorted into bins by the generator or mingled in a single-collection container or bag. If they are presorted, the driver is generally responsible for placing the recyclables into individual bins on the collection vehicle.

- *Material Specific Drop-Boxes* (dumpster-type boxes) located around the community. Residents are responsible for transporting recyclables to the location and separating the waste into material-specific drop-boxes.

- *Redemption Centers* are centralized facilities that pay for recyclable materials. The operation of these facilities varies, but typically the generator is required to collect and transport the recyclables to the center where the recyclables are weighed and the generator is paid on a per-pound basis.

- *Commingled Collection* is a process where recyclables are placed in colored-coded bags according to waste type and collected as part of the solid waste stream. At a processing facility, the colored bags are manually removed from the remainder of the waste and processed for markets. A concern with this type of system is that compaction in the collection vehicle may break glass containers and "contaminate" the other recyclables.

Collection rates vary with the type of collection system used, the effectiveness of public education efforts, the economic status of residents, the educational levels of residents, and the locale — rural, suburban, or urban.

Processing the recyclables follows collection. First, the recyclables are separated by material type, e.g., glass, aluminum, paper, corrugated cardboard, etc. Once separated, each material is processed depending on the demands of the market, the collection method, and the materials involved. For example, with a Source-Separated Collection system, processing is limited to the baling of certain materials and separation of glass by color. But with a Commingled Collection system involving a wide number of materials, much more extensive separation processing is required.

Mechanical methods for processing MSW and source-separated and commingled recyclables to recover marketable materials have been extensively studied since the 1970s. While there have been some successes, e.g., magnetic devices for ferrous materials and air classifiers to segregate light-weight fractions, most mechanical devices have been unable to reliably perform in the harsh environment inherent in waste handling and at the same time produce market-quality materials. As a consequence, materials recovery relies to a large degree on low-skilled manual laborers.

With Commingled Collection wastes, the first step is to open the bags. In mechanical processing this is accomplished by a shredder or flail mill. Typically, a processing facility will employ a combination of mechanical and manual methods. Following is a summary of processing requirements/equipment for selected materials.

Glass — Material should be color-separated to enhance value. Breaking the glass may or may not be required depending on the market.

Aluminum — Material may be crushed and baled to faciliate handling during shipping.

Ferrous metals — A magnet is used to separate ferrous metals from other materials. Stockpiles are typically crushed and baled prior to shipping.

Newsprint — Typically baled. In some areas, newsprint is shredded and sold as animal bedding.

Mixed paper — Paper may be separated by grade of paper and color (white versus colored paper) and may be shipped either loose in large bins or baled.

Plastics — Material should be separated by polymer. Typically, plastics are baled to reduce the volume for shipping.

Corrugated cardboard- Typically baled prior to shipping.

Marketing is the most critical step in the recycling process. A lack of stable markets has hindered the development of recycling programs across the United States. As recycling programs gain popularity, materials markets become overloaded, driving prices down, resulting in making recycling less competitive compared to the relatively low cost of other disposal options. As word of this condition spreads, some consumers and operators become disillusioned thereby reducing the supply. At the same time, a drop in price of recycled materials encourages industries to use them in new ways.

As this uncertain interplay of market forces proceeds, materials markets face the classic conundrum of "which comes first, the chicken or the egg?" Manufacturers are unwilling to make large capital investments in equipment to handle recyclables in their production process unless there is a stable supply of raw material available and the demand for goods made from recyclables increases. As a means of increasing both the supply of recyclables and the demand for recycled materials, a number of states have adopted legislation requiring recycling programs as a condition to the continued landfilling of waste and others have offered tax and financial incentives to encourage industries to set up recycling facilities. Governments have also required their agencies and contractors to use goods made from recycled materials. These actions have helped stabilize and increase the market (and price) for recyclables.

One of the main problems with marketing recyclables is the quality of the materials. Many industries have quality standards for their raw materials. Generally, the more guarantees that can be made about the quality of a material, the higher price the marketer can get for the material. Coincidentally, higher quality requirements create more problems and cost for the recycler.

Marketing of recyclables is complex. Large, national waste disposal firms with their own recycling programs can collect a large amount of material. This stable source of materials has enabled these firms to negotiate higher prices for the materials they recycle. Smaller individual recycling programs have less material to market and are more subject to market fluctuations. In addition, some markets set a minimum volume of materials that they will accept at one time, making it even more difficult for the smaller recycling program to operate. To offset these limitations, materials brokers have emerged to contract with a number of small recycling programs and provide a stable supply of materials. While brokers can produce better prices for the recyclables, this extra revenue is accompanied by an increase in handling and transportation costs.

An alternative approach to marketing recyclables is to look for new uses for recyclables near the source. For example, tires are being used as a rubber source, and glass is being used as a replacement for aggregate in asphalt. Local markets may not pay as much as larger markets, but the reduction in transportation costs can offset the lost revenue.

9.6.4 Composting
Composting is the controlled biological decomposition of organic wastes. Typically, composting systems are aerobic. Solid waste composting has been very popular in Europe for a number of years, but has only recently gained in popularity in the United States.

- *Yard waste composting* involves composting leaves, grass, and other yard waste. It is relatively common, and its use is supported by bans on the disposal of yard waste in landfills.

- *Municipal solid waste (MSW) composting* involves composting some or all of the MSW waste stream. This form of composting is less common due to the high levels of impurities in the waste (e.g., glass and plastics) and the lack of stable markets.

- *Co-composting* consists of composting the organic fraction of MSW or other organic solid wastes with sludge from municipal or industrial wastewater treatment facilities.

No matter what the waste stream, there are a number of factors that control the decomposition process in composting. The primary factors are:

- *Aerobic conditions.* The bacteria involved in the decomposition process require oxygen to survive. If the compost pile becomes anaerobic, the aerobic bacteria die and anaerobic bacteria dominate. While anaerobic composting is possible, it is typically less efficient and results in more odors. For this reason, aerobic decomposition processes are preferred.

- *Moisture content.* For the bateria to remain active in the compost, a moisture content of 40 to 60% is required. If there is too little moisture, the decomposition process will stop. If there is too much moisture, the process becomes anaerobic and causes odors.

- *Nutrients.* The ratio between the amount of carbon, nitrogen, and phosporus controls the rate of decomposition. The ideal carbon-nitrogen ratio for raw compost is between 25 to 30:1, while initial ratios between 20 to 40:1 consistently provide for rapid composting (NRAES.54, 1992).

- *Heat.* Decomposition is an exothermic process that releases heat to the atmosphere. To optimize the decomposition process, the temperature is typically maintained in the mesophillic range from 45°C to 55°C.

A typical composting process consists of the following steps:

Step 1. *Waste acceptance.* Waste is received at the composting site. Large noncompostible items or other contaminants should be removed from the waste stream.

Step 2. *Waste processing.* Waste processing typically consists of some form of volume reduction. This may include shredding or hand sorting operations to remove large items, or other mechanical processes to improve the compost. Some communities run the waste through recycling centers prior to composting to reduce the number of noncompostible materials in the waste stream.

Step 3. *Composting.* Various composting processes exist. Some of the commonly used processes include:

Drum systems
Static pile
Windrows
Biological digestors

Step 4. *Curing.* Following the initial composting stage, the compost is allowed to digest over a 6- to 12-week period.

Step 5. *Screening.* To improve the quality and marketability of the compost, it is common to screen the final composted material. Large pieces and other impurities are removed.

Step 6. *Marketing.* As with recyclables, marketing is essential, otherwise the compost will have to be landfilled.

The most common problems associated with composting programs include odors and the presence of contaminants such as heavy metals or noncompostible materials in the final compost. Odors are difficult to address because each person has a different opinion as to what odor is acceptable and at what level. Odors can be caused by a variety of factors including lack of air, too much moisture, too much or too little heat, and contaminants in the waste stream which are toxic to the process. Chemical contaminants, such as heavy metals, may appear in some compost. Originally, it was thought that MSW compost would have high concentrations of heavy metals. However, a number of studies have shown variable levels of heavy metals in MSW compost, depending on the source of the waste. Therefore, testing is required to ascertain the extent of metals in the compost. If metal concentrations are too high, sales of the compost are adversely affected because application rates can be restricted and markets limited. Noncompostible materials in the compost can be a significant problem. Many buyers set purity standards for compost limiting the amount of noncompostible materials in the final product. Excess glass and plastic are the two most common contaminants.

Example 9-7 — Waste recycling and composting

A community of 100,000 people proposes to implement a curbside collection program for recyclables and yard waste within the community. City planners anticipate collecting the following materials:

Yard waste	50%
Aluminum	75%
Ferrous metals	20%
Paper	35%
Glass	25%
Plastics	15%

Assume the average per capita waste generation rate is 4 pounds, including yard waste. Use the information in Tables 9-5 through 9-9.

   a. How many tons of recyclables and compost will be collected annually?

     Solution:

Waste generated = 100,000 people • 4 lb / capita day • 365 day / yr • 1 ton / 2,000 lb
                   = 73,000 ton / yr

From Table 9-5

yard waste	73,000 • 18.5% • 50%	= 6,753 ton/yr
+ aluminum	73,000 • 0.5% • 75%	= 274 ton/yr
+ ferrous	73,000 • 6.0% • 20%	= 876 ton/yr
+ paper	73,000 • 34.0% • 35%	= 8,687 ton/yr
+ glass	73,000 • 8.0% • 25%	= 1,460 ton/yr
+ plastics	73,000 • 7.0% • 15%	= 767
Total		= 18,817 tons/yr collected

   b. What will be the percent reduction in the volume of materials sent to the landfill, assuming there is not a significant change in the density of solid waste as a result of the recycling program?

Solution:

(18,817 ton / yr) / (73,000 ton / yr) = 26% reduction

c. If the remaining waste will be composted, with 25% of material (by volume) being rejected as unsuitable for composting, 10% of the composted material (by volume) being screened out at the end of the composting process, and there being a 50% reduction in the volume of material due to the composting process, how many cubic yards of final compost will be generated each year? Assume a solid waste density of 500 lb/yd^3 (See Table 9-6) and that rejected screenings have the same density as the incoming solid waste stream.

Solution:

73,000 ton/yr - 18,817 ton/yr = 54,183 ton/yr to composting facility

54,183 ton/yr • 2,000 lb/ton • 1 yd^3/500 lb = 216,732 yd^3/yr

216,732 yd^3/yr • 25% rejects = 54,183 yd^3/yr sent to landfill as rejects

216,732 yd^3/yr - 54,183 yd^3/yr = 162,549 yd^3/yr • 50%
= 81,275 yd^3/yr compost generated

81,275 yd^3/yr • 10% screened out = 8,128 yd^3 to landfill

81,275 yd^3/yr - 8,128 yd^3/yr = 73,147 yd^3 of final compost

d. If the community implements the curbside collection program and the composting system discussed above, what will be the total volume of material sent to the landfill? What is the final percent reduction, by volume, of material sent to landfill? Assume a solid waste density of 500 lb/yd^3 (Table 9-7) and that there is no significant volume of rejects at recycling plant.

73,000 ton/yr • 2,000 lb/ton • 1 yd^3/500 lb = 292,000 yd^3/yr

54,183 yd^3/yr rejects + 8,128 yd^3/yr screened
= 62,311 yd^3/yr to landfill

292,000 yd^3/yr - 62,311 yd^3/yr = 229,689 yd^3/yr reduction in volume

229,689 yd^3/yr/292,000 yd^3/yr = 79% reduction

9.7 Hazardous Waste Treatment

9.7.1 Introduction As discussed previously in Section 9.2, the Hazardous and Solid Waste Amendments of 1984 (HSWA) and accompanying rules require that hazardous waste be treated to meet treatment standards some of which were set using Best Demonstrated Available Technology (BDAT) to reduce the concentra-

TIP FOR THE P.E. EXAMINATION CANDIDATE

The candidate needs to know how to solve a hazardous waste treatment problem by correctly identifying and justifying the use of the applicable unit processes. In addition, the candidate must be able to correctly configure a treatment process train; apply correct design principles and assumptions to determine the size of the treatment train; identify and calculate the necessary energy and material inputs and any associated health and safety concerns; and identify and calculate all wastes and emission discharges including environmental or public health hazards associated with these discharges.

tion of certain hazardous constituents below the EPA-determined land disposal restriction (LDR) standards. The generator of the waste must identify the point of generation of the waste when determining whether the waste is subject to the LDR standards. The purpose of this identification is to restrict subsequent dilution of the waste as a means of reducing the levels of hazardous constituents below the LDR concentrations.

Treatment process trains (a combination of unit processes) are often necessary to treat hazardous waste from industrial sources and the remediation of uncontrolled hazardous waste disposal sites. A wide variety of treatment technologies can be used to treat hazardous waste. Table 9-14 provides a listing of several treatments commonly used to process industrial waste and waste generated from remediation of uncontrolled hazardous waste disposal sites (Martin & Johnson (1987); Anderson Vol. 1 (1993); Berokowitz, Funkhouse, Stevens (1978)). Most of the technologies listed in Table 9-14 are applicable to liquid wastes. Fewer technologies exist for treatment of hazardous sludges and solids.

Table 9-14
Commonly used hazardous waste treatment technologies[1]

Physical Media[3]	Application[2]	Physical
Activated Carbon Adsorption	wide variety of organics and inorganics, metals, phenols, cyanide	L, G
Air Stripping	volatile organics, ammonia	L
Dissolution	solid wastes containing metallics and organo-metallics	S, SL
Evaporation	concentrating radioactive, heavy metal, hydrocarbons, and wide variety of organic compounds from wastes	L, S, SL
Filtration/Ultrafiltration	heavy metals, colloidal and suspended materials, oil/water mixtures	L, SL, G
Ion Exchange	metals, halides, sulfates, nitrates, cyanides, organic acids, organic amines	L, G
Gravity Separation	two phased liquids, oil/water mixtures	L

Table 9-14 (cont.)
Commonly used hazardous waste treatment technologies[1]

Reverse Osmosis	dilute solutions of metals and some organics	L
Solidification/Stabilization	heavy metals, acids, oxidizers, sulfates, halides low level radioactive materials	S, SL
Steam Stripping	volatile organics, hydrogen sulfide, ammonia, phenol	L L
Solvent Extration/Distillation	complex mixtures of organic solvents, halgenated organics and radioactive compounds	L, SL, G
Chemical		
Calcination and Sintering	radioactive materials, organic and organo-metallic compounds, hydroxides, sulfites, sulfates, nitrates, carbonates	L, S, SL
Neutralization	acids, bases, heavy metal containing wastes	L, S, SL, G
Oxidation	sulfites, sulfides, phenols, cyanide, mercaptans, some pesticides, ferrous iron, lead, formaldehyde	L, G
Ozonation	cyanides, phenols, aromic hydrocarbons, pesticides	L, G
Precipitation, Flocculation Sedimentation	soluble heavy metals, particulates	L
Reduction	hexavalent chromium, mercury containing wastes	L, G
Wet-air Oxidation	halogenated aliphatics, halogenated and non-halogenated aromatic hydrocarbons	L, SL
Biological[4]		
Aerobic	petroleum hydrocarbons, creosote, low molecular weight alcohols, ketones, esters, less chlorinated aliphatics, chlorinated aromatics	L, S, SL, G
Nitrate Reducing	BTEX, creosote, less chlorinated aliphatics	L, S, SL, G
Other anaerobic	BTEX, low molecular weight gasoline and #2 fuel oil, creosote, low molecular weight alcohols, ketones, esters, chlorinated aliphatics, highly chlorinated aromatics	L, S, SL, G

1) Not all technologies used for treatment of hazardous wastes are listed. Table identifies most commonly used methods.
2) Common uses of the technologies are listed. There may be factors (such as waste mixtures, treatment requirements, regulatory setting) that minimize the effectiveness of the technology treat the materials identified. Futhermore there may be special circumstances where the technologies may have useful application in treatment of wastes other than those listed.
3) Technologies most commonly used for the media are shown.
   L = liquid   S = solid   SL = sludge or slurry   G = gas
4) Biological conditions are listed instead of specific treatment processed. Refer to Table 9-16 for summary of treatment processed.
   Adapted from Anderson, 1993, Berkowitz, 1978, Martin, 1987.

9.7.2 Carbon Adsorption Adsorption using granular or powdered activated carbon is a proven technology for treatment of a wide variety of organic and inorganic hazardous constituents. Soluble constituents are removed from the liquid solution by the attachment of molecules to the surface and pores of the carbon as a result of intermolecular attraction.

Laboratory tests provide useful information regarding the effectiveness of activated carbon and the quantity needed to provide the degree of treatment required for a specific application. Laboratory tests are also helpful in determining the amount of time until the adsorptive capacity or availability of the carbon is spent, commonly referred to as "time to break through." Bed-depth/service time (BDST) analyses use breakthrough curves to develop bed-depth/service time curves for given effluent concentrations.

TIP FOR THE P.E. EXAMINATION CANDIDATE

Although there are many innovative technologies under field demonstration, only the more commonly used and proven technologies are covered on the P.E. examination. The American Academy of Environmental Engineers has published a set of monographs presenting design and application data for a variety of innovative treatment technologies for site remediation. The reference section of this chapter provides more detail on texts for further study of innovative technologies.

The adsorptive capacity of the carbon in a column is first exhausted closest to the influent, often leaving a large quantity of partially spent carbon in the column. Multiple columns of carbon are normally used in series to maximize the useful life of the carbon. Breakthrough is measured at the influent to the final column. When the first column is exhausted, it is removed from the series for regeneration or disposal, and a new column is added. In a well-designed carbon system, the lead column is exhausted just prior to breakthrough of the last column. Weber (1972), and Crittendon (1987) provide design procedures incorporating BDST curves.

TIP FOR THE P.E. EXAMINATION CANDIDATE

The candidate should be familiar with the key design parameters and design formulas of a packed air stripping tower to remove volatile constituents from an aqueous waste stream and be able to calculate the treatment efficiency of the unit, determine the quantity of volatile constituents discharged to the air, and evaluate technologies for treatment of the air discharge.

Any chemical sorbed to the carbon may require hazardous waste treatment, depending on the results of a Toxicity Characteristic Leaching Procedure (TCLP) test of the exhausted carbon. Exhausted carbon can be thermally regenerated; however, the air exhaust from the furnace may be hazardous and require proper control and treatment.

9.7.3 Air Stripping Air stripping is an effective method to remove volatile constituents from a liquid or solid waste. Air stripping a liquid waste is accomplished by allowing the liquid to flow down through a tower filled with a packed medium or through a system of trays. Air is blown counter-current or cross-current to the liquid flow. As the air passes along the liquid adhering to the medium, gas-phase transfer of the volatile constituents occurs. Volatile constituents with a Henry's coefficient greater than 0.003 are most readily stripped. The height of the packing in a stripping tower is determined using the following formula:

$$Z = \frac{ln[(X_2/X_1)(1-A)+A]L}{KL_aC(1-A)}$$ [9-12]

where:
- Z = height of packing, ft
- L = water velocity, lb-mole/hr/ft^2
- X_2 = influent concentration of constituent to be removed, mole fraction
- X_1 = effluent concentration of constituent, mole fraction
- KL_a = mass transfer coefficient, gal/hr; determined from pilot-scale treatability studies; function of compound being removed, liquid temperature, type of packing and tower geometry
- C = molar density of liquid; water = 3.47 lb-mole/ft^3
- H = Henry's law constant, mole fraction in air per mole fraction in water
- G = air velocity, lb-mole/hr/ft^2
- A = L/HG

In recent years, computer modeling has provided a convenient tool to generate the packing height for a variety of tower dimensions.

9.7.4 Gravity Separation Gravity separation is applicable for two-phased liquid wastes such as PCB-laden oil and water. Emulsion-breaking chemicals may be necessary to enhance the effectiveness of the treatment process. Extracted waste must be collected and further treated or disposed of as a hazardous waste. This process is often used as part of a treatment train to remove oils that may interfere with later processing of the waste or to reduce the quantity of hazardous waste requiring further treatment or disposal.

9.7.5 Chemical Reduction The hazardous constituents in many hazardous wastes can be rendered non-hazardous through chemical reduction. This treatment process is most often used to reduce hexavalent chromium to its nonhazardous trivalent form, as demonstrated in the following reaction using sulfur dioxide as the reducing agent:

$$2H_2CrO_4 + 3SO_2 + 3H_2O \rightarrow Cr_2(SO_4)_3 + 5H_2O$$ [9-13]

The resulting solution will be acidic and may require neutralization prior to further treatment or disposal.

Example 9.8 — Chemical reduction

A plating facility generates 500 gal/day of a plating-line wastewater containing 10% by weight CrO_3 in a sulfuric acid solution. Assume all the chromium is hexavalent and must be reduced prior to disposal. The plant proposes to use sulfur dioxide as a reducing agent followed by lime for neutralization. Assuming the plant operates 300 days a year, what is the annual quantity of sulfur dioxide required to treat the wastewater?

Step 1. Plating facility chrome reduction

Weight of CrO_3 = 500 gal / day • 365 day / yr • 8.34 lb / gal • 10
= 152,000 lb / yr

Step 2. Determine the molar ratios

CrO_3	H_2CrO4	SO_2
Cr - 52	H_2 -2	S - 32
O_3 - 48	Cr - 52	O_2 - 32
	O_4 - 64	
100	118	64

Step 3. Determine the weight of H_2CrO_4
H_2CO_4 = 118 / 100 • 152,000 lb / yr = 179,000 lb / yr

Step 4. Calculate SO_2 usage
SO_2 = 3.64 / 2.118 • 179,000 = 146,000 lb / yr

9.7.6 Neutralization Neutralization is required to increase or decrease the pH of a hazardous wastewater, sludge or gas to meet the hazardous characteristic limitations. In addition, it is often used to precipitate heavy metals, to prevent corrosion of wastewater handling equipment, to condition the wastewater for biological treatment, and to facilitate process water reuse.

Neutralization is accomplished by adding an acid to a base, or vice versa, in a carefully controlled and designed system until the desired effluent pH is achieved. Strong acids such as sulfuric and hydrochloric acids and bases such as caustic soda and sodium hydroxide are commonly used within industry to treat hazardous waste streams. The chemistry involved in this process is discussed is Chapter 3.

Example 9.9 — Base neutralization

A groundwater leachate from a mining site has a pH of 12.5. An NPDES discharge permit requires the mining company to treat 100,000 gal/day of the groundwater to a pH of 7.0 prior to discharging to a neighboring stream. Using the titration curve presented below, determine the annual quantity of neutralizing chemical required to treat the waste for surface water to the required pH.

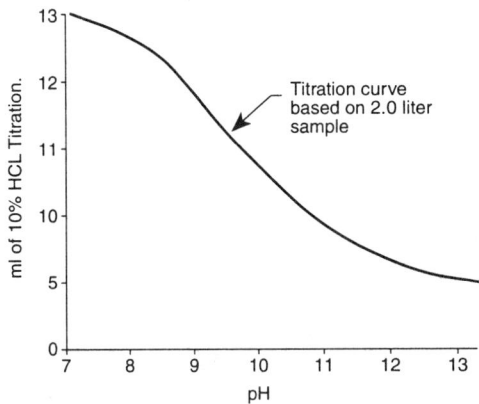

Solution:

Amount of titrant = 13 - 5 = 8 mL

100,00 gal / day • 365 day / yr • 8 mL / 2,000 mL = 146,000 gal / yr of 10% HCL

9.7.7 Chemical Precipitation/Flocculation/Sedimentation
Chemical precipitation, flocculation, and sedimentation are well-proven technologies often used together as a treatment train for the removal of hazardous particulates and heavy metals from wastewaters. Chemical precipitation is a physiochemical process whereby the solubility of the targeted hazardous constituent is changed so that the constituent becomes less soluble. Flocculation is the process whereby small, unsettleable particles group together to form larger, more settleable solids. Sedimentation uses gravitational and inertial forces to settle out suspended solids. Refer to Chapter 6 for a more detailed discussion of these processes.

The most common hazardous constituents removed from liquids and slurries using chemical precipitation, flocculation, and sedimentation are heavy metals such as arsenic, cadmium, chromium, copper, lead, mercury and nickel. Chemical precipitation is often initiated by adjusting the pH of the waste stream. Many metals become less soluble at a higher pH. A solubility curve is essential in determining the correct pH required to maximize the insolubility of the heavy metal constituents. Common inorganic flocculants include alum, lime and iron salts. These flocculants form large, gelatinous precipitates, which surround and enmesh the small particles formed during chemical precipitation, thus creating larger, more settleable solids. These solids are then removed by sedimentation. Often polymers are used to enhance the speed of settling.

The environmental engineer needs to understand the stoichiometric relationships involved in chemical precipitation in order to determine the most economical material usage and the residence time required to achieve the level of treatment desired. It is also necessary to ascertain whether volatile, hazardous or toxic constituents may be created or released during the chemical mixing process, potentially causing negative impact to the environment or harmful human exposure.

TIP FOR THE P.E. EXAMINATION CANDIDATE

The candidate should know that the sludge resulting from chemical precipitation, flocculation, and sedimentation processes may be a hazardous waste and require additional treatment or hazardous waste disposal.

9.7.8 Oxidation

Chemical oxidation is the most common treatment technology used to render a cyanide-containing waste nonhazardous. Oxidation is also used for treatment of waste streams that contain phenols and organic sulfur compounds. One of the limitations of chemical oxidation is that the oxidation reactions are often incomplete, creating the potential for production of other toxic compounds. When iron and other transition metal ions are present in a waste stream, metal cyanide sludges such as ferrocyanide can develop during the oxidation process, limiting the availability of the cyanide to be oxidized. Oxidation is most economically suited to waste streams with few and well-defined constituents. Table 9-15 lists common oxidizing agents and the associated stoichiometry.

9.7.9 Ion Exchange

Ion exchange is a reversible process whereby ions from a hazardous constituent are interchanged with ions from a resin medium. Ion exchange has been used successfully for aqueous and slurry waste streams to remove a broad range of ionic species, including metals, halides, sulfates, nitrates, cyanides, organic acids and organic amines. Ion exchange systems are subject to exhaustion and require regeneration of the medium. The regeneration solution may require hazardous waste disposal.

TIP FOR THE P.E. EXAMINATION CANDIDATE

The candidate needs to understand the stoichiometry associated with chemical oxidation; be able to compute the quantities of reactants necessary to complete an oxidation reaction in a given waste stream; and understand the environmental and human health impacts associated with the chemicals used in the oxidation process and the appropriate protective measures.

Table 9-15
Common commercially-available oxidants used for cyanide treatment

Conversion to cyanate
potassium permanganate
$$2NaCN + 2KMnO_4 + KOH \oslash 2K_2MnO_4 + NaCNO + H_2O$$
hydrogen peroxide
$$NaCN + H_2O_2 \oslash NaCNO + H_2O$$
sodium hypochlorite
$$NaCN + NaOCl \oslash NaCNO + NaCl$$
chlorine gas
$$NaCN + Cl_2 + 2NaOH \oslash NaCNO + 2NaCl + H_2O$$

Cyanide Destruction
calcium hypochlorite
$$4NaCN + 5Ca(OCl)_2 + 2H_2O \oslash 2N_2 + 2Ca(HCO_3)_2 + 3CaCl_2 + 4NaCl$$
sodium hypochlorite
$$2NaCN + 5NaOCl + H_2O \oslash N_2 + 2NaHCO_3 + 5NaCl$$
chlorine gas
$$2NaCN + 5Cl_2 + 12NaOH \oslash N_2 + 2Na_2CO_3 + 10NaCl + 6H_2O$$

TIP FOR THE P.E. EXAMINATION CANDIDATE

The candidate should know when ion exchange may be the preferred treatment system and the practical limitations of implementing an ion exchange system.

Influent concentrations higher than 2,500 to 4,000 mg/L may result in a high rate of media exhaustion, thus requiring frequent regeneration. The frequency of regeneration has a significant impact on the cost of ion exchange. To prevent plugging of the ion exchange medium, the suspended solids concentration should be less than 50 mg/L. Oxidants can also limit the effectiveness of ion exchange.

9.7.10 Reverse Osmosis Reverse osmosis (RO) provides an extremely high level of treatment of wastewaters and is often the preferred treatment system when water reuse is desired. The process involves the application of pressure sufficient to force a fluid through a semipermeable membrane, overcoming osmotic pressure, and causing the constituents of concern to be separated out of solution on the upstream side of the membrane. RO processes are subject to membrane chemical incompatibilities, fouling and plugging of the membrane caused by microbiological organisms, high concentrations of organic solvents, oxidizing materials, particulate matter, oil and grease, and high or low pH. When RO is used to treat a hazardous waste stream, the

TIP FOR THE P.E. EXAMINATION CANDIDATE:

The candidate needs to understand the applicability of reverse osmosis for hazardous waste treatment and have a practical knowledge of the design limitations of the process.

material that is rejected from the RO system, i.e., the material that does not pass through the membrane, must be considered a hazardous waste.

9.7.11 Solvent Extraction/Distillation Distillation is often used in industry to recover economically-viable organic solvents from aqueous and air waste streams. Distillation is becoming a common method of treating "difficult-to-treat" waste, such as radioactive materials and diverse mixtures of organic compounds, for which chemical treatment or other technologies may be too expensive because of operational, maintenance, and waste-disposal costs. Distillation works by bringing the waste stream to the identified boiling point of the organic compound to be recovered and condensing the organic vapors that are generated. Still bottoms may require hazardous waste characterization and appropriate hazardous waste disposal.

TIP TO THE P.E. EXAMINATION CANDIDATE

The candidate needs to understand the applicability, limitations, and design parameters of the various common biological treatment methods identified in Table 9-16.

9.7.12 Biological Treatment Several biological treatment methods may be used to treat aqueous, solid, semi-solid, and gaseous organic and some inorganic hazardous wastes. The applicability of biological treatment is highly dependent on the waste mixture, biodegradable organic content, toxicity of the waste mixture to micro-organisms, and physical and chemical limitations associated with location of the material to be treated (in situ or ex situ). Table 9-16 presents a summary of the more commonly used biological treatment processes and their applications to certain hazardous wastes. Many of these technologies are still in developmental stages; treatability studies are essential prior to application of a biological treatment technology to a particular waste stream. The candidate is referred to the Innovative Site Remediation Technology Monograph Volume 1, Bioremediation (Anderson, 1993) for design considerations and limitations.

9.7.13 Stabilization/Solidification Stabilization and solidification treatment processes cover a wide range of treatment technologies that generally apply to sludges and solid hazardous metal and radioactive wastes. The primary objectives of stabilization and solidification are to reduce the potential for leaching and to improve the physical handling characteristics of the waste. Depending on the application, hazardous solids and sludges can often be rendered nonhazardous or inert, thus greatly reducing the cost of hazardous waste disposal. Table 9-17 lists several of the proven technologies and their applicability to various wastes.

Solidification involves mixing a solidifying material with the waste to form a material that has reduced permeability and improved handling characteristics. Solidification dramatically reduces of the available surface area of the waste and mechanically encapsulates the contaminants within the solidified matrix. Thus,

Table 9-16
Biological technologies for hazardous waste treatment and remediation

Technology	Typical Application
In Situ Techniques	
Land Treatment	remediation of soils/sludges contaminated with oils, metals, wood-preserving wastes, petroleum hydrocarbons, PAHs; wastes from coal gasification, food processing, leather tanning, pulp and paper production, wood-preserving, and petroleum refining.
Bioventing	remediation of soils contaminated with petroleum hydrocarbons
Air Sparging in the Unsaturated Zone	remediation of soils contaminated with petroleum hydrocarbons
Air Sparging in the Saturated Zone	remediation of groundwater/soils contaminated with petroleum hydrocarbons
Liquid Delivery Systems (Aerobic)	remediation of groundwater/soils contaminated with petroleum hydrocarbons, nonchlorinated solvents
Ex Situ Techniques for Aqueous Waste	
Suspended-growth Reactors, Activated Sludge, Aerated Lagoon, Extended Aeration, Contact Stabilization	aromatic and aliphatic hydrocarbons, mono- and dichlorinated-aromatic and aliphatic hydrocarbons, wastes from petroleum refineries, pulp and paper mills, textile, pharmaceutical and food processing facilities
Fixed-film Reactors, Trickling Filters Rotating Biological Contactors	petroleum hydrocarbons, wastes from petroleum refineries, pulp and paper mills, textile, pharmaceutical and food processing facilities
Submerged-film Reactors	aromatic and aliphatic hydrocarbons
Activated Carbon Reactors	pesticides, volatile organic compounds, low concentrations of chlorinated organics
Ex Situ Techniques for Soils, Slurries, Sludges	
Land Treatment	petroleum hydrocarbons, polynuclear aromatic hydrocarbons, wood-preserving wastes
Soil-pile Treatment	petroleum hydrocarbons, polynuclear aromatic hydrocarbons, solvents, chlorobenzenes
Composting	sewage sludge, soils contaminated with petroleum products, wastes and residues from food processing, brewing, antibiotic, fermentation, mineral oil and munitions operations
Anaerobic Digestion	simple organic waste streams such as alcohols and organic acids
Ex Situ Techniques for Air	
Biofiltration	removal of odors, hydrocarbons, some chlorinated organics from gas streams

Adapted from Anderson, 1993, and Berkowitz, 1978.

Table 9-17
Applicability of common solidification and stabilization techniques to various wastes

Techniques	Applicability
Vitrification	organics, acids, oxidizers, sulfates, halides, heavy metals, radioactive materials
Cement based	organic resins and tars, acids, oxidizers, heavy metals, radioactive materials
Lime based	organic resins and tars, acids, oxidizers, sulfates, heavy metals, radioactive materials
Sorbants/Surfactants	oily wastes, inorganic contaminated soils and sludges
Soluble Phosphates	heavy metals
Bituminization	low to medium level radioactive waste solutions, heavy metal sludges
Emulsified Asphalt	metal contaminated solids, petroleum contaminated soils
Thermoplastics	organics, acids, heavy metals, radioactive materials

Adapted from Anderson, 1993, and Martin, 1987.

contaminant loss to the environment is greatly reduced. Vitrification, a form of solidification, involves the application of high levels of electrical energy to a waste mass causing the mass to form a molten glass-like material that resists leaching.

Stabilization may be achieved by two different approaches: pH solubility control and chemical fixation.

- *pH solubility control* involves mixing a highly alkaline material such as Portland cement with the waste material, thereby reducing the pH to a point at or near the minimum solubility pH of the metal of concern.

- *Chemical fixation* involves adding a chemical to form insoluble or less soluble chemical compounds, thereby preventing the constituents from leaching.

Important design limitations for the more frequently used solidification and stabilization technologies are discussed below:

- Cement-based solidification is suitable for immobilizing many metals because of the high pH of the cement. However, this process works only for metals that form insoluble hydroxides or carbonates at a high pH and it does not work well for amphoteric metals such as lead. In addition, the solidified material is not considered inert since it can leach contaminants under the acidic conditions of the EPA's Toxicity Characteristic Leaching Procedure (TCLP) test. When analyzing the use of cement for solidification, the environmental engineer must determine the extent of organic matter, silt and clay, and certain metal and sodium salts in the waste, all which can interfere with the setting, curing, and permanence of the cement-waste mixture. Although cement is relatively inexpensive compared to other treatment additives, the volume of the waste expands twofold, and the weight of the cement-waste mixture increases significantly, thus creating higher disposal costs. For these reasons, cement is often used as a setting agent prior to further solidification and stabilization using other materials.

- Silicate-based processes involve mixing a siliceous material with lime, cement, gypsum or other setting agent to provide stabilization of a broader range of compounds, including organic solvents. Use of a silicate-based process should be made on a site-specific basis with a thorough knowledge of the mixture of materials contained within the hazardous waste. An important limitation of silicate-based processes is the large amount of water that remains after solidification. This liquid can leach contaminants and must be collected and treated or disposed.

- Sorbents such as fly ash, cement kiln dust, bentonite and activated carbon can be used to eliminate free liquids in a hazardous waste. It is important to note that, although sorbents prevent movement of the free liquids, they may not inhibit leaching of hazardous constituents. Laboratory and/or field scale studies must be conducted to determine the leachability of the final stabilized product.

- Vitrification offers the highest degree of containment achievable with solidification. The main limitations of this method are its high cost and the need for secondary treatment for vapors emitted during the vitrification process.

In most applications, a mixture of materials and processes is used to overcome the limitations of a single process.

9.7.14 Wet-Air Oxidation Wet-air oxidation can be applied to a wastewater or sludge to effectively destroy halogenated aliphatic compounds, aromatic hydrocarbons, and halogenated aromatic hydrocarbons with at least one nonhalogen group on the aromatic ring. The process consists of high-speed mixing, after which the waste is passed through a heat exchanger and then into a reactor tank where, under atmospheric conditions, the organic constituents are oxidized. Catalysts are sometimes used in the reactor tank to remove more resistant organic constituents. Certain organics may oxidize into other organic compounds that are as hazardous or more hazardous than the original constituent. Vapor emissions are generated during the process and require control and treatment prior to discharge.

9.8 Site Remediation

9.8.1 Introduction Remediation of waste disposal and other contaminated sites is a complex subject and cannot be covered in detail in a single reference. For details, consult the references listed at the end of this section. This section will provide an overview of the fundamental procedures and technologies employed in site remediation.

As with any engineering activity, the first and most important step in site remediation is to thoroughly define the problem. Every possible piece of historical and record information should be gathered to begin the process. An outline of the common steps in conducting a remedial investigation and feasibility study is contained in the National Contingency Plan regulations at 40 CFR 300.66. This outline identifies activities that are required on federally-funded projects. It also serves as a good checklist for projects that are not federally-funded. Conducting work in "substantial accordance" with these regulations also helps to ensure that the costs of the remedial activity are "recoverable" when the time comes for cost-recovery litigation. Detailed guidance on conducting remedial investigations is also available in the publication *Guidance for Conducting Remedial Investigations and Feasibility Studies Under CERCLA* (U.S. EPA (1988)).

The following sources of site information are invaluable in initially defining the nature and extent of the problem:

- Interviews with persons familiar with the history and operation of the site.
- Record documents including photographs and prior studies that may be available from:
    - the site's owners and operators
    - local, state and federal regulatory and public-service agencies
    - the local register of deeds
    - the local newspaper
    - the local library
- Aerial photographs from the U.S. Geological Survey and the U.S. Soil Conservation Service, satellite photos, and other photos from local and regional planning agencies, libraries and local universities.

Following collection and evaluation of the materials listed above, field investigations can begin. The first step in a field investigation is the preparation of a Health and Safety Plan (HSP) and a Quality Assurance and Protection Plan (QAPP) in accordance with, respectively, applicable Occupational Safety and Health Administration (OSHA) regulations (29 CFR 1910.120) and the National Contingency Plan. These plans are intended to ensure the safety of onsite workers and the quality of the data gathered. Also, a well-prepared workplan that summarizes the site information, the goals and objectives of the investigation, the data to be gathered and the appropriate methodologies, the timeline for the work, the anticipated output and the estimated cost is essential to the reliability and cost-effectiveness of the investigation.

Field investigation techniques can be generally divided into two categories: intrusive and nonintrusive. Nonintrusive investigations can provide a wealth of information about the site at low cost and little risk. The simplest nonintrusive technique is a site walkover by experienced personnel after the HSP is prepared. OSHA's Hazardous Waste Operations and Emergency Response ("HAZWOPER") requirements (29 CFR 1910.120) may require, depending on the characteristics of the site, that field personnel and their supervisors be properly-trained and enrolled in a medical monitoring program prior to entering a hazardous waste disposal site. Every professional involved in environmental site investigation and remediation should be thoroughly familiar with the 29 CFR 1910.120 regulations.

During a site walkover, a thorough visual examination of the site and surrounding properties should be conducted. Evidence of grade changes, settlement, vegetation types and stress, stained soils, surface debris and other surface features can help define the affected area that requires further investigation. Geophysical surveys may also be conducted early in the investigation to help delineate affected areas and to screen for buried metallic objects and other anomalies prior to intrusive activities. Electromagnetic (EM) ground-penetrating radar and resistivity surveys are commonly used to define subsurface anomalies at a site. EM surveys are particularly useful and inexpensive ways of defining subsurface metallic anomalies and/or waste deposits before drilling. They can also be useful in defining groundwater contamination plumes where significant electrolytic variations exist in the groundwater.

Only after the nonintrusive measures have been conducted, or at least thoroughly considered, should intrusive investigations take place. Intrusive activities can vary from simple surface soil sampling and hand-boring probes to deep bedrock drilling using reverse-air rotary and multiple-casing techniques to avoid cross-contamination. Drilling programs should be designed and implemented only by experienced professionals. The dangers involved in drilling on a site include worker exposure and health and safety risks, penetration of underground utilities and waste deposits, and exacerbation of vertical contaminant migration by drilling through low-permeability strata. On some sites, careful backhoe excavations can be used to quickly define the nature and extent of waste deposits, although this work must be conducted only by trained and experienced personnel. Many buried wastes can create safety hazards for equipment operators.

Small-diameter monitoring wells are commonly installed at waste disposal or environmentally-contaminated sites to evaluate the direction of groundwater flow and to monitor the quality of the groundwater. The workplan for the investigation should estimate the probable direction of groundwater flow based on the topographic and hydrologic information available for the site. A minimum of one upgradient and two downgradient wells placed in a triangular orientation are typically required to define an approximate direction of groundwater flow and to assess the impact on groundwater quality. The upgradient wells must be placed far enough upgradient to define the quality of groundwater flowing onto the site (background water quality). The two downgradient wells should be placed sufficiently far enough apart from each other and from the upgradient well to measure the difference in the static water levels across the piezometric surface of the groundwater so as to allow discernable differences in the static elevations when referenced to a common datum. Groundwater flow direction can then be determined by contouring between the data points, plotted in three dimensions, and by drawing a directional arrow perpendicular to the contours. Typically, several sets of static elevation groundwater data are collected from multiple wells in order to observe differences in water elevation and flow direction over time and with the season.

TIP FOR THE P.E. EXAMINATION CANDIDATE

The candidate should be able to design a simple groundwater monitoring system and interpret the data obtained from the system to determine groundwater flow direction.

Example 9-10 — Groundwater flow and direction

Determine the hydraulic gradient, groundwater flow direction, and groundwater flow velocity from the groundwater elevations at three wells, given the information below, through the solution of a three-point problem. The hydraulic conductivity of the soils is $4 \cdot 10^{-3}$ cm/sec, and the porosity is 0.2.

Well	Casing Elevation	Depth to Water
MW-1	1072.34	20.78
MW-2	1074.98	23.94
MW-3	1073.23	22.43

Step 1. Determine the groundwater elevations at each of the wells by subtracting the depth to groundwater from the casing elevation.

Groundwater Elevation
1051.56
1051.04
1050.80

Step 2. Plot the well locations (Figure 1), and identify the well with the intermediate groundwater elevation (MW-2).

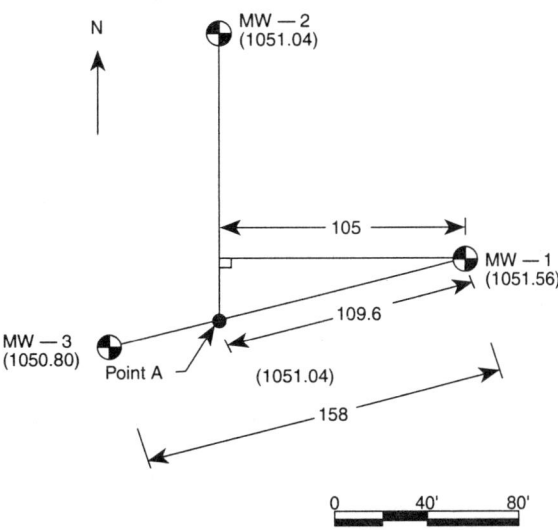

Step 3. Calculate the position between the well with the highest groundwater elevation (MW-1) and the well with the lowest groundwater elevation (MW-3) which has the same groundwater elevation as the intermediate well (Point A). This can be accomplished using the proportional relationship between groundwater elevation and distance along the line between MW-1 and MW-3.

$$\frac{\text{(Distance from MW}-1\text{ to Point A)}}{\text{(El. change from MW}-1\text{ to Point A)}} = \frac{\text{(Distance from MW}-1\text{ to MW}-3)}{\text{(El. change from MW}-1\text{ to Point3)}} \quad [9\text{-}14]$$

or

$$= \frac{\text{(Distance from MW}-1\text{ to Point A)}}{(1051.56 - 1051.04)} = \frac{(158)}{(1051.56 - 1050.80)} \quad [9\text{-}15]$$

or

$$\text{(Distance from MW}-1\text{ to Point A)} = \frac{(158)(0.52)}{(0.76)} = 109.6 \quad [9\text{-}16]$$

Thus Point A is 109.6 units from MW-3 along the line between MW-1 and MW-3.

Step 4. Draw a line of equal groundwater elevation between MW-2 and Point A. Draw and measure the line perpendicular to the line of equal groundwater elevation and MW-1. This line gives the groundwater flow direction which is always perpendicular to lines of equal groundwater elevation and goes in the direction of lower head (due West).

Step 5. The hydraulic gradient is calculated from the change in elevation between MW-1 and the line of equal groundwater elevation divided by the shortest distance from MW-1 to the line of equal groundwater elevation.

Hydraulic gradient (I) = dh / dx = (1051.56 – 1051.04) / 105 = 0.005

Step 6. The groundwater velocity can now be calculated:

$V = KI / n = (4 \bullet 10^{-3} \text{cm} / \text{sec}) \bullet (0.005) / 0.2 = 0.0001 \text{ cm} / \text{sec}$

After the groundwater flow direction has been verified, groundwater samples are taken and analyzed to permit comparison of the upgradient well water quality with the quality of the water in the downgradient wells. Often, several to several dozen wells are required on complex sites to define patterns of a groundwater flow and quality variations. Care must be taken in sampling and analyzing water samples from monitoring wells to ensure good-quality data. The QAPP plan for the sampling program needs to be carefully prepared, considering the hydrogeology of the site and the chemical parameters of interest. Sampling and analysis techniques should generally conform with the guidelines in EPA's publication SW 846 and/or other applicable guidelines and requirements, depending on the nature and location of the site.

Some sites contain multimedia contamination which must be adequately defined in order to undertake appropriate risk assessment and, if necessary, site remediation. Solid waste may exist in bulk deposits, in containers, or integrated with site soils. Contaminated soils and/or ground and surface waters may have dispersed far from the disposal site. Furthermore, the waste material may volatize and cause airborne migration of contaminants. Good references for defining the nature and extent of contamination in different media include USEPA (1986, 1995) and Nielsen and Johnson (1989).

9.8.2 Groundwater Remediation Extraction and treatment of contaminated groundwater, commonly known as "pump-and-treat," has been used widely in the United States since Love Canal first focused public attention on groundwater contamination in the early 1980s. Pump-and-treat technology uses vertical or horizontal wells to remove contaminated groundwater from an aquifer. Understanding the hydrogeology and geochemistry of the site is essential to design a groundwater extraction system. The degree of interbedding of zones of relatively high and relatively low permeability soils must be carefully defined or the extraction system may be ineffective. Downhole geophysical methods such as gamma-ray and resistivity logging are especially useful in defining and confirming variations in soil character in porous media aquifers and aquitards. Even a thin layer of soil with a high clay or silt content can compromise extraction system performance.

Similarly, the geochemistry of a contaminant plume must also be well understood. Certain dense nonaqueous phase liquids (DNAPLs) create density gradients, which can dramatically effect their migration patterns. Perchloroethylene (common dry cleaning fluid), with a specific gravity of 1.6, typifies this type of fluid. When released to the environment in pure product form, this material can pool at the bottom of an aquifer and act as a long-term subsurface source. Light nonaqueous phase liquids (LNAPLs) such as gasoline and fuel oil can pool in pure product form at the surface of groundwater, requiring surface skimming for removal. References by Schwille (1988), Cohen (1993) and American Petroleum Institute (1981) are recommended to provide detailed design information of remediation systems for high- and low- density fluids in groundwater.

The most common groundwater pump-and-treat system involves the purging of dissolved contaminants flowing with the groundwater by applying the accepted principles of advection and dispersion. Once the horizontal and vertical extent of such a plume is defined and the physical character of the interbedded aquifers and aquitards are reasonably well understood, design of an extraction system can begin. Typically, in very shallow groundwater environments where saturated thicknesses are limited, horizontal wells or interceptors are used to capture contaminated groundwater. When the depth to an underlying aquitard is less than about 20 ft, especially when permeability is poor or there is a high degree of interbedding, a trench system with a horizontal perforated interceptor pipe will often be most effective in collecting and containing groundwater flow. In these settings, vertical wells are limited by the available drawdown and flow rates, and the resulting capture zones are small, thereby requiring a large number of relatively ineffective, low-flow wells. Typically, once the flow rates drop below about 5 to 10 gal/min per well, alternative technologies should be evaluated.

In very permeable aquifers, vertical wells can produce large flow rates (300 to 500 gal/min or more per well) and create large capture zones. If the available drawdown and aquifer yield are sufficiently large, deep wells can be aligned along the central longitudinal axis to capture and contain the plume. Where there is less available well yield because of drawdown or permeability limitations, multiple wells across the width of the plume may be required to create overlapping capture zones to control the full plume. Normally, at least one vertical well or horizontal interceptor is used at the extreme downgradient end of the plume to prevent further downgradient migration. Depending on the age and three-dimensional configuration of the plume, additional capture wells are placed in the areas of maximum contaminant concentration to help remove the maximum mass of contaminant in the shortest possible time. Wells are also often placed near the source of the contamination to capture the contaminants before they can migrate downgradient and adsorb to the soils.

Technologies now under development, with some field use, involve directional drilling of horizontal or variable-elevation wells. In special circumstances, e.g., under an active manufacturing facility or within an old landfill, these techniques can be very cost-effective. Innovative uses of geosynthetic materials to create both physical and hydraulic containment within the same trench, while minimizing or eliminating the need for very costly offsite disposal of contaminated trench spoils, are also becoming more common. Also, barrier technologies and *in-situ* treatment barriers are often used in conjunction with pump-and-treat technologies to contain and treat contaminant plumes.

In recent years, it has become increasingly apparent that, for a variety of reasons, pump-and-treat systems are not very effective in "cleaning up" an aquifer. The effects of aquifer heterogeneity and anisotropy, adsorption of contaminants onto soil particles, and the complexities of aquifer geochemistry combine with a variety of other factors to limit the efficiency of even the best-designed pump and treat systems. At best, pump-and-treat systems take a very long time to completely remediate an aquifer. Over these time periods, natural attenuation processes such as hydrodynamic dispersion, soil adsorption, volatilization and biological activity have significant impacts on the plume characteristics.

The key to design of an extraction system is understanding how groundwater moves through the soil and applying these principles (refer to discussion of Darcy's Law on page in Section 9.4.3). A breakthrough curve forms as contaminated groundwater moves through a porous medium because of the processes of molecular diffusion and hydrodynamic dispersion. Molecular diffusion is a mixing process caused by Brownian motion, whereby molecules from locations of high concentration tend to migrate to locations of low concentrations, and vice versa. Hydrodynamic dispersion includes mixing caused by the creation of turbulent eddies as fluids move through pore spaces and by the channeling of flow through tortuous passages. The effect in both cases is a smearing of the front of a plume, causing a gradual rather than instantaneous breakthrough at a fixed point that is downgradient of the source.

Except at extremely low flow rates, hydrodynamic dispersion always dominates molecular diffusion by orders of magnitude. The broadening of a plume as it advances is also determined by dispersion. Coefficients of longitudinal and transverse dispersion can be measured empirically. Longitudinal dispersion typically exceeds transverse dispersion, and coefficients measured in the field typically exceed those measured in laboratory columns by orders of magnitude. The shapes and positions of breakthrough curves can be determined as functions of time by formulating and solving advection-dispersion equations, including known or assumed values of average linear velocity, retardation coefficients, and dispersion coefficients.

An equation that is often used in understanding the effect of retardation on the movement of contaminant in porous media is

$$\frac{v}{v_c} = 1 + \left(\frac{\rho_b}{n}\right) \bullet K_d \qquad [9\text{-}17]$$

where: v = average linear velocity of the groundwater
   v_c = velocity of the breakthrough curve of the retarded constituent

The right side of the equation is the retardation factor:

where: n = porosity
   ρ_b = bulk mass density
   K_d = distribution coefficient

The distribution coefficient can be described as the adsorbed mass of the solute species per unit of solids, divided by the dissolved solute concentration. Measured K_d values are normally reported as milliliters per gram (ml/g). This treatment implicitly assumes the adsorption isotherm to be linear, a traditional approach that may be valid over a narrow range of concentrations, but invalid over a wide range. For this reason, the use of published values of K_d to predict plume velocity is not advisable.

An advection-dispersion equation is a differential equation that expresses the mass balance of the contaminant as it is transported by advection or dispersion from one volume of pore space to another, or as it leaves solution and becomes adsorbed to solids. A retardation factor, as shown above, is incorporated to reflect adsorption. Chemical- and biologically-mediated reactions can also be incorporated into advection-dispersion equations, which are presented algebraically in Freeze and Cherry (1979) and other standard texts.

After the conceptual design of a groundwater capture system has been completed and the approximate locations, effective capture areas, and flow rates of the capture well system have been tentatively established based upon available information, field trials are conducted to define both the hydraulic characteristics of the aquifer and the chemical characteristics of the captured groundwater. Typically, this is done by placing a test well in the area of expected highest concentration of contaminants and pumping this well for a period of one to two weeks to actively draw contaminants into the well from a reasonably significant distance away from the well in order to define the chemical characteristics of the water and the hydraulic characteristics of the aquifer. Treatment of the water purged from the ground during the testing is generally required, depending upon the discharge point, and provides an opportunity to gather data on treatability characteristics using likely treatment technologies. It is important during this phase to get a complete baseline characterization of the groundwater, including full scans for volatile and organic and inorganic compounds and to define the characteristics of the water for purposes of assessing treatability and discharge options. These data are extremely useful for defining baseline characteristics of the groundwater and its associated contaminants, projecting mass balances around future treatment systems, preparing discharge permits, and assessing residual disposal options and costs. Given the uncertainty inherent in hydrogeologic characterizations (including the often used but fundamentally erroneous assumptions of homogeneity and isotropic conditions in the aquifer), basing a multi-million dollar remediation system on office-derived estimates of capture rates, flow rates and water quality can be a serious error in the design of remediation systems. Field and laboratory treatability studies should be the minimum foundation upon which every remediation design is built.

There are many technologies available to treat contaminated groundwater. Typically, there are three or four treatment technologies available for each general category of groundwater contaminant. For volatile organic contaminants, such as benzene, toluene, xylene, and the chlorinated aliphatics, such as perchloroethylene (tetrachloroethene) and its degradation products, the two most common treatment technologies are air stripping and carbon absorption. The references by Weber (1972), Kavanaugh and Trussel (1980), Gossett et al. (1985) American Petroleum Institute (1986), Crittenden et al. (1987), USEPA/ORD (1988), and Damle and Rogers (1990) offer good guidance on the application of these technologies.

TIP FOR THE P.E. EXAMINATION CANDIDATE

The candidate should understand these technologies and be able to use simple kinetic calculations to size such facilities.

Typically, for volatile organic compounds with high Henry's coefficients, air stripping is a very economical method for removing contaminants from groundwater. Once removed from groundwater, volatile contaminants in the resulting offgas can be addressed in a number of ways. The most economical alternative is discharge to the atmosphere if ambient conditions and receptor risks permit. However, in many cases it is necessary to treat the offgases with either carbon, incineration or ultraviolet technologies. Carbon absorption is typically employed to treat these types of organic compounds. However, a careful examination of the potential interferences by inorganic species and the effect of biological growth on the carbon material must be made to ensure reliable operation. Whenever carbon is used, either in liquid-phase or vapor-phase treatment, the carbon must be either regenerated or disposed. For small-scale systems, offsite disposal or regeneration is common.

Inorganic compounds such as metals are removed using conventional water treatment technologies, with various forms of precipitation and ion exchange technology being the most common. Both of these technologies require offsite disposal of concentrated contaminants in either slurry or sludge form. These factors need to be taken into account in assessing alternative treatment technologies.

9.8.3 Landfill Remediation A variety of techniques are available to remediate or stabilize old industrial and mixed-waste landfills depending on the nature of the hydrogeologic setting. A fundamental step in assessing a landfill remediation program is to conduct thorough geologic and hydrologic studies of the site and its surrounding environment, combined with a risk assessment to verify the migration pathways that must be controlled to reduce risks to acceptable levels.

TIP FOR THE P.E. EXAMINATION CANDIDATE

The candidate should be able to perform the sizing and cost-benefit analysis for life-cycle evaluation of air-stripping and carbon absorption for a variety of chemical compounds using standard reference texts during the examination.

The most common problem associated with older landfills is the production of leachate caused by the infiltration of rainwater through the landfill surface and underlying waste. In highly permeable sites, leachate will migrate down or away from the landfill to contaminate area groundwater resources. In areas with low-permeability soils, the landfill refuse mass becomes saturated and the leachate seeps to the surface and runs off to contaminate surface waters. In either situation, the remedy most often used in closing old landfill sites, now being termed a "presumptive remedy," is to construct a relatively impermeable barrier or "cap" across the old landfill surface. Additionally, lateral containment barriers consisting of trenches or physical barriers can be constructed around the perimeter of the landfill, depending upon the hydrogeologic setting, to further minimize the offsite contamination potential.

Once the need for a landfill remediation program has been established in the remedial investigation by defining the impacts on groundwater and/or surface water at the landfill site, the next step is to define the areal and vertical extent of the landfill deposits. The same types of investigation strategies discussed in Section 9.8.1 are often used for this purpose. After the hydrogeologic setting, the landfill history, and physical characteristics have been defined, alternative remediation strategies can be developed and evaluated.

A cap is often the sole remedy for closing an old landfill site, followed by long-term groundwater monitoring and maintenance of the cap. Creating a cap on the landfill involves preparing a stable sub-grade with adequate slopes to promote runoff from the newly-installed, low-permeability surface barrier to infiltration. Once the sub-grade has been constructed, a gas transmission layer is sometimes installed to

TIP FOR THE P.E. EXAMINATION CANDIDATE

The candidate should understand the sequencing and uses of these various layers in designing a final landfill cap and be able to calculate the velocity and quantity of flow through the various layers and the runoff from the closed landfill surface.

control and direct gas migration toward either passive or active collection points. A barrier layer is then constructed above the gas migration layer. This barrier layer typically consists of at least 2 to 3 ft of clay and/or a synthetic membrane liner. Above the barrier layer, another drainage layer is normally installed to direct infiltration to lateral drainage facilities around the perimeter of the site. Above the drainage layer, another layer of soil is required in northern climates to prevent freeze-thaw cracking and other physical damage to the underlying barrier layer. Above the frost protection layer, a minimum of 4 to 6 in. of top soil is placed and seeded with shallow-root type vegetation. Adequate drainage facilities are installed around the landfill perimeter to ensure that the runoff, which is now dramatically increased by the low-permeability cap, is adequately managed. Additional discussion of final cover design is included in Section 9.4.6.

When a landfill is capped, provisions for long-term control of gases must be constructed on the site if decomposition gases are expected. The simplest gas control device is a passive venting well, which is drilled through the cap and into the refuse deposits. Multiple venting wells on the site vent the decomposition gases through the cap and into the atmosphere, thereby minimizing the potential for lateral migration. In circumstances where economics are favorable or where explosion risks must be mitigated, active gas collection systems are installed, consisting of blowers to create negative pressure on extraction wells or lateral collection pipes constructed within the gas collection layers. Additional discussion of decomposition gas control is included in Section 9.4.8.

TIP FOR THE P.E. EXAMINATION CANDIDATE

The candidate should understand the fundamentals of the design of the gas systems, including the need to drain condensate from the lines, the relative percentages of landfill gases that commonly exist in these collective gases, and the fire and explosion hazards that may exist in decomposition gas systems.

On very small municipal or industrial landfill sites, excavation and transport of all wastes to a nearby licensed landfill is often the most economical remedial option. The cost of long-term maintenance and monitoring of an encapsulated landfill often involves life-cycle costs well in excess of the initial cost to excavate and haul the waste deposits to a controlled facility. However, in large municipal or industrial landfills, the costs of excavation and the ancillary concerns regarding dust, traffic impacts, and the human health risks associated with excavating and transporting the material, far outweigh the costs and risks of simply capping and controlling the wastes in place. Site workers employed in excavation and removal activity must be adequately trained according to 29 CFR 1910.120. Adequate health and safety plans must be in place prior to excavation, and perimeter air-monitoring systems often must be installed to monitor offsite impacts of the gases that are frequently present in such an excavation. Moveover, a very thorough understanding of the nature and extent of the waste must be acquired long before excavation commenc.

On some sites, the hydrogeologic setting is such that perimeter containment systems can be installed economically to create a lateral barrier to either leachate or gas migration or both. This approach is especially viable where the wastes have been placed in a porous soil stratum above an underlying low-permeability stratum that acts as a lower barrier to migration. When the lower barrier layer is within 50 to 70 ft of the ground surface, a variety of barrier techniques can be used. Sheet-steel walls, slurry walls, collection trenches, flexible membrane liner material, and treatment barriers all can be constructed, depending upon the depth of the underlying barrier layer and the nature of the contaminants encountered. Typically, a partially penetrating barrier system is less effective in these circumstances. This type of system, sometimes referred to as a "hanging curtain" barrier system, often results in underflow below the lower limits of the barrier unless additional hydraulic contaminant systems are used. If a contaminated groundwater plume has migrated from the landfill site, the groundwater remediation technologies previously discussed can be used to contain, capture, and treat the groundwater contamination.

Once a landfill has been capped or otherwise contained in-place, groundwater monitoring is typically required if there are aquifers that have been or may be affected. In these circumstances, the initial hydrogeologic monitoring facilities are operated (possibly with added wells) to monitor the effectiveness of the landfill cap/containment system over time. An additional important element of a long-term maintenance program is the care and maintenance of the constructed landfill cap. Typically, during the first year after construction, erosion channels must be repaired at the gathering locations of surface runoff. Also, the landfill needs to be mowed at least twice a year to prevent the intrusion of volunteer, deep-rooted species through the cap and barrier layer. Cottonwood trees, for instance, will often start to appear within one or two years after a landfill cap is constructed. Unless these and other deep-rooted species are controlled, the landfill cap can be compromised early in its life. Any erosion or other damage of the vegetation layer must be repaired.

9.9 Standard Solid and Hazardous Waste Notations

GEOSYNTHETICS STABILITY

Symbol	Definition	Units
W	weight of material or applied load (per unit width of slope)	F/L
Wn	normal component of weight or applied load	F/L
Ws	shear component of weight or applied load	F/L
Fb	friction force acting on bottom surface of geosynthetic	F/L
Ft	friction force acting on top surface of geosynthetic	F/L
T	tensile force generated within a geosynthetic	F/L
Ty	grab tensile yield strength of geosynthetic	F/L
Fa	force generated by geosynthetic anchor system	F/L
Fe	force generated by equipment weight and operations	F/L
FS	factor of safety	none
α	slope angle	deg
β	interface surface friction angle	deg
σ	strain induced in geosynthetic	L/L

HYDROGEOLOGY

Symbol	Definition	Units
A	area	L^2
A_d	drainage area	acres
B	aquifer width	L
C_p	rational runoff coefficient	—
C_r	retardance coefficient	—
d	drawdown, or distance between stations	L
E	evaporation	L/T
F	frequency of occurence	L/T
I	rainfall intensity	L/T
K	a constant (general usage)	none
K_p	constant of permeability	L^3/TL^2
L_c	centroidal stream length	L
L_o	overland flow path length	L

Symbol	Definition	Units
L_s	main stream length	L
N	average percipitation per year, or time from peak to end or runoff	L or T
P	precipitation over a short period	L
Q	flow quantity	$L^3/T7$
r	radial distance from well	L
R	runoff	L
s	slope of hydraulic gradient	L/L
S_c	storage constant	—
t	time since pumping	T
t_c	storm duration (time of concentration)	T
t_e	overland flow time	T
t_p	time from start of storm to peak runoff	T
t_r	rain storm duration	T
T	transmissivity	L^3/TL
v	flow velocity	L/T
V	volume	L^3
W(u)	well function	none
y	aquifer thickness after drawdown	L
Y	original aquifer thickness	L

OPEN CHANNEL FLOW

Symbol	Definition	Units
A	area	L^2
b	weir width	L
C	coefficient (e.g. orifice flow)	none
d	depth	L
d_H	hydraulic diameter	L
D	pipe diameter	L
E	specific energy	LM/M
f	Darcy friction factor	none
g	acceleration due to gravity (32.2)	L/T^2
h	head	L
H	head	L

K	minor loss coefficient	none
L	channel length	L
m	Bazen coefficient	none
n	Manning roughness coefficient	none
N	number of end contractions	none
p	pressure	F/L^2
P	wetted perimeter, or weir height	L
Q	flow quantity	L^3/T
r_H	hydraulic radius	L
S	slope of energy line (energy gradient)	none
S_o	channel slope	none
v	velocity	L/T
w	channel width	L
Y	weir height	L
z	height above datum	L
ρ	density	m/L^3

LEACHATE PUMPS/LFG BLOWERS (HYDRAULIC MACHINES)

Symbol	Definition	Units
bhp	brake horsepower	hp
c	specific heat	btu/lb-°F
C_v	coefficient of velocity	none
D	diameter	L
ehp	electrical horsepower	hp
E	energy	ML^2/T^2
f	Darcy friction factor	none
fhp	friction horsepower	hp
g	acceleration due to gravity (32.2)	L/T^2
h	static head	L
H	dynamic head	L
L	pipe length	L

Chapter 9: Solid and Hazardous Waste Management 387

n	rotational speed	1/T
n_s	specific speed	1/T
n_{ss}	suction specific speed	1/T
NPSHA	net positive suction head available	L
NPSHR	net positive suction head required	L
p	pressure	F/L^2
Q	flow quantity	L^3/T
S.G.	specific gravity	none
T	temperature	deg
v	velocity	L/T
w	weight	F
whp	hydraulic ('water') horsepower	hp
z	height above datum	L

Symbols

η	efficiency	none
ε	specific pipe roughness	L
ρ	density	M/L^3
β	blade angle	deg

HYDRAULICS

Symbol	Definition	Units
a	speed of sound in an elastic pipe	L/T
A	area	L^2
c	speed of sound in a pure medium	L/T
C	compressibility, or Hazen-Williams constant	L^2/F
d	diameter, or depth	L
D	diameter	L
e	pipe wall thickness	L
E	modulus of elasticity or bulk modulus	F/L^2
f	Darcy friction factor	none
F	factor, or force	F
g	acceleration due to gravity (32.2)	L/T^2
h	height, or head	L
I	moment of inertia	L^4

Symbol	Definition	Units
K	fitting loss coefficient	none
L	length	L
m	mass flow rate	M/T
n	Manning roughness constant, or friction exponent	—
N_{Re}	Reynolds number	—
p	pressure	F/L^2
P	power	LF/T
Q	flow quantity	L^3/T
r_H	hydraulic radius	L
R	resultant force	F
S	slope	L/L
SG	specific gravity	none
t	time	T
v	velocity	L/T
V	volume	L^3
w	weight flow	F/T
z	height above datum	L
υ	specific volume	L^3/F
ρ	density	M/L^3
μ	absolute viscosity	FT/L^2
τ	shear stress	F/L^2
ν	kinematic viscosity	L^2
ε	specific roughness	L
δ	Hardy Cross correction	L^3/T
β	ratio of small diameter to large diameter	L/L

SOILS AND WASTE

Symbol	Definition	Units
a	area	L^2
A	area	L^2
C_u	uniformity coefficient	none
C_z	coefficient of curvature	none
CBR	California bearing ratio	none

Chapter 9: Solid and Hazardous Waste Management

D	diameter	L
e	void ratio	none
F	percent passing through the sieve	none
G_H	hydraulic gradient	L/L
h	head	L
I_c	compression index	none
I_d	density index	none
I_g	group index	none
I_l	liquid index	none
I_p	plasticity index	none
k	coefficient of permeability	L/T
L	flow path length	L
LS	linear shrinkage	F
M	mass	M
n	porosity	none
p	pressure	F/L^2
P	load	F
PPS	percent pore space	none
Q	flow quantity	L^3/T
R	overconsolidation ratio, or Hveem's resistance	none
s	degree of saturation	none
S	strength	F/L^2
SG	specific gravity	none
SL	shrinkage linit	none
SR	shrinkage ratio	none
t	time	T
v	velocity	L/T
V	volume	L^3
vs	volumetric shrinkage	none
w	water content	none
W	weight	F
w_l	liquid limit	none
w_p	plastic limit	none

x	gravimetric fraction	none
ρ	density	M/L^3
σ	normal stress	F/L^2
φ	angle of internal friction	deg
τ	shear stress	F/L^2
ε	strain	L/L
θ	angle of principal stress plane	deg

9.10 References

American Petroleum Institute. Cost Model for Selected Technologies for Removal of Gasoline Components from Groundwater. Washington, D.C.: API, 1986.

---. Guide to the Assessment and Remediation of Underground Petroleum Releases. 2nd ed. Publication 1628. Washington, D.C.: API, 1989.

American Society for Testing and Materials. ASTM Standards on Ground Water and Vadose Zone Investigations. 2nd ed. Philadelphia: ASTM, 1994.

Anderson, William C., ed. Innovative Site Remediation Technology Monograph Series. Annapolis: American Academy of Environmental Engineers, 1993. 8 vols.

---. Remediation Technologies Screening Matrix and Reference Guide. Annapolis: American Academy of Environmental Engineers, 1995.

Bagchi, A. Design Construction, and Monitoring of Landfills, 2d. John Wiley. 1994.

Berkowitz, Joan B., et al. Unit Operations for Treatment of Hazardous Industrial Wastes. Park Ridge: Noyes, 1978.

Bonner, T., et al. Hazardous Waste Incineration Engineering. Park Ridge: Noyes, 1981.

Corbitt, R. Standard Handbook of Environmental Engineering. McGraw-Hill. 1990

Crittenden, John C. et al. "Design Considerations for GAC Treatment of Organics Chemicals." Journal of American Water Works Association. 19.1 (1987): 74-82.

Freeze, R. Allan, and John A. Cherry. Groundwater. Englewood Cliffs: Prentice, 1979.

Glysson, E. A., J. R. Packard, and C. H. Barnes. The Problem of Solid Waste Disposal. Ingenor Series No. 9. Ann Arbor: U. of Michigan P., 1972.

Gossett, J.M., et al. Mass Transfer Coefficients and Henry's Constants for Packed Tower Air Stripping of Volatile Organics: Measurements and Correlations. Final Report ESL-TR-85-18. Tyndall AFB: Engineering and Services Laboratory, 1985.

Government Institutes. Environmental Statutes. 1994 ed. Rockville: Government Institutes, 1994

Kavanaugh, Michael C. and R. Rhodes Trussel. "Design of Aeration Towers to Strip Volatile Contaminants from Drinking Water." Journal of American Water Works Association. 72 (1980): 684-692.

Martin, Edward J. and James H. Johnson, Jr., eds. Hazardous Waste Management Engineering. New York: Van Nostrand Reinhold, 1987.

Moore. 1980, 1989. USEPA. Washington, D.C.: GPO.

Niessen, Walter R. Combustion and Incineration Processes. 2nd ed. New York: Marcel Dekker, Inc., 1995.

Norris, Robert D., et al. Handbook of Bioremediation. Boca Raton: Lewis, 1994.

NRAES, (Northwest Region Agricultural Engineering Services), On Farm Composting Services, NRAES.54, 1992.

"RCRA Land Disposal Restrictions: A Guide to Compliance." Hazardous Waste Consultant. 12.6 (1994).

Schwille, Friedrich. Dense Chlorinated Solvents in Porous and Fractured Media. Chelsea: Lewis, 1988.

Sincero, A., and G. Sincero. Envronmental Engineering A Design Approach. Prentice Hall. 1996.

Tchobanoglous, G., H. Thiesen, and S. Virgil. Integrated Solid Waste Management. New York: McGraw, 1993.

United States Environmental Protection Agency. Cleanup of Releases from Petroleum USTS. EPA/530/UST-88/001. Washington, D.C.: GPO, 1988.

---. Groundwater. EPA/625/6-87/016. Washington, D.C.: GPO.

---. Office of Air Quality, Planning, and Standards. Air Stripper Design Manual. By Ashok S. Damle and Tony N. Rogers. Research Triangle Park: GPO, 1990.

---. Office of Emergency and Remedial Response. Guidance for Conducting Remedial Investigations and Feasibility Studies under CERCLA. EPA/540/G-89/004. Washington, D.C.: GPO, 1988.

---. Office of Research and Development. Performance of Air Stripping and GAC (Granular Activated Carbon) for SOC (Sythetic Organic Chemical) and VOC (Volatile Organic Chemical) Removal from Groundwater. Cincinnati: GPO, 1988.

---. Office of Research and Development. Stabilization/Solidification of CERCLA RCRA Wastes, Physical Tests, Chemical Testing Procedures, Technology Screening and Field Activities. EPA/625/6-89/022. Washington, D.C.: GPO, 1989.

---. Office of Research and Development. Stabilization Technologies for RCRA Corrective Action. EPA 625/6-91/026. Washington, D.C.: GPO, 1991.

---. Remedial Action at Waste Disposal Sites. EPA/ 625/6-85/006. Washington, D.C.: GPO, 1985.

---. Risk Reduction Engineering Laboratory, Office of Research and Development. Technical Guidance Document: Quality Control for Waste Management Facilities. By David E. Daniel and Robert M. Koerner. EPA/600/R-93/182. Cincinnati: GPO, 1993.

---. Robert S. Kerr Environmental Research Laboratory. DNAPL Site Evaluation. By Robert M. Cohen, James W. Mercer, and C.K. Smoley. EPA/600/R-93/022. Springfield: GPO, 1993.

--. Technical Handbook for Stabilization/Solidification of Hazardous Wastes. By M. J. Cullaine and L. W. Jones. EPA/540/2-86/001. Cincinnati: GPO, 1985.

---. Test Methods for Evaluating Solid Waste. [SW486]. Washington D.C.: GPO, 1995.

---. United States Department of Health and Human Services, et al. Occupational Safety and Health Guidance Manual for Hazardous Waste Site Activities. Washington, D.C.: GPO, 1985.

---. United States Government. National Archives and Records Administration. Code of Federal Regulations. V. 40, parts 141-149. Washington, D.C.: GPO, 1994.

---. Weber, Walter J., Jr. Physiochemical Processes for Water Quality Control. New York: Wiley-Interscience, 1972.

Wagner, T. The Complete Guide to the Hazardous Waste Management Regulations, 2d. Van Nostrand Reinhold, 1991.

Chapter 10

HEALTH, SAFETY, AND ENVIRONMENTAL PROTECTION

by Ronald L. Kathren, P.E., DEE

10.1 Introduction

Health, safety and environmental protection encompass a broad and important area of environmental engineering. The primary concern is protection of humans from workplace hazards including control of airborne toxic chemicals in the form of dusts, gases, mists and vapors; extremes of temperature and pressure; noise; and ionizing and nonionizing radiation. The professional discipline known as industrial hygiene deals with all of the above, although the specialized area known as health physics or radiological engineering relates specifically to radiation protection, and more specifically to ionizing radiations. Also included is risk assessment, the determination of the extent and nature of a response to a given dose. While principally applied in the context of human exposures, the technique can also be applied to other components of the ecosystem.

10.2 Legal Bases and References

The legal bases for the control of health, safety and environmental hazards are many and diverse. The environmental engineer, irrespective of the state(s) in which engineering is practiced, needs a basic understanding of the federal laws and regulations that apply. Federal regulatory agencies such as the Occupational Safety and Health Administration (OSHA), the Environmental Protection Agency (EPA), and Nuclear Regulatory Commission (NRC) have been granted specific responsibilities related to the control of health, safety and environmental hazards, and, through agreements with a number of states, have delegated enforcement responsibilities to state officials. In those states without suitable regulatory programs, enforcement falls to the relevant federal agency.

The primary regulatory foundation for the control of occupational hazards, including chemical and other hazards in the workplace, is the Occupational Safety and Health Act of 1970, which authorized the federal government to set mandatory occupational safety and health standards. The act created OSHA within the Department of Labor and also established the National Institute of Occupational Safety and Health (NIOSH) to conduct research that would serve as the basis for standards setting. NIOSH, it should be noted, is not a regulatory body. OSHA standards provide, perhaps, the single most important reference source with respect to the control of health, safety and environmental hazards.

A second important legal foundation is the Toxic Substances Control Act of 1976 (TSCA), which required industry to provide data on the production, use, and health and environmental effects of chemicals. In addition, TSCA provides the means for employees to receive information on the nature of potential exposures in the workplace.

Detailed regulations for ionizing radiation are promulgated by the NRC and are found in Title 10 of the Code of Federal Regulations, Part 20. These regulations are not comprehensive in that they do not generally apply to natural radioactive sources or to the control of machine-produced sources of radiation such as x-rays. Control of these sources of ionizing radiation is a state regulatory responsibility. In addition to the regulations, guidance is published by the National Council on Radiation Protection and Measurements and by the International Commission on Radiological Protection.

The Clean Air Act and the Clean Water Act, as amended over the past two decades, provide another important legal foundation for control of health, safety, and environmental hazards. In general, these laws govern the release of pollutants to the environment and fall under the purview of the EPA. The EPA is also concerned with indoor air quality and has promulgated standards for radon, a naturally-occurring radioactive gas associated with the decay of radium.

Since its inception in 1938, the American Conference of Governmental Industrial Hygienists (ACGIH) has played an important role in developing and tabulating limits for exposures in the workplace. The ACGIH has promulgated workplace exposure limits known as Threshold Limit Values (TLVs) for more than 600 chemicals. Many of these TLVs have been incorporated into regulations. In addition, the ACGIH has begun to provide biological exposure indices for a number of chemicals. It also publishes a number of basic guidance documents relating to air sampling and the design of exhaust ventilation systems. Key references published by ACGIH are listed in Section 10.7 and are published on a regular basis. The most recent available edition is recommended. In addition, the ACGIH publishes *Documentation of Threshold Limit Values for Substances in the Workroom Air,* along with supplements, at periodic and irregular intervals. This publication provides the underlying bases for the TLVs.

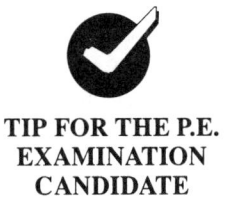

TIP FOR THE P.E. EXAMINATION CANDIDATE

The P.E. Examination does not include test items requiring the candidate to demonstrate detailed knowledge of the applicable laws and regulations. However, the candidate should know the statutory basis for control of health and safety.

10.3 Indoor Air Quality

10.3.1 Basic Control Strategies In general, indoor air quality relates to the control of airborne contaminants such as dusts, fumes, gases, mists or vapors, whether chemical, biological, or radioactive, and to assurance that the atmosphere does not contain explosive mixtures and has sufficient oxygen to sustain life. Although the basic route of entry for most airborne contaminant is inhalation, consideration must always be given to whether the contaminants can be absorbed via the skin, or whether it will result in illness or irritation after a suitable latent period. Numerous examples of airborne contaminants can be given: evaporation of solvents from cleaning or degreasing operations that result in vapors; generation of dusts by mechanical processes, such as grinding, or even from apparently innocuous operations, such as loading of materials; fume production from welding operations; and gases such as carbon monoxide produced by internal combustion engines.

Once the hazard has been recognized, there are a number of basic strategies for the control of indoor air pollutants. These are briefly outlined below in decreasing order of desirability. The most desirable strategies provide for control of the hazard at the point of origin and can be characterized as follows:

1. *Elimination.* Control of airborne contaminants in the workplace is best accomplished by elimination of the toxic material or the process that generates it. This may be accomplished either by use of a substitute

process, such as the use of battery-powered fork lift trucks in lieu of fork lifts powered by internal combustion engines. In general, elimination is the most effective and simplest form of control and should be favored over other control strategies.

2. *Substitution.* Substitution of a material of lesser toxicity or hazard is another means of controlling pollutants. An example is the use of toluene-based paint strippers in place of those that contain the more toxic benzene. Substitution is often the most practical and reliable solution to a potential indoor air pollution problem and can be a highly-effective means of achieving control.

3. *Process and Equipment Modification.* Control of potential airborne hazards at the source of generation can be achieved through suitable engineering of the industrial processes likely to generate air contaminants. Wet processes are effective at reduction of generation of airborne contaminants. Continuous flow operations that rely on automatic or mechanical introduction of process feed materials typically provide less opportunity for release of air contaminants than do batch operations, or operations in which chemicals are manually introduced. Leakage paths for chemicals should be eliminated by paying proper attention to equipment design. Enclosures and ventilation, discussed below, can be used on a process line. Effluent control for industrial processes can be achieved in various ways. Filters, cyclones or inertial separation devices and electrostatic precipitation are typically utilized for dust control. Scrubbers may be needed for removal of gases and vapors (Table 10-1a). The design and operation of air quality control devices is discussed in Chapter 8. In general, process and equipment modifications are far more expensive to install and maintain than the simpler solutions of elimination or substitution described above, but may be dictated out of necessity.

4. *Isolation or Enclosure.* Related to process and equipment modification is process isolation or enclosure. Isolation or total enclosure may be a

Table 10-1a
Recommended minimum design velocities for various materials

Material collected	Minimum conveying velocity (ft/min)
Fumes, vapors, gases, mists, very fine dusts	1500–2000
Fine dusts (dry)	1000
Asbestos fibers	3000
Typical industrial dusts	3500
Coarse particles and dusts	3500–4500
Large particles, wet materials	≥4500

necessity for operations involving highly or acutely toxic chemicals or radioactive materials. In its most sophisticated forms, isolation can involve robotics to permit remote operations. Generally, however, a process is isolated by a physical barrier of some type. It may be necessary to totally isolate large areas and volumes. An extreme example of isolation is the complete physical isolation and containment of a process in a room or building from which personnel are excluded; the containment of a nuclear reactor or of various chemical engineering process reactors are examples. At the other end of the spectrum is the use of pressure differentials to isolate a process by having personnel located in areas that have positive pressure with respect to the location of the pollutant-generating process. Small processes may be contained in a glove box that is kept under negative pressure with respect to the room in which it is located.

5. *Exhaust Ventilation.* Exhaust ventilation provides a means of removing the pollutant or contaminant from the point of generation before it has the opportunity to become airborne in the general room environment. The key element is that the airflow must be sufficient to provide adequate capture of the particulates or gases that need to be removed, and this requires skillful design of the total ventilation system. General dilution ventilation is suitable only for control of such factors as temperature and humidity, or for small amounts of low-toxicity contaminants. Hoods—sometimes called fume hoods, although they can be applied to dust-, gas- and vapor-generating operations as well—can be used to provide local exhaust or removal of potential air contaminants. Hoods and associated ductwork and air movers must be suitably designed to assure proper operation (Table 10-1b). The principles discussed in Chapter 4, Fluid Mechanics, apply.

TIP FOR THE P.E. EXAMINATION CANDIDATE

The priority of basic control stategies should be followed in answering indoor air quality test items.

In addition to the five basic control strategies outlined above, that control potentially toxic indoor air pollutants at their point of origin to a greater or lesser degree, a sixth control strategy can be applied to the workers themselves. This is accomplished through the use of personal protective equipment (PPE). PPE basically isolates the worker, rather than the process. It includes respirators of various types, ranging from positive-pressure oxygen-demand type equipment to so-called half masks equipped with cartridges to remove simple dusts of low toxicity. PPE, in the broader sense, includes gloves and protective clothing and eyewear of all types, which prevent contact with the airborne contaminant. It should be stressed that the use of PPE is not nearly as desirable as removal of the toxic materials at the point

Table 10-1b
Exhaust hood flow rate equations

Hood type	Minimum flow rate (Q)
Plain opening, round, square or rectangular	$(10X^2 + A)V$
Flanged opening, round, square or rectangular	$0.75(10X^2 + A)V$
Flanged opening, resting on bench or table	$0.75(5X^2 + A)V$
Plain slot, aspect ratio ≤ 0.1	$3.7L_s XV$
Flanged slot, aspect ratio ≤ 0.1	$2.8L_s XV$
Unflanged tank	$2.3W^{1.15} L^{0.85} V$
Flanged tank	$1.9W^{1.3} L^{0.7} V$
Canopy	$1.4 PHV_C$

A = face area of hood (ft^2)
H = height of canopy above tank (ft)
L = tank length (ft)
L$_4$ = slot length (ft)
P = tank perimeter (ft)
V = capture velocity at distance X (ft/min)
V$_C$ = velocity at edges of canopy (ft/min)
X = distance from point of capture to hood face (ft)

of generation and should, therefore, be considered as the last or least desirable control strategy.

10.3.2 Air Concentrations and TLVs Concentrations of airborne contaminants are typically expressed in parts per million (ppm) or milligrams per cubic meter (mg/m^3). When making unitary conversions, one should bear in mind that ppm is based on volume, while mg/m^3 is a weight-per-volume relationship. Dusts are generally expressed in terms of mg/m^3, although the older unit of millions of particles per cubic foot of air (MPPCF) is sometimes still used. The relationship of ppm to mg/m^3 is given by

$$\text{ppm} = \frac{24.45}{\text{gram} - \text{molecular weight}} \qquad [10\text{-}1]$$

TLVs are based on the toxicity of the specific chemical or material, with due consideration being given to effects such as irritant potential. The general underlying concept is that the TLV represents the level at which a person may be continuously exposed without demonstrable ill effect (Table 10-2). The TLV is expressed in terms of an eight-hour-time weighted average; that is, the TLV is stated in terms of the average or mean concentration permitted over an eight-hour period. Thus, for some substances, the TLV can be exceeded for short periods, provided the overall eight-hour average exposure does not exceed the TLV. An excursion TLV factor, which is a function of TLV value, is applied to determine the upper limit of allowable concentration and ranges from 1.25 for chemicals with the highest TLVs to 3

Chapter 10: Health, Safety, and Environmental Protection

Table 10–2
TLV's for common chemicals in the work place

Substance	Time Weighted Average (TWA) ppm	mg/m^3	Short Term Exposure Limit (STEL) ppm	mg/m^3
Acetone	750	1780	1000	2380
Acrolein	0.1	0.23	0.3	0.69
Ammonia	25	17	35	24
Benzene	0.3	0.96	—	—
Bromine	0.1	0.66	0.2	1.3
Carbon dioxide	5000	9000	30,000	54,000
Carbon monoxide	25	29	—	—
Carbon tetrachloride	5	31	10	63
Chlorobenzene	10	46	—	—
Chloroform	10	49	—	—
Cyclohexane	300	1030	—	—
DDT	—	1	—	—
Dichloroethyl ether	5	29	10	58
Dichlorotetra–fluoroethane	1000	6990	—	—
Ethanol (ethyl alcohol)	1000	1880	—	—
Ethyl butyl ketone	50	234	—	—
Fluorine	1	1.6	2	3.1
Formaldehyde	—	—	0.3	0.37
Formic acid	5	9.4	10	19
Furfuryl alcohol	10	40	15	60
Gasoline	300	890	500	1480
Heptane	400	1640	500	2050
Hexane (normal)	50	176	—	—
Hexane (isomeric)	500	1760	1000	3500
Hydrogen peroxide	1	1.4	—	—
Hydrogen sulfide	10	14	15	21
Iron oxide dust and fume	2	5	—	—
Isoamyl alcohol	100	361	125	452
Isopropyl alcohol	400	983	500	1230
Isopropyl ether	250	1040	310	1300
Lead	—	0.15	—	—
LPG	1000	1800	—	—
Malathion	—	10	—	—
Mercury (metallic)	—	0.025	—	—
Methanol	200	262	250	328
Methyl chloroform	350	1910	450	2560
Methyl isobutyl ketone (MIBK)	50	205	75	307
Methyl isopropyl ketone	200	705	—	—
Methyl methacrylate	100	410	—	—
Napthalene	10	52	15	79
Nicotine	—	0.5	—	—
Nitric acid	2	5.2	4	10
Nitric oxide	25	31	—	—
p–Nitrochlorobenzene	0.1	0.64	—	—
Nitrogen dioxide	3	5.6	5	9.4
Nitroglycerin	0.05	0.46	—	—

Table 10–2 (cont.)
TLV's for common chemicals in the work place

Substance	Time Weighted ppm	Average (TWA) mg/m^3	Short Term Exposure Limit (STEL) ppm	mg/m^3
Nitrous oxide	50	90	—	—
Ozone	—	—	0.1	0.2
Parathion	—	0.1	—	—
Particulates (not otherwise classified)	—	10	—	—
Pentachlorophenol	—	0.5	—	—
Perchloroethylene	25	170	100	685
Phenol	5	19	—	—
Phenyl ether	1	7	2	14
Phosgene	0.1	0.40	—	—
Phosphorus (yellow)	0.02	0.1	—	—
Propylene dichloride	75	347	110	508
Rouge	—	10	—	—
Silica, amorphous				
Diatomaceous earth	—	10	—	—
Fume	—	2	—	—
Fused	—	0.1	—	—
Silica, crystalline				
Cristobalite	—	0.05	—	—
Quartz	—	0.1	—	—
Stoddard solvent	100	525	—	—
Styrene monomer	50	213	100	425
Sulfuric acid	—	1	—	3
1,1,1,2–Tetrachloro–2,2–difluoroethane	500	4170	—	—
1,1,1,2–Tetrachloro–1,2–difluoroethane	500	4170	—	—
1,1,2,2–Tetrachloroethane	1	6.9	—	—
Toluene	50	188	—	—
Toluene–2,4–diisocyanate	0.005	0.036	0.02	0.14
Trichloroethylene	50	269	100	537
Trifluorobromomethane	1000	6090	—	—
Turpentine	100	556	—	—
Vinyl chloride	5	13	—	—
Welding fumes	—	5	—	—
Xylene	100	434	—	—

Source: American Conference of Governmental Industrial Hygienists 1995–96 Threshold Limit Values for Chemical Substances and Physical Agents and Biological Expsure Indices

for those with the lowest values. For a number of chemical substances, the TLV must not be exceeded under any circumstances; such substances are designated with a "c" or ceiling designation. Exposure concentrations in excess of the TLV have also been defined in the American National Standard Z.37 series in terms of an Acceptable Ceiling Concentration which establishes the maximum concentration permissible during the period of exposure. A Maximum Peak Permissible Concentration has also been defined. This permits the Acceptable Ceiling Concentration to be exceeded under certain conditions related to concentration, duration of exposure, and the number of times such peaks may occur in an eight-hour work day.

10.3.3 Exhaust Ventilation

Control of indoor air contaminants is often achieved through exhaust ventilation. The basic principle is expressed by the simple formula:

$$Q = AV \qquad [10\text{-}2]$$

in which Q refers to volume of air moved through a duct of cross-sectional area A at a velocity V. Air movement results from the pressure differential, which is characterized in terms of the total pressure (TP). The TP is the sum of the velocity pressure (VP), or pressure created by moving air, and the static pressure (SP) (refer to Equation 4-15). Pressures are normally measured in inches of water. The velocity is related to VP by the relationship

$$V = 4005\sqrt{VP} \qquad [10\text{-}3]$$

This equation is used in all hood-design calculations. Velocity must be sufficient to capture the particles or gases being removed. For effective removal, the velocity of air across the face of a hood must be at least 100 feet per minute (FPM), with 150-200 FPM recommended for high-toxicity or radioactive materials. A minimum transport velocity of 2000 to 4500 FPM is the "rule of thumb" for particles depending on the fineness of the dust. For gases and vapors, duct velocities are lower, but typically 1500 to 2500 FPM. The specific design of the hood will determine its airflow characteristics. The coefficient of entry is an expression of how efficient the hood is with respect to air intake and is directly proportional to the velocity pressure in the ductwork downstream from the hood.

Example 10-1 — Sample ventilation problem

A flanged rectangular hood is used to remove very fine particles generated at the hood face. The hood dimensions are 3 ft by 4 ft. Determine: a) minimum acceptable hood flow rate; b) hood face velocity, (neglect all losses); c) the hood entry loss if the velocity pressure is 0.6 in. of water; d) the hood static pressure.

Solution:

  a) Required transport velocity for fine particles is 2,000 FPM. Hood area = 3 • 4 = 12 ft². Distance • from point of generation = 0
For flanged hood:

$$Q = 0.75\left(10X^2 + A\right)$$

$$= 0.75\left(10(0)^2 + 12\right)2,000$$

$$= 18,000 \, \text{CFM}$$

b) Face velocity = $\dfrac{Q}{A}$

$$= \dfrac{18{,}000}{12} = 150 \text{ FPM}$$

c) Coefficient of entry = C_e = 0.82 for flanged opening hood entry loss,

$$h_e = \dfrac{1 - C_e^2}{C_e^2} VP = 0.49 VP$$

$$= \dfrac{1 - (0.82)^2}{(0.82)^2} \cdot 0.6 = 0.3 \text{ in. of water}$$

d) The hood coefficient of entry is related to the pressure as follows:

$$C_e = \sqrt{\dfrac{VP}{SP_h}}$$

$$0.82 = \sqrt{\dfrac{0.6}{SP_h}}$$

$$SP_h = \dfrac{0.6}{(0.82)^2} = 0.9 \text{ in. of water}$$

The contaminant-bearing air, once collected by a hood or other entry device, is directed through ductwork to its ultimate destination. Ducts provide resistance to flow through friction and other mechanisms. In general, pressure losses are least for large-diameter straight ducts. Impediments to laminar flow increase losses. Bends provide considerable resistance to flow, and to minimize such resistance and associated turbulence, the joining angles between ducts is kept small. Right-angle bends in ducts should be avoided; elbows should have a radius greater than two duct diameters. Air-cleaning devices by their nature typically provide considerable resistance to flow. This resistance needs to be carefully considered along with each bend, change in duct cross-section, and other factors when sizing the air-moving device (AMD)—fan or blower—for a system. Another consideration relates to weather caps on the discharge stack. A rule of thumb is that the weathercap be located no closer than one duct diameter to the stack.

TIP FOR THE P.E. EXAMINATION CANDIDATE

The engineering of indoor air quality control management devices apply fluid principles discussed in Chapter 4.

Example 10-2 — Sample duct design problem

A 20 ft run of a 2 ft diameter duct is to be added to an existing hood. Duct velocity is 2,000 FPM. What is the duct loss?

Solution:

The equation for duct loss per 100 ft of run is

$$h_d = \frac{2.74\left[\dfrac{V}{1,000}\right]^{1.9}}{D^{1.22}}$$

Assume:

1. Straight run of duct
2. Duct loss proportional to length

then,

$$h_d = \frac{2.74\left[\dfrac{2,000}{1,000}\right]^{1.9}}{(24)^{1.22}}$$

$$h_d = \frac{2.74[2]^{1.9}}{48.3} = 0.21 \text{ in. of water for } 100 \text{ ft}$$

$$= 0.04 \text{ in. of water for the } 20 \text{ ft run}$$

Note: Duct diameter has been converted to inches before inserting in the headlloss formula.

The AMD is typically located downstream of the collection point at least six discharge stack diameters from the point of discharge to assure adequate mixing. The air may pass through an air-cleaning device first. Particulate removal is best achieved by a high-efficiency particulate air (HEPA) filter, which, according to standard, must have a removal efficiency of 99.97% for particles 0.3 μm in diameter and a pressure drop of no more than one inch of water when new. Pressure drop will increase as filter loading occurs. Two or more filters can be used in series to increase particle removal. The particle removal factor (also known as the decontamination factor) for a HEPA filter is 3300.

Example 10-3 — Filter problem

A fabric bag filter is to be used to remove fine dry particulates. Filter data given are:

> Design collector system loss = 4 in. of water
> Initial filter resistance coefficient = 0.0005 in. H$_2$O per FPM
> Dust deposit resistance = 0.0007 in. /lb of dust per square foot of filter
> Area of collector = 1.7 ft^2 = duct area

What is the filter capacity in pounds of dust?

Solution:

For a bag filter, the system pressure loss is related to the collection capability by

$$DP = K_o V + K_1 VW$$

For fine dust, the minimum design velocity is 3,000 FPM (Table 10-1a). Therefore:

$$4 = 0.0005 \bullet 3,000 + 0.0007 \bullet 3,000 \bullet W$$

and solving for W, the filter capacity per ft^2 is

$$W = \frac{4 - (0.0005 \bullet 3,000)}{0.0007 \bullet 3,000} = 1.2 \, lbs / ft^2$$

or, for 1.7 ft^2 of filter area:

$$1.2 \, lbs / ft^2 \bullet 1.7 \, ft^2 = 2 \, lbs$$

10.3.4 Monitoring The specific contaminants that may be released by a process will determine the selection of monitoring methods for indoor air and stack effluents. Direct-reading instruments can be used for general room air measurement of a number of aerosols, gases and vapors. A simple means of measuring ambient air concentrations of various chemicals is direct-reading colorimetric devices, in which a quantity of air is drawn through a transparent tube containing a suitable chemical mix. They are specific for individual contaminants and provide a grab sample only. Their results may be suspect if potential interferences are present. General room air measurements can be made with a variety of sampling devices, with the sample subjected to laboratory analysis to determine the amount of contaminant present. Paper, fiberglass or molecular-sieve type filters are used

Chapter 10: Health, Safety, and Environmental Protection

for collection of dusts, while impregnated papers or bubblers of various types can be used to sample for chemical contaminants.

More useful, particularly for dust sampling, is the lapel or personal air sampler (PAS). The PAS is worn by the worker in such a manner as to draw air from the breathing zone of the worker, thereby providing a more representative sample of the actual exposure. Such samplers determine only average concentration, which is determined by dividing the amount of contaminant measured in the laboratory by the volume of air that passed through the sampler. They cannot provide a measure of peak exposure concentration.

Example 10-4 — Monitoring

An air sampler collects 16 mg of methyl chloride during six-hour work shift. The sampler has the following characteristics:

Average flow rate = 1.2 LPM
Collection efficiency = 0.9

a) What is the 8 hour TWA of methyl chloride?

Solution:

The total air collected was

$$1.2 \text{ LPM} \bullet 6 \text{ hr} \bullet 60 \text{ min/hr} = 432 \text{ L}$$

$$\frac{432 \text{ L}}{1,000 \text{ L/m}^3} = 0.432 \text{ m}^3$$

Sampler content was 16 mg or $\frac{16}{0.9} = 17.8$ mg in the air sampled.

Average air concentration during the collection time

$$= 41.2 \frac{\text{mg}}{\text{m}^3} \bullet \frac{6 \text{ hr}}{8 \text{ hr}} = 31 \frac{\text{mg}}{\text{m}^3}$$

b) Was the TLV exceeded?

TWA for methyl chloride is 103 mg/m³ and was not exceeded.

TIP FOR THE P.E. EXAMINATION CANDIDATE

Indoor air quality measurements are often related to a specific time period in regulations. Control designs must be designed accordingly.

Continuous monitors that provide a real-time measurement can be used to measure peak concentrations and provide the eight-hour time-weighted average concentration. These instruments generally measure ambient, and not breathing zone air. The advantages of these instruments are many, including real-time readout of concentration, alarm capability, ability to provide a continuous permanent record, elimination or reduction of the number of laboratory analyses required, and reduced personnel requirements. Disadvantages include high initial and maintenance costs, low sensitivity, lack of portability, and lack of specificity or effects produced by interferences.

Example 10-5 — Continuous monitoring

Lead fumes are produced in a lead-burning operation. A continuous monitor in the work room is operated for 5 consecutive days. The following are known:

   Initial sample rate: 1CFM
   Final sample rate: 0.7 CFM
   Filter efficiency: 0.999 for particulates ≥ 20 μm AMD
     0.83 for particulates < 20 μm AMD
   Mass of dust collected: 1.7 g
   % of lead in dust: 23

Solution:

Determine the TWA lead concentration in the work room air.

   quantity of air sampled = average sampling rate • sampling time

Assume approximately linear reduction in sampling rate as filter loads. Hence, average sampling rate is

$$\frac{1+0.7}{2} = 0.85 \text{CFM}$$

and total air sampled is

   $0.85 \text{CFM} \bullet 5 \text{days} \bullet 1,440 \text{min} / \text{day} = 4,180 \text{ft}^3$

Total dust collected = 1.7 g or 1.7 • 0.23 = 0.39 g of lead, since 23% of dust is lead. Since lead was present as a fume, AMD <20 μm and total lead collected is

$$\frac{0.39}{0.83} = 0.47 \, g$$

Average lead concentration is the total amount divided by sample volume or

$$\frac{0.47 \, g}{4,180 \, ft^3} = \frac{470 \, mg}{4,180 \, ft^3} \bullet \frac{1}{35.2 \, m^3 / ft^3}$$

$$= 0.0032 \, mg / m^3$$

Assuming lead generation occurs only during 8 hour work day, the TWA is

$$TWA = \frac{0.0032 \, mg}{m^3} \bullet \frac{24}{8} = 0.01 \, mg / m^3$$

Dusts pose a particular problem with respect to determining acceptable air concentrations. The particle size distribution needs to be taken into account, and the specific respirable fractions, as published in the literature, must be used to determine exposure. Dust particle sizing can be accomplished by size-selective samplers, typically multistage impactors.

Example 10-6 — Dust monitoring

An air sample collected in the vicinity of an operation mechanically generating diatomaceous earth dust was found to contain 17.3 mg/m³ of diatomaceous earth dust. Particle size analysis of the total sample yielded the following:

AMD(μm)	Mass Collected (mg)
>20	173
>10–20	61
>8–10	34
>6–8	27
5	21
4	12
3	9
2	6
1	1.5
<1	0.5

Calculate the respirable particulate mass concentration and determine if it exceeds the 8 hr TWA for a worker exposed 6.5 hr per day.

Solution:

Using the ACGIH table of respirable particulate mass and linear interpolation, the above table can be expanded as follows:

I AMD(μm)	II Mass Collected (mg)	III Mass Fraction	IV ACGIH RPM (%)	V II • IV
>20	173	0.50	—	0
>10–20	61	0.18	—	0
>8–10	34	0.10	1	.03′ .3
>6–8	27	0.08	11	3.0
5	21	0.06	30	6.3
4	12	0.03	50	6.0
3	9	0.03	74	6.7
2	6	0.02	91	5.5
1	1.5	0.005	97	1.5
<1	0.5	0.002	100	0.5
Total	348			29.8

Note: RPM = Respirable particulate mass

By inspection, 68% of the mass is outside the respirable range; therefore no more than $(1-0.68)\ 17.3 = 5.5$ mg/m^3 can be within the respirable range. Time weighted, this is

$$5.5 \bullet \frac{6.5}{8} = 4.5 \frac{mg}{m^3} \text{ or}$$

below the TLV of 10 mg/m^3.

The exact value of respirable mass concentration is calculated by determining the respirable mass fraction,

$$\frac{\sum_i (\text{mass collected} \bullet \text{RPM})}{\text{total mass}} = 8.6\% \text{ of total mass}$$

and multiplying it by the total dust mass concentration and the time weighting factor:

$$0.086 \bullet 17.3 \frac{mg}{M^3} \bullet \frac{6.5\,hr}{8\,hr\,day} = 1.2 \frac{mg}{M^3}$$

this amount is also less than the TLV of 10 mg/m^3.

10.4 Noise

10.4.1 Physical Principles Sound is the physical manifestation of variations in pressure in a medium such as air. Noise is unwanted sound. The frequency of sound refers to the rate at which a complete cycle of high- and low-pressure regions is produced by the source. This is measured in hertz (Hz); 1 Hz is one complete cycle per second. The range of audible frequencies for a young person is from 20 Hz to 20 KHz. As a person ages or is exposed to excessive noise, the ability to hear the higher frequencies becomes reduced. Older adults may have an effective cutoff of 10 KHz or less.

Sound intensity (I) is the average rate at which sound energy is transmitted through a unit area normal to the direction of sound propagation. It is expressed as an energy fluence rate, normally in terms of decibels (dB). The decibel has been defined as 10^{-12} watts per square meter. Sound intensity is a function of the sound power (P) of the source. The sound power is the total sound energy radiated by the source per unit of time and is normally expressed in *watts*. The dB is also used to quantify sound power level. Under free-field ideal conditions, sound intensity will drop as the square of the distance from the source. The relationship of sound intensity to sound power is given by:

$$I = \frac{P}{4\pi d^2} \qquad [10\text{-}4]$$

Sound power should not be confused with sound pressure (p). Sound pressure refers to the root mean square (rms) value of the pressure changes above ambient atmospheric level produced by a sound source. It is typically measured in units of pascals, newton per square meter (N/m^2) or dB. The dB, as used for sound pressure, is referenced to 20μPa or $2 \cdot 10^{-5}$ N/m^2, which is considered the threshold of hearing for a normal person. Peak sinusoidal pressure is typically 3 dB greater than the rms value, although for complex waveforms associated with common noises, the differences

TIP FOR THE P.E. EXAMINATION CANDIDATE

Sound pressure is not the same as sound power. Sound pressure is measured in dB which must be added logrithmically.

may be much greater and approach values as large as 25 dB. Sound pressure is also measured in microbars, although this unit is now obsolete. The range of sound pressures encountered in everyday life is quite broad. Representative levels are shown in Table 10-3.

Table 10-3 also serves to illustrate the relationship between sound pressure level expressed in dB and sound pressure. As measured by an instrument, sound pressure level, L_p, is characterized in units of dB and is expressed mathematically in dB as:

Table 10-3
Representative sound pressure levels

Activity	dB	N/m^2	Watts
Minimum Perception Level	1	0.00002	10^{-12}
Night, quiet rural location	20	0.0002	10^{-10}
Quiet room	40	0.002	10^{-8}
Normal conversation (3 ft)	60	0.0210^{-6}	
Heavy vehicular freeway traffic (automobiles, 50 ft)	80	0.2	10^{-4}
Inside helicopter	90	0.7	10^{-3}
Power lawn mower	95	1	0.04
Jet aircraft (100 ft)	110	7	0.1
Pain threshold	120	20	1

$$L_p = 20 \log \frac{p}{p_o} \qquad [10\text{-}5]$$

where p is the measured rms sound pressure and p_o is the reference value, viz., 0.00002 N/m².

Sound pressure levels cannot be added arithmetically because, as shown in the equation above, the dB is a logarithmic quantity. Sound pressures must be added logarithmically using sound intensities. The relationship between sound pressure, L_I, and sound intensity I is:

$$L_I = 10 \log \left(\frac{I}{I_o} \right) \qquad [10\text{-}6]$$

The reference intensity, I_o, is 10^{-12} watts/m². Intensity is proportional to the square of the rms sound pressure and inversely proportional to the speed of sound and the density of the medium. As a rule of thumb, addition of sound pressure levels that differ by 20 dB or more will not increase the overall sound pressure level. For example, adding an 80 dB noise to already existent 100 dB level will still result in a sound pressure level of 100 dB. Adding two equal sound pressure levels increases the total sound pressure level by 3 dB. For example, adding an 80 dB noise to an already present 80 dB sound pressure level will result in a total of 83 dB.

Example 10-7 —Sound pressure/intensity relationship

A new noise source with a sound level of 83 dBA is installed in a plant with a background noise level of 75 dBA. Will the combined level exceed the 8 hr daily duration TLV for noise?

Solution:

The ACGIH guide gives the 8 hr noise TLV as 85 dBA. The combined level of noise in the plant is determined by converting the A weighted sound pressure levels to sound intensity, adding the levels and reconverting to dBA. Alternatively tables from various publications (e.g., Patty, Vol. 1, Parta, p. 947) can be used. The table shows that for a difference in 8 PL of 83–75=8 dBA, an additional 0.6 dBA should be added to the higher value, or 83 + 0.6 = 83.6 dBA. The calculation can be done by noting intensity, L.

$$L = 10\log\left(\frac{I}{I_o}\right) dBA$$

The reference value $I_o = 10^{-12}$ watt/m² and calculating pressure from the new source:

$$L = 83 = 10\log\left(\frac{I}{10^{-12}}\right)$$

$$\frac{83}{10} = \log\left(\frac{I}{10^{-12}}\right)$$

$$8.3 = \log\left(\frac{I}{10^{-12}}\right)$$

Taking antilogs and rearranging gives

$$2 \bullet 10^8 \bullet 10^{-12} = I$$

$$I = 2 \bullet 10^{-4}$$

Performing the same operation for the background SPL gives

$$75 = 10\log\left(\frac{I}{10^{-12}}\right)$$

$$3.2 \bullet 10^7 \bullet 10^{-12} = I$$

$$I = 3.2 \bullet 10^{-5}$$

Adding the two sound intensities gives $2 \cdot 10^{-4} + 3.2 \cdot 10^{-5} = 2.32 \cdot 10^{-4}$ which can then be used to determine the combined SPL

$$L = 10\log\left(\frac{2.32 \cdot 10^{-4}}{10^{-12}}\right)$$
$$= 10\log(2.32 \cdot 10^{8})$$
$$= 10 \cdot 8.37 = 83.7 \, dBA$$

The perceived loudness of a sound is related not only to its intensity, but to the individual's ability to hear certain frequencies. The normal human ear does not have the same efficiency with respect to sounds of all frequencies; a pure tone with a frequency of 20 Hz, for example, would not seem as loud to an observer as a pure tone of 400 Hz, even though both sounds produced the same sound pressure level at the ear. For this reason, sound pressure levels are sometimes expressed in a weighted form to take into account the ear's ability to perceive sounds of different frequencies. The A-weighting scale accomplishes this purpose. Instruments are designed to specifically measure the A-weighted scale, which is reported as dBA; the 'A' signifies that the scale has been appropriately weighted. Tabulations of corrections for various frequencies are given in the literature. The corrections range from -44.7 dB at 25 Hz, to +1.3 at 2,500 Hz.

10.4.2 Noise Exposure and Noise Control For continuous exposure to noise over an eight-hour day, OSHA has set a limit of 90 dBA as an eight-hour time-weighted average. Employers are required to provide a hearing conservation program, noise monitoring, annual audiograms and hearing protection for employees whose average daily eight-hour exposure exceeds 85 dBA. Higher levels are permitted for shorter intervals; a 5 dBA increase is allowed up to a limit of 115 dBA for each halving of exposure time. Thus, 95 dBA would be permitted for 4 hours per day, 100 dBA for 2 hours, and so on. The maximum level of 115 dBA would be permitted for only one-quarter hour daily. In general, noise levels above 85 dBA make telephone use impossible and voice communication very difficult and limited, requiring shouting at a distance of only two to three feet.

Example 10-8 — Noise exposure

A worker is exposed to a noise level of 94 dBA for 30 minutes and to 88 dBA for 3 hours each day. Does this exposure exceed the TLV for noise?

Solution:

The TLV for 94 dBA noise is 1 hour
The TLV for 88 dBA is 4 hours

The combined noise TLV can be expressed by

$$\frac{C_1}{T_1} + \frac{C_2}{T_2} \cdots \frac{C_n}{T_n} = \text{Fraction of TLV}$$

For the two noises

$$\frac{0.5}{1} + \frac{3}{4} = 1.25$$

the TLV is exceeded.

Noise control is effected in a number of ways. Machinery can be designed to reduce noise. The basic principle is similar to that described above for control of indoor air pollutants: eliminating the source of the noise. This can be accomplished through improved machinery design and maintenance or by damping vibrations by increasing stiffness or using rubber or plastic bumpers or shock absorbers. Larger, slower machines typically produce less noise than smaller and faster devices. Belt drives can be used to replace gears and cams, and hydraulic or electrical devices can be substituted for mechanical ones.

Noise screens of various types can be designed and constructed. Noise baffles are effective, particularly for higher frequencies. Cosmetic corrections, such as tree plantings or visual screens, may provide better acceptance of noise levels, but do little to reduce actual sound pressure levels. Distance is a very effective noise control strategy. It is noted that higher sound frequencies are more easily attenuated than lower frequencies.

Example 10-9 — Noise control

A turbine has an acoustic power output of 10 watts. A workstation is to be located 10 feet from the turbine in an acoustically shielded booth providing a 29 dB reduction. What will the sound pressure level be in the booth? Will it be sufficiently low to permit conversation?

Solution:

For a free-field, omnidirectional source, use Equation 10-4 to calculate the intensity at d = 10 ft

$$I = \frac{P}{4\pi(305)^2}$$

Converting d to meters:

10 ft • 0.305 M/ft = 3.05 M and, then

$$I = \frac{10}{4\pi(305)^2} = 0.086 \text{ watts} / M^2$$

Converting to sound pressure:

$$L_I = 10\log\left(\frac{I}{I_o}\right) = 10\log\left(\frac{0.086}{10^{-12}}\right) = 99.3 \text{ dB}$$

with no attenuation.

Level inside booth = 99.3 − 29 = 70.3 dB

70 dB (unweighted) will not interfere with normal speech.

Personnel protection can be achieved through the design of quiet booths or spaces and the use of ear protection. Ear plugs are small devices that fit inside the ear canal and mold to the shape of the canal. Good quality, well-fitting plugs can achieve a reduction in sound pressure levels of 20 to 45 dB. The greatest attenuation is achieved in the high frequencies. Ear muffs fit over the ear and can achieve somewhat better attenuation. Muffs and plugs can be used together, but the total attenuation is the logarithmic sum of the dB attenuation values for each. For example, an ear plug that reduces the sound pressure level by 30 dB in combination with a muff that reduces the level by 35 dB would provide a total reduction of 36.2

dB. Personal ear protective devices are not the preferred means of noise control, but may be necessary if adequate engineered control cannot be achieved.

10.4.3 Noise Monitoring The basic instrument for noise monitoring is the sound-level meter, which is equipped with a suitable microphone, amplifier, and calibrated volume control and measures the rms sound pressure level in dB. Sound-level meters should meet the standards for these devices promulgated by the American National Standards Institute and OSHA. A sound-level meter may measure the unweighted sound pressure level or may be equipped with appropriate circuitry to provide a weighted response. Special, fast instruments are required to measure impulse noises or peak sound levels.

It is important to know the specific frequency distribution of the sound energy in order to properly evaluate the hazard from the noise. This is accomplished with frequency analyzers of various types. Octave-band analyzers are most commonly used for this purpose. They measure the sound pressure levels in the various octaves, normally expressed in terms of the center frequency. Octave bands typically measured are 31.5 Hz, 63 Hz, 125 Hz, and so on to a maximum of 31.5 KHz. Weightings are provided for the various octave lands so that conversion can be made to a weighted decibels, as shown in Example 10-10.

Example 10-10 — Noise monitoring

Determine the A weighted SPL for a noise with the following octave and analysis:

Frequency (Hz)	Level (dB)
31.5	89
63	80
125	74
250	62
500	60
1,000	65
2,000	61
4,000	50
8,000	50

Applying the weighted factors yields the following expanded table:

Frequency (Hz)	Measured Level (dB)	Weighted Response (dB)	A Weighted Level (dBA)
31.5	89	-39	50
63	80	-26	56
125	74	-16	58
250	62	-9	53
500	60	-3	57
1,000	65	0	65
2,000	61	+1	62
4,000	50	+1	51
8,000	50	-1	49

The A weighted values for each octave band are then added logarithmically (see Example 10-7) taking the highest value at 1,000 Hz as reference.

Frequency (Hz)	A Weighted Level (dB)	Difference from Reference Value (dBA)	Additional Contribution to Reference Value (dBA)*
31.5	50	15	0.1
63	56	9	0.5
125	58	7	0.8
250	53	12	0.3
500	57	8	0.6
1,000	65	0	0
2,000	62	3	1.8
4,000	51	14	0.2
8,000	49	16	<u>0.1</u>

Total dBA to be added = 4.4

* Obtained from tabulated values (published)

A weighted sound level is 65 + 4.4 = 69.4 dBA

Noise dosimeter or noise hazard meters are used to assess the hazard from noise exposure. These may be worn by personnel and integrate the noise exposure received over the time the dosimeter is worn, providing a measure of the average sound pressure level.

10.5 Ionizing Radiation

10.5.1. Physical Principles Ionizing radiation is *corpuscular* or electromagnetic radiation capable of producing ion pairs as it traverses matter. Radioactive elements are those with unstable nuclei that emit particles and electromagnetic radiations as they undergo radioactive decay, converting or transmuting to a new element in the process. All isotopes of all elements with atomic numbers greater than 82 (lead) are radioactive, except perhaps for the 209 isotope of bismuth, whose half-life (the time required for half of the atoms of a radioactive substance to undergo radioactive decay) is long enough for it to be considered stable.

TIP FOR THE P.E. EXAMINATION CANDIDATE

Every competent environmental engineer should be, at minimum, conversant with the terms and principles described in this section.

The heavy elements typically decay by emission of an *alpha* particle from the nucleus. An alpha particle is the same as a helium nucleus. Although alpha particles may have a considerable amount of associated kinetic energy, they have very limited penetrating power because of their relatively large atomic mass. Alpha particles pose a problem if taken into the body, however. In particular, long-lived alpha emitters such as plutonium-239 and radium-226 are significant radiological hazards because these elements tend to accumulate in the skeleton and remain there for many years because of their long physical and biological half-lives.

Beta-emitting radioactive elements include the fission products and a number of useful isotopes such as cobalt-60 or molybdenum-99 made by neutron activation in a reactor. Beta particles are electrons with a range of as much as 30 ft in air, depending upon their energy. They are easily shielded by a few millimeters of most common materials. Most beta radioactive nuclides emit negatrons or electrons with a negative charge. There are some positron emitters, however, that emit positively charged electrons (antimatter) from the nucleus. These interact quickly and convert to photons of electromagnetic radiation having a specific energy of 511 keV.

Electromagnetic radiation in the form of *x-rays* and *gamma rays* (photons) is associated with the decay of both alpha and beta radioactive elements. X-rays and gamma rays are both electromagnetic radiations, differing only in their mode of production. X-rays are produced outside the nucleus. Gamma rays are emitted directly from the nucleus as a means by which an excited nucleus can drop to a lower energy state. Typically, x-rays have lower photon energies than do gamma rays. X-ray energies associated with radioactive decay ordinarily have energies of a few tens of kiloelectron volts, while gamma ray energies typically range up to several million electron volts.

X-rays are also produced, either purposefully or incidentally, by a number of devices. Purposeful production provides x-rays for medical diagnosis and therapy and industrial radiography. Typical diagnostic x-ray units produce x-rays with peak energies of 100 to 150 keV, while therapy machines may operate in the 250 to 300 keV range. Television sets, cathode ray tubes, radar sets and similar electronic devices operating at voltages in excess of 10,000 V may incidentally produce x-rays when energized.

Neutrons are uncharged particles associated with the fission of heavy elements and are produced as a byproduct of high-energy particle accelerators. *Protons* are hydrogen nuclei and, hence, positively charged. Like alpha particles, they have limited penetrating power and are found only in special situations, such as around particle accelerators.

All radionuclides, i.e., alpha, beta, etc., are characterized by a specific physical or radiological half-life, the time required for half of the radioactive nuclei to decay. Radioactive decay is an exponential process and is described by the fundamental decay law, which is expressed mathematically as:

$$A = A_o e^{-\lambda t} \quad [10\text{-}7]$$

where A refers to the activity at any time t, and A_o refers to the activity at zero time. The decay constant l is expressed in units of reciprocal time and is an indication of the rate or probability of decay. The larger the value of l, the shorter the half-life. The relationship between the decay constant and the half-life, $T_{1/2}$, is expressed by:

$$T_{1/2} = \frac{0.693}{\lambda} \quad [10\text{-}8]$$

A rule of thumb is that after 7 half-lives, the activity is reduced to less than 1% of the original amount, and after 10 half-lives, to less than 0.1%. Activity refers to the number of radioactive transformations per unit of time. In the old system of units, activity was expressed in curie (Ci) units (or multiples or submultiples thereof); one Ci was defined as $3.7 \cdot 10^{10}$ disintegrations per second. The comparable unit in the System Internationale (SI) is the becquerel (Bq), which is simply one disintegration per second.

Example 10-11 — Radionuclide decay

How long will it take for 750 Bq of a radionuclide to decay to 200Bq if the decay constant is 0.086 day^{-1}? What is the half life?

Chapter 10: Health, Safety, and Environmental Protection

Solution:

$A_o = 750$ Bq
$A = 200$ Bq
$l = 0.086$ day^{-1}

Time to decay:

$$A = A_o e^{-\lambda t} \qquad [10\text{-}7]$$

$$200 = 75 e^{0.086t}$$

$$\ln\left(\frac{250}{750}\right) = 00.086 t$$

$$t = 15.4 \text{ days}$$

Half Life,

$$T_{1/2} = \frac{0.693}{\lambda} \qquad [10\text{-}8]$$

$$= \frac{0.693}{0.086}$$

$$= 8.06 \text{ days}$$

Radioactive materials taken into the body are removed by two mechanisms: physical decay and biological processes, each with its own characteristic half-life. The combined removal by these two mechanisms is known as the *effective half life*, T_{eff}, which is determined by:

$$T_{eff} = \frac{T_{rad} \bullet T_{biol}}{T_{rad} + T_{biol}} \qquad [10\text{-}9]$$

The effective half life is always smaller than either the biological or radiological half-life, although if either the radiological or biological half-life is very much smaller than the other, it will approximate the effective half life.

The fundamental decay law described above has the same form as the shielding equation for x-rays and gamma rays. The basic shielding equation is:

$$I = I_o e^{-\mu x} \qquad [10\text{-}10]$$

where I refers to the intensity or dose of radiation passing through a shield of thickness x, given an original intensity of I_0. The linear attenuation coefficient, μ, is a specific function of the energy of the photons and the shielding material, and is usually given in units of reciprocal cm. Instead of the linear attenuation coefficient, sometimes the mass attenuation coefficient is used. This is the linear attenuation coefficient divided by the density and symbolized by μ/r. If the mass attenuation coefficient is used, the thickness of the shielding material has to be expressed in comparable reciprocal units, which are obtained by multiplying the linear thickness by the density.

Example 10-12 — Biological half life

A worker inhales a radionuclide in a single acute exposure. Urine bioassays show a half life of 6.1 days. The nuclide has a decay constant of 0.08 per day. What is the biological half life in the worker?

Solution:

The radiological half life is determined by Equation 10-8:

$$T_{1/2} = \frac{0.693}{l} = \frac{0.693}{0.08} = 8.7 \text{ days}$$

The biological half life is obtained form Equation 10-9:

$$T_{eff} = \frac{T_{rad} \bullet T_{biol}}{T_{rad} + T_{biol}}$$

$$6.1 = \frac{8.7 \bullet T_{biol}}{8.7 + T_{biol}}$$

$$6.1(8.7) + 6.1 T_{biol} = 8.7 T_{biol}$$

$$53.1 = 8.7 T_{biol} - 6.1 T_{biol}$$

$$T_{biol} = 20.4 \text{ days}$$

Radiation doses are a function of the energy absorbed. The older dose units were the *rad*, which was numerically equal to the absorption of 100 ergs of energy per gram (0.01 J/kg) and the corresponding dose equivalent unit the *rem*, which was the rad multiplied by a quality factor to account for the different effectiveness of various radiations at producing biological damage. The SI units are is now becoming widely used. The comparable units are the *gray* (Gy) and the *Seivert* (Sv). The

Gy is numerically equal to 1 J/kg and, hence, equal to 100 rad. The factor of 100 holds for the rem and Sv relationship as well. The Sv is obtained by multiplying the absorbed dose in Gy by a radiation weighting factor; analogous to the older quality factor.

The newer radiation quantities include an effective dose equivalent, which utilizes weighting factors to account for partial body irradiation, such as when an organ receives a high dose relative to the rest of the body from an internally-deposited radionuclide. To obtain the effective dose, the dose equivalent is multiplied by appropriate tissue weighting factors. Often the expression is in terms of the 50-yr committed effective dose equivalent, in which the dose from a radioactive substance deposited is integrated over a period of 50 yrs. The units for all these new quantities are the Sv.

Finally, mention is made of *collective dose*, which is a concept of expressing the dose to a population in terms of a single number. This is done by summing the doses to all members of the population and expressing it as person-rad, person-rem, person-Gy or person-Sv, as appropriate.

10.5.2 Protection Principles The basic radiation exposure limit is 50 mSv (5 rem) annually, usually referred to as the *maximum permissible dose* (MPD). Doses from all sources, whether from internal or external radiations, are included. Radiation exposure from sources external to the body is from electromagnetic radiations, beta rays and neutrons. Alpha particles do not penetrate the dead layer of the skin and hence produce no external dose. However, when taken into the body, alpha-emitting radionuclides provide significant doses because of the relatively large amount energy associated with their decay and the short range of the alpha particles in tissue. There are specific Annual Limits on Intake (ALI) established for various radionuclides. Intake of one ALI will produce a maximum permissible annual equivalent effective dose of 50 mSv (5 rem). Permissible air concentrations for specific radionuclides are obtained by dividing the ALI by the annual breathing rate, and converted to activity per volume of air. This is known as the Derived Air Concentration or DAC which is commonly expressed in units of μCi/ml in the old system or Bq/m^3 in the SI.

Internal radiation dose is calculated by the amount of energy deposited by the decay of a radioactive substance in a specific tissue or organ. The deposition of radioactivity in a tissue or organ can be determined by *in vivo* counting techniques if there is electromagnetic radiation of sufficient energy to exit the body associated with the decay of the alpha- or beta-emitting nuclide. Measurement of the radioactivity excreted in urine can be used along with suitable mathematical excretion models to calculate the deposition.

With the exception of radon, tritium, and radioiodines, airborne radioactivity is generally in the form of particulates. Air samples are collected on filters of various types and analyzed in the laboratory for their radioactivity content. Radioactive decay is a statistical process, following Gaussian statistics. Thus, the standard deviation of a count will be equal to the square root of the count. Since all counting systems have some background associated with them, the total count must be sufficiently large to provide statistical significance. A conservative rule of thumb is that a sample is positive for radioactivity if it exceeds the background count by 2 standard deviations, although there are more precise formulations in most textbooks.

Example 10-13 — Airborne radioactivity

The following data are obtained from the laboratory for a room air sample:

Total sampling time: 15 min
Sampling rate: 30 CFM
Counter background: 83 counts in 100 min
Sample count rate: 603 counts in 10 min
Detector efficiency: 30%

What is the airborne radioactivity concentration (Bq/m^3) and the standard deviation of the concentration?

For Gaussian distribution $SD = \sqrt{Count}$

$$SD = \sqrt{603} = 24.6 \text{ or } \frac{24.6}{10} = 2.46 \, CPM$$

$$CV = \frac{2.46}{603/10} = 0.04 = 4\% \text{ of count rate}$$

Net count rate:

$$\frac{603}{10} - \frac{83}{100} = 60.3 - 0.83 = 59.5 \, CPM$$

$$Activity = \frac{count\ rate}{efficiency} = \frac{59.5}{0.3} = 198 \, dis/min$$

Converting to becquerel:

$$\frac{198 \, dis/min}{60 \, sec/min} = 36 \, DPS = 36 \, Bq$$

Volume of air sampled

$$15 \text{ min} \bullet 30 \text{ CFM} = \frac{450 \text{ ft}^3}{35.2 \text{ ft}^3 / \text{m}^3} = 12.8 \text{ m}^3$$

$$\text{concentration is} \frac{36 \text{ Bq}}{12.8 \text{ m}^3} = 2.8 \text{ Bq} / \text{m}^3$$

Standard deviation of concentration:

Since CV=4% of count rate, it will also be 4% of activity. Therefore, $0.04 \bullet 2.8 = 0.11$ Bq.

Radon is a naturally-occurring radioactive gas that is present everywhere in our environment. Homes and other buildings in areas of high radium content in the soil, or those built from brick or other materials containing radium, will show elevated levels of radon. Remediation measures include ventilation of the structure, sealing of cracks in foundations and elsewhere that may serve as pathways for radon into the living spaces, and lining of crawl spaces with plastic vapor barriers to minimize the escape of radon into the living spaces.

Protection from external radiations is achieved by reducing the time of exposure, increasing the distance from the source, and by shielding. Radiation levels from a point or small source follow the inverse-square law. Thus, doubling the distance from the source reduces the exposure rate to one fourth of that at the closer distance. Shielding is frequently necessary. Beta particles should be shielded with material that has a low atomic number since high-Z material may produce photons from bremsstrahlung interactions, much as x-rays are produced in an x-ray tube. High-density materials provide the best shielding for electromagnetic radiations, with lead, iron and concrete most commonly used. The effectiveness of a shield is sometimes expressed in terms of the half-value thickness, which is the thickness of the material required to reduce the radiation intensity by one-half. The tenth-value layer is the thickness required to reduce the intensity by a factor of ten.

Although radiation shielding follows the exponential attenuation principle described above, under broad-beam conditions, and particularly with thick shields, attenuation will be less than predicted by the exponential attenuation equation. This is attributable to radiation scattering within the shield and is a phenomenon known as buildup. Buildup can be rule-of-thumb estimated by the factor $(1 + \mu x)$ which is added as a multiplier to the right side of the attenuation equation as shown:

$$I = I_o B e^{-\mu x} \qquad [10\text{-}11]$$

$$B = 1 + \mu x \qquad [10\text{-}12]$$

The term μx is simply the number of half-value layers.

Example 10-14 — Radiation shielding

A stored ^{137}Cs source produces an exposure rate of 100 μGy per hour in areas accessible to personnel. How much addition lead shielding is needed to reduce the dose rate to 20 μGy per hour? The source is isotropic and stored in a spherical cask.

This is a case of broad beam geometry, and buildup must be considered. The buildup factor B is approximated by $1+\mu x$, where μx is equivalent to a half value thickness. To obtain a 5-fold reduction in dose rate, (i.e., from 100 to 20 μGy per hour), requires somewhere between 2 and 3 half value layers, or, more precisely

$$5 = 2^n$$
$$n = 2.3$$

Hence $B = 1 + 2.3 = 3.3$

The basic shielding equation, including buildup, is

$$I = I_o B e^{-\mu x}$$

For the 0.662 MeV photons from ^{137}Cs, $\mu = 1.24$ per cm. Solving for the thickness required

$$20 = 100 \bullet 3.3 e^{-1.24 x}$$
$$x = 2.26 \, cm$$

10.5.3 Monitoring Personnel monitoring for external radiation exposure from beta rays, electromagnetic radiation and neutrons is required if the exposure is anticipated to exceed one-quarter of the MPD. Film badges and thermoluminescent dosimeters are used for this purpose. Since the dosimeter does not respond to or absorb radiation in the same manner as the human body, specific shields or absorbers are used to correct for the energy dependence of the dosimeter.

Since radiation cannot be detected by the human senses, it is necessary to use various instruments to detect and measure it. Survey meters of various types are used to monitor external radiation fields. Ionization chamber instruments typically provide the best and most accurate measurements in the field, but they are not as sensitive as the familiar Geiger-Mueller counter, which may show unacceptable energy dependence characteristics. Constant air monitors are used to ensure that airborne radioactivity levels do not exceed permissible levels. These typically draw air continuously through a filter that is continuously examined by a suitable detec-

10.5 Risk Assessment

Risk assessment is built on the principle that small exposures carry with them some risk of an untoward health effect such as development of a malignant tumor or leukemia at some time in the future. Such risks are generally considered to be stochastic or probabilistic in nature and are expressed in terms of a risk coefficient, which is basically an expression of the probability or chance of the specific health effect occurring per unit of exposure. Although response to ionizing radiation is generally considered to be linear with dose, the risk from chemical carcinogens may not be; thus, risk evaluation needs to consider such effects as the existence of a threshold dose below which the effect does not occur, as well as the shape of the dose-response curve. Also, the latency period—the time between the exposure and the onset of disease—needs to be taken into account. For example, if the latency period is 25 yrs and exposure occurs at age 50, the individual may die from other causes prior to developing the specific health effect under consideration. Even if the individual lives long enough for the health effect to be manifested, the number of years of life lost, and productive years of life lost, will be less than if the exposure had occurred at an earlier age. Health and safety risks are sometimes expressed in terms of the number of years of productive (or healthy) life lost per thousand worker-years.

Example 10-15 — Cancer risk assessment

a) Estimate the total number of fatal leukemias in 50 yrs to a population of 100,000 exposed to an average annual dose of 500 µGy if the risk coefficient is 0.03 Gy^{-1}.

$$\text{Fatal Leukemias} = \frac{0.03}{\text{Gy}} \cdot \frac{5 \bullet 10^{-4} \text{Gy}}{\text{person} - \text{year}} \bullet 50 \text{ years} \bullet 10^5 \text{ persons}$$
$$= 75$$

b) What would the probability of causation be in a person exposed to 2 µGy if the natural incidence is 0.0003?

$$PC = \frac{\text{induced risk}}{\text{natural frequency} + \text{induced risk}}$$

$$= \frac{2 \cdot 10^{-3} \text{Gy} \cdot 0.03 \text{Gy}^{-1}}{2 \cdot 10^{-3} \text{Gy} \cdot 0.03 \text{Gy}^{-1} + 0.0003}$$

$$= \frac{6 \cdot 10^{-5}}{3.6 \cdot 10^{-4}} = 0.17$$

TIP FOR THE P.E. EXAMINATION CANDIDATE

The fundamental principle underlying risk assessment is the relationship of benefit to cost. Accordingly, the candidate should be prepared for any type of environmental scenario, not just cancer risk assessment as illustrated.

The fundamental principle underlying risk analysis is related to benefit-cost. If the legally permissible levels for exposure are not exceeded, the presumption is that the benefit will exceed the cost or risk. Risk, however, is not the same as cost; risk refers to the likelihood that an untoward event will occur, while cost refers to result of the event should it occur. Similarly, risk and hazard are not identical; for each hazard there is a risk of one or more untoward events (e.g., detrimental health effects). The risk analysis process provides an estimate of the number and likelihood of these events occurring referenced to the exposure. This process produces a basis for ascertaining the acceptability of the risk. Acceptability goes beyond the technical considerations used to carry out the risk analysis and includes societal and economic factors.

In addition to the risk coefficients, many parameters need to be evaluated to achieve a proper and complete risk analysis, including the effect that factors such as age, gender, and life style have on risk. The latency period and the effects of other exposures also need to be given along with assessment of the value of a life or years of life lost and other costs associated with the potential health effects. This has been done for radiation exposure; a value of $1000 to $2000 per person-rem ($100,000 to $200,000 per person-Sv) is often used as the basis for determining whether dose-reduction measures should be implemented.

Chapter 10: Health, Safety, and Environmental Protection 427

Example 10-16 — Integration of risk and cost

What is the value of a human life inherent in an alara monitization of $200,000 Sv^{-1} if the risk coefficient is 0.02 Sv^{-1}? If an individual affected would lose 20 yrs of life, what is the cost per year of life lost?

$$\text{Value of life} = \frac{\$200,000\,\text{Sv}^{-1}}{0.02\,\text{Sv}^{-1}} = \$10,000,000$$

If the life reduction is 20 yrs, this corresponds to $\frac{\$10,000,000}{20} = \$500,000$ per year.

10.6 Standard Notations

AIR

Symbol	Definition	Units
A	Duct cross-sectional area	[ft^2]
φ	Volume of air	[ft^3]
TP	Total pressure in duct	[in. of water]
VP	Velocity pressure in duct	[in. of water]
TLV	Threshold limit value	ppm, mg/m^3, others depending on hazard
C	Concentration of airborne contaminant	[ppm or mg/m^3]
V	Velocity	[ft/min]

NOISE

Symbol	Definition	Units
dB	Decibels	[10^{-12} watts/m^2]
dBA	Scale-weighted decibel	dBA
Hz	Sound frequency	Hertz
I	Sound intensity	watts/m^2
L	Sound pressure	[n/m^2 or dBA]
P	Sound power	watts

RADIATION

Symbol	Definition	Units
B	Buildup factor	dimensionless
A$_o$	Initial activity	Bq or Ci
λ	Decay constant	reciprocal time
T$_{1/2}$	Radiological half life	time
T$_{eff}$	Effective half life	time

I	Intensity	dose rate
J_o	Initial intensity	dose rate
μ	Attenuation coefficient	cm^{-1}
x	Thickness (shielding)	cm
μ/p	Mass attenuation coefficient	cm^2/g
rad	Absorbed dose	=100 erg/g
		=0.01 Jkg
rem	Dose equivalent	=100 erg/g
		=0.01 Jkg
Gy	Absorbed dose	=1J/kg
Sv	Equivalent dose, effective dose	=1J/kg
Q	Quality factor	dimensionless
W_R	Radiation weighting factor	dimensionless
W_T	Tissue weighting factor	dimensionless

10.7 References

American Conference of Governmental Industrial Hygienists. 1995. *Industrial Ventilation: A Manual of Recommended Practice*. 22nd Edition, Cincinnati.

American Conference of Governmental Industrial Hygienists. 1983. *Air Sampling Instruments for Evaluation of Atmospheric Contaminants*. Sixth Edition. Cincinnati: ACGIH.

American Industrial Hygiene Association. 1989. *Engineering Field Reference Manual*. Akron: AIHA.

American Industrial Hygiene Association. 1994. *Industrial Noise Manual*. Akron: AIHA.

American Society of Heating, Refrigerating and Air Conditioning Engineers (ASHRAE). 1989. *1989 Fundamentals Volume*. Atlanta: ASHRAE.

American Society of Heating, Refrigerating and Air Conditioning Engineers (ASHRAE). 1995. *Heating, Ventilating Air Conditioning Guide*. Atlanta: ASHRAE.

American Society of Heating, Refrigerating and Air Conditioning Engineers (ASHRAE). 1985. *Methods of Testing the Performance of Laboratory Fume Hoods*. ANSI/ ASHRAE Standard 110–1985, Atlanta:ASHRAE.

Baturin, V. V. 1973 *Fundamentals of Industrial Ventilation*. New York: Pergamon Press.

Buonicore, A. J. and W. T. Davis, Eds. 1992. *Air Pollution Engineering Manual*. New York: McGraw–Hill.

Cember, H. 1995. *Introduction to Health Physics*. Third Edition. New York: McGraw Hill.

Cheremisinoff, N. P., J. A. King and R. Boyko, Eds. 1994. *Dangerous Properties of Industrial and Consumer Chemicals*. New York: Marcel Dekker.

Clayton, G. D. and F. Clayton, Eds. 1991. *Patty's Industrial Hygiene and Toxicology*, Fourth Edition, Volume I, Parts A and B, *General Principles*; Volume II, Parts A, B, and C, *Toxicology*; Volume III, Parts A and B, *Theory and Rationale of Industrial Hygiene Practice*. New York: John Wiley and Sons..

Cralley, L. V and L. J. Cralley, Eds. 1986. *Industrial Hygiene Aspects of Plant Operations*. 3 volumes. Melbourne, FL: Kreiger.

Goetsch, D. L. *Occupational Health and Safety*. Second Edition. New York: Prentice Hall, 1996.

Kathren, R. L. 1995. *Radiation Protection*. Bristol: Adam Hilger.

Killuru, R., S. Bartell, R. Pitblado and S. Stricoff. 1996. *Risk Assessment and Management Handbook for Environmental, Health and Safety Professionals*.

Koren, H. 1995. *Illustrated Dictionary of Environmental Health and Occupational Safety*.

Kryter, K. D. 1970. *The Effects of Noise on Man*. New York: Academic Press.

Petersen, J., Ed. 1991. *Industrial Health*. American Conference of Governmental Industrial Hygienists: Cincinnati.

Peterson, A. P. G. and E. E. Gross, Jr. 1972. *Handbook of Noise Measurement*. Seventh Edition. Concord: General Radio.

Porteus, A. 1992. *Dictionary of Environmental Science and Technology*. (Revised). New York: Wiley.

Shapiro, J. 1991. *Radiation Protection*. Third Edition. Cambridge: Harvard.

Sheehan, A. 1995. *Industrial Hygiene Practice*. Boca Raton: Lewis Publishers.

Turner, D. B. and A. C. Stern, Eds. 1984. *Fundamentals of Air Pollution*. New York: Academic Press.

U.S Department of Health, Education and Welfare, PHS, CDC, NIOSH. 1973. *The Industrial Environment — Its Evaluation and Control*. Washington, DC: NIOSH.

U. S. Public Health Service. 1973. *Air Pollution Engineering Manual.* USPHS Publication No. 999– AP–40, Washington: USPHS.

Wald, P. and G. M. Stave. 1994. *Physical and Biological Hazards in the Workplace.* Van Nostrand Reinhold.

Wang, C. K. 1993. *OSHA Compliance and Management Handbook.* Noyes Publishers.

Wight, G. D. 1994. *Fundamentals of Air Sampling.* Boca Raton: Lewis Publishers.

Appendix

CONVERSION FACTORS

UNIT CONVERSIONS

in.	=	inch	lb	=	pound
ft	=	foot/feet	kg	=	kilogram
m	=	meter	sec	=	second
cm	=	centimeter	Mgal	=	million gallons
min	=	minute(s)	mph	=	miles per hour
mi	=	mile	kW	=	kilowatt
ha	=	hectare	kJ	=	kilojoule
ac	=	acre	cal	=	calories
km	=	kilometer	Btu	=	British thermal units
oz	=	ounce	atm	=	atmosphere
L	=	liter	Pa	=	Paschal
gal	=	gallon	psi	=	pounds force per in.^{2}
g	=	gram			

LENGTH

1 in. = 2.54 cm
1 ft = 12 in = 0.3048 m = 1/3 yd
1 mi = 5,280 ft = 2,760 yd = 1.609 km

AREA

1 ft = 0.929 m^2
1 ac = 34,560 ft^2 = 0.4047 ha
1 mi = 640 ac = 2.59 km^2
1 yd = 0.8361 m^2

VOLUME

1 oz = 0.02957 L
1 gal = 3.785 L = 0.003785 m^3 = 128 oz
1 ft^3 = 7.48 gal = 28.3 L = 0.0283 m^3 = 1,728 in^3
1 yd^3 = 0.765 m^3
1 m^3 = 264.17 gal

MASS

1 oz = 28.35 g
1 lb = 435.6 g = 16 oz = 0.454 kg = 7,000 grains
1 ton = 2,000 lb = 1.016 metric tons
1 metric ton = 1,000 kg = 2,200 lb

FLOWRATE

1 gal/day = 1.547 • 10^{-6} ft^3/sec = 3.785 • 10^{-3} m^3/day = 4.381 • 10^{-5} L/sec
1 Mgal/day = 1.547 ft^3/sec = 4,381 m^3/s
1 ft^3/sec = 2,446.7 m^3/day

CONCENTRATIONS

lb/ft^3 = 16.02 kg/m^3
lb/mg = 0.1198 mg/L
grains/ft^3 = 2.288 g/m^3

VELOCITY

V ft/sec = 0.3048 m/sec
1 mph = 1.466 ft/sec = 0.447 m/sec

FORCE

$$1N = 1 \text{ kg m/sec}^2 = 1 \cdot 10^5 \text{ dynes} = 1 \cdot 10^5 \frac{\text{g} \cdot \text{cm}}{\text{sec}^2} = 0.2248 \text{ lb}_f$$

ENERGY

$1J = 1N \cdot M = 1 \cdot 10^7 \text{ ergs} = 1 \cdot 10^7 \text{ dynes} \cdot \text{cm}$
$= 2.778 \cdot 10^{-7} \text{ kw} \cdot \text{hr} = 0.24 \text{ cal} = 7376 \text{ ft} \cdot \text{lb}_f$
$= 9.486 \cdot 10^{-4} \text{ Btu}$

POWER

$1 \text{ w} = 1\text{J/sec} = 0.239 \text{ cal/sec} = 0.7376 \text{ ft} \cdot \text{lb}_f/\text{sec}$
$= 9.486 \cdot 10^{-4} \text{ Btu/sec} = 1.341 \cdot 10^{-3} \text{ HP}$

PRESSURE

$1 \text{ atm} = 1.013 \cdot 10^5 \text{ N/M}^2 \text{ (Pa)} = 1.013 \text{ bars}$
$= 1.013 \cdot 10^6 \text{ dynes/cm}^2$
$= 760 \text{ mm Hg @ } 0°C$
$= 10.33 \text{ m H}_2\text{O @ } 4°C$
$= 144.696 \text{ psi}$
$= 33.9 \text{ ft H}_2\text{O @ } 4°C$
$= 29.92 \text{ in Hg @ } 0°C$

INDEX

A

Acids 62–64
Aeration basin design 224
Aerobic sludge digestion 245
Air pollution regulations 264–267
Air stripping 255, 362
Ambient air quality 261, 264, 267–269
 air quality control region 265
 national ambient air quality standards 265
 opacity 269
American Conference of Governmental Industrial Hyg 395
Ash management 348
 bottom ash 348
 fly ash 348
Attached-growth processes 215

B

Baghouses 283
Bases 62–64
Bernoulli equation 73
Biological oxygen demand 183
Biological treatment 207–227
 attached-growth processes 215
 process kinetics 208
 suspended-growth processes 213

C

Capitalized costs 46. *See also* Engineering economics
Carbon adsorption 256, 362
Channel properties 104
 hydraulic depth 105
 hydraulic radius 105
 most efficient 105
Characterization of wastewaters 190
Chemical equation
 balanced 52

Chemical kinetics 54
Chemical precipitation 365
Chemical reduction 363
Chezy equation 112
Chlorination basin design 237
Chlorine requirements 238
Clarifier design 165
Clean Air Act 264, 394
Clean Air Act Amendments 304
Clean Water Act 304, 394
Coagulation 55
Coagulation/flocculation 165–167
  unit design 166
Composition of air 263
Comprehensive Environmental
    Response, Compensation and
    Liability Act 302
Compression settling 163
Conduit flow 74–85
  closed 74
  Darcey-Weisbach resistance
    coefficient 75
  energy loss 74, 79
  fully turbulent 74
  laminar 74
  Moody diagram 75
  Reynolds number 74
  transitional 74
Cone of depression 141
Conservation of
  energy 72
    Bernoulli equation 73
  mass 70
Corrosion control 177
Credentials
  accreditation 24
  certification 24
  licensing 24
  registration 24
  specialty certification 25
Critical flow 109, 119
CT value 169
Cyclone collectors 273

D

Darcey-Weisbach resistance coefficient 75
Darcy-Weisbach friction factor 112
Darcy's Law 138
Dense nonaqueous phase liquids 377

Dewatering 252
Diffused aeration 231
Dimensional analysis 36
Disinfection 169–172, 236–238
  breakpoint chlorination 236
  CT value 169
Disposal 255
Domestic wastewater characteristics
  191

E

Electrostatic precipitators 279
Energy grade line 79
Energy Loss
  major 74
  minor 79
Engineering economics 39–42
  annual cash flow 44
  annual cost analysis 44
  capitalized costs 46
  present worth analysis 42
  rate of return 46
Environmental engineering
  history 13
Environmental Protection Agency
  394
Equivalent weight 50, 52

F

Filtration 167–168, 234–236
  backwashing 167
  straining 167
Fine particulate removal 174
  reverse osmosis 174
  ion exchange 174
Flexible membrane liners 333
Flocculation 55, 365
Flow properties 67–69
Flow regimes 108
Fluid properties 67–69
Fluids
  compressible 67
  incompressible 67
  Newtonian 68
  non-Newtonian 68
Fluoridation 178
Free surface flow 102–123
  Channel properties 104
  critical flow 119

Index

Flow regimes 108
  gradually-varied 104
  one-dimensional 103
  rapidly-varied 104
  spatially varied 104
  Specific energy 115
  steady 104
  uniform 104
    Chezy equation 112
    Darcy-Weisbach friction factor 112
    Manning's equation 112
Froude number 109

G

Gas transfer 228
  diffused aeration 231
  mechanical aeration 229
  oxygen supply requirements 228
  nitrification applications 229
Gravity separation 363
Grit chamber design 201
Groundwater hydrology 135–142
  aquifers 135
  flow in porus media 136
  Flow nets 139
  hydraulic conductivity 137

H

Hazard management 320
Hazardous and Solid Waste Amendments 302
Hazardous wastes 304
  characteristic 304
  listed 304
Hazardous waste treatment 359–371
  Air stripping 363
  biological treatment 368
  carbon adsorption 362
  chemical precipitation 365
  chemical reduction 363
  distillation 368
  flocculation 365
  gravity separation 363
  ion exchange 366
  neutralization 364
  oxidation 366
  reverse osmosis 367
  sedimentation 365
  solidification 368
  solvent extraction 368
  stabilization 368
  wet-air oxidation 371
Hazen-Williams pipe flow analysis 89–91
Hook's law 68
Hydraulic grade line 79
Hydrographs 128
Hydrology 123–135
  hydrologic cycle 123
  linear storage equation 128
  rainfall 125
    probable maximum precipitation 125
  rational equation 134
  surface runoff 126
Hydrostatics 69–70
  Newton's Second Law of Motion 69

I

Incineration 193
Incineration energy recovery 344
  air pollution control 346
  energy production 349
  mass burning 344
  refuse derived fuel (RDF) furnaces 345
Indoor air quality 261, 395–408
  air concentrations and TLVs 398
  exhaust ventilation 401
  monitoring 404
Ion exchange 255
Ionizing radiation 417–425
  alpha particle 417
  beta-emitting radioactive elements 417
  biological half life 420
  half-life 418
  maximum permissible dose 421
  neutrons 418
  protons 418
  radionuclide decay 418
  radon 423

L

Landfilling 321–343
  composite liner systems 335

design 321
final cover system 336
flexible membrane liners 333
gas management 341
leachate management 337
siting 324
soil liner systems 327
Laplace equation 140
Leachate management 336
Light nonaqueous phase liquids 377

M

Manning's equation 112
Mass balance 37
Math
  basics 33
  exponentials 34
  logarithms 34
Maximum contaminant levels 149
Mechanical aeration 229
Membrane processes 256
Mole 50
Molecular weight 50
  calculation 50
Momentum 71
  Newton's Second Law of Motion 71
Moody diagram 75
  zone of transition 78

N

National Emission Standards for Hazardous Air Poll 267
National Environmental Policy Act 303
National Research Council 216
Neutralization 255, 364
Newton's Second Law of Motion 69
Nitrogen 292
Noise 409–416
  control 412
  exposure 412
  monitoring 415
  physical principles 409
  sound intensity 409
  sound power 409

O

Occupational Safety and Health Administration 394
Open channel flow. *See* Free surface flow
Oxidation 52, 255
Oxygen supply requirements 228

P

Particulate control 273–285
  baghouses 283
  cyclone collectors 273
  electrostatic precipitators 279
Physical chemical treatment 255
  air stripping 255
  carbon adsorption 256
  ion exchange 255
  membrane processes 256
  neutralization 255
  oxidation 255
  precipitation 256
Piezometric head 79
Pipe networks 91–92
Porous media flow
  one-dimensional 140
  radial 141
    cone of depression 141
Precipitate 53
Precipitation 256
Preliminary treatment 199–201
  grit chamber design 201
Pressure
  considerations 95
Pressure conversions 35
Primary clarification 203
Principles and practice (P.E.) examination
  development 27
  examination format 28
  scoring 27
Process kinetics 208
Process selection 194
Public Utility Regulation and Policy Act 303
Pumps 92–101
  centrifugal 93
  classification and characteristics 92
  installation 96
    parallel 96
    series 96
  NPSH considerations 95

Pressure considerations 95
   cavitation 95
   Specific speed 94

R

Radon 423
Rate of return 46. *See also* Engineering economics
Rational equation 134
Reaction equilibrium 59–60
   solubility 60
Reaction kinetics 60–62
Receiving stream water quality 184–190
   deoxygenation coefficient 186
   maximum dissolved oxygen deficit 185
   oxygen deficit 186
   reaeration coefficient 186
Recycling 193
Reduction 52
Regulations 301
   Clean Air Act Amendments 304
   Clean Water Act 304
   Comprehensive Environmental Response, Compensation 302
   Hazardous and Solid Waste Amendments 302
   National Environmental Policy Act 303
   Public Utility Regulation and Policy Act 303
   Resource Conservation and Recovery Act 302
   Toxic Substances Control Act 304
Resource Conservation and Recovery Act 302
Resource recovery systems 350–359
   composting 356
   recycling 352
   waste minimization 350
Reynolds number 74, 109
Risk assessment 425–427

S

Scaling control 177
Scrubbers 285–292
   particulate control 290
   venturi scrubber 290

Secondary clarification 204
Secure land disposal 193
Sedimentation 202, 365
   objectives and design criteria 202
   primary clarification 203
   secondary clarification 204
Sedimentation types 157
   type I 158
   type II 160
   type III 162
   type IV settling 163
Settleable solids 153
Settling 155–165
   clarifier design 165
   detention time 155
   overflow rate 155
   Stoke's Law 157
   type I 158
   type II 160
   type III 162
   type IV 163
Site remediation 372–383
   geophysical surveys 373
   groundwater remediation 376
      dense nonaqueous phase liquids 377
      light nonaqueous phase liquids 377
      retardation 378
   landfill remediation 381
Siting 324
Sludge generation rates 240
Sludge pumping 244
Soil liner systems 327
Solid and hazardous waste 301
   characterization of solid waste 312
Solids management 172
Source reduction 193
Specific speed 94
Specific weight and moisture 314
Stabilization 245
Stationary air pollution regulations 267
   National Emission Standards for Hazardous Air Nati 267
   new source performance standards 267
Stationary source air pollution 269–273
   aerodynamic diameter 269

particle size distribution 269
Stoichiometric
  calculations 56–57
    molar approach 56
    valence 56
  chemical equation 52
Stoke's Law 157
Suspended-growth processes 213

T

Thickening 249
Threshold Limit Values 395
Total dissolved solids 153
Total Kjeldahl nitrogen 183
Total suspended solids 153
Total trihalomethanes control 178
Toxic Substances Control Act
    304, 394
Treatment standards 182
  biological oxygen demand 183
  total Kjeldahl nitrogen 183
  Water pollution control act 182
Trickling filter design 218
TSD facility 305

U

Uniform flow 110
  Darcy-Weisbach friction factor
    112
Units of environmental engineering
    34

V

Valence 50
Volatile organic gases 293
  excess air 294
  volatile organic compounds 293

W

Waste solids management 239
  conditioning 248
  dewatering 252
  disposal 255
  sludge generation rates 240
  sludge pumping 244
  solids characterization 239
    sludges 239

stabilization 245
thickening 249
Wastewater treatment 190–199
  characterization of wastewaters
    190
  domestic wastewater characteristics 191
  key design factors 195
  process selection 194
Water Pollution Control Act 182
Water quality 148–154
  parameters 151
    alkalinity 154
    hardness 154
    settleable solids 153
    total dissolved solids 153
    total suspended solids 153
    turbidity 154
  regulations 149
    maximum contaminant levels
      149
Water softening 53, 174
Water treatment 154–155

Z

Zone of transition 78
Zone settling 162, 163